Genetics, Environment, and Behavior

Contributors

V. ELVING ANDERSON
ERNST CASPARI
J. C. De FRIES
THEODOSIUS DOBZHANSKY
BRUCE E. ECKLAND
LEE EHRMAN
JOHN L. FULLER
BENSON E. GINSBURG
I. I. GOTTESMAN
LEONARD L. HESTON
ARTHUR R. JENSEN
L. ERLENMEYER-KIMLING
GERALD E. McCLEARN
AUBREY MANNING
NEWTON E. MORTON
ARNO G. MOTULSKY
GILBERT S. OMENN
P. A. PARSONS
SATYA PRAKASH
CLAUDINE PETIT
WILLIAM S. POLLITZER
W. R. THOMPSON
E. TOBACH
S. G. VANDENBERG
PETER L. WORKMAN

Genetics, Environment, and Behavior

Implications for Educational Policy

Edited by

LEE EHRMAN
Division of Natural Sciences
State University of New York
College at Purchase
Purchase, New York

GILBERT S. OMENN
Division of Medical Genetics
School of Medicine
University of Washington
Seattle, Washington

ERNST CASPARI
Department of Biology
University of Rochester
Rochester, New York

1972

ACADEMIC PRESS   New York and London

COPYRIGHT © 1972, BY ACADEMIC PRESS, INC.
ALL RIGHTS RESERVED.
NO PART OF THIS PUBLICATION MAY BE REPRODUCED OR
TRANSMITTED IN ANY FORM OR BY ANY MEANS, ELECTRONIC
OR MECHANICAL, INCLUDING PHOTOCOPY, RECORDING, OR ANY
INFORMATION STORAGE AND RETRIEVAL SYSTEM, WITHOUT
PERMISSION IN WRITING FROM THE PUBLISHER.

ACADEMIC PRESS, INC.
111 Fifth Avenue, New York, New York 10003

United Kingdom Edition published by
ACADEMIC PRESS, INC. (LONDON) LTD.
24/28 Oval Road, London NW1

LIBRARY OF CONGRESS CATALOG CARD NUMBER: 72-84369

PRINTED IN THE UNITED STATES OF AMERICA

Contents

LIST OF CONTRIBUTORS	xi
FOREWORD	xiii
PREFACE	xvii
ACKNOWLEDGMENTS	xix

CHAPTER 1   INTRODUCTORY REMARKS

ERNST W. CASPARI	1
References	4

CHAPTER 2   QUANTITATIVE ASPECTS OF GENETICS AND ENVIRONMENT IN THE DETERMINATION OF BEHAVIOR

J. C. DEFRIES

The Heritable Nature of Group Differences	6
Quantitative Aspects of Environmental Determination	11
Summary	15
References	16

Discussion

JOHN L. FULLER	17
References	21
Reply to Professor Fuller: J. C. DeFries	21

v

Comment

ARTHUR R. JENSEN	23
Reply to Professor Jensen: J. C. DeFries	24

Comment

PETER L. WORKMAN	26

CHAPTER 3   **QUALITATIVE ASPECTS OF GENETICS AND ENVIRONMENT IN THE DETERMINATION OF BEHAVIOR**

CLAUDINE PETIT

Genetic Determinism and the Influence of Physical Environment on Some Types of Behavior	28
Influence of Physiological Factors, Including Hormones, on Genes	34
Influence of Biological and Social Environment	36
Conclusion	42
References	43

Discussion

AUBREY MANNING	48
References	52

CHAPTER 4   **GENETIC DETERMINATION OF BEHAVIOR (ANIMAL)**

GERALD E. MCCLEARN

The Comparative Method	55
Animal Behavioral Genetics	58
Genetics of Learning	59
Summary and Conclusions	64
References	65

Discussion

SATYA PRAKASH	68
References	71

Comment

THEODOSIUS DOBZHANSKY	72

CONTENTS

CHAPTER 5   **GENETIC DETERMINATION OF BEHAVIOR (MICE AND MEN)**

P. A. PARSONS

Introduction	75
The Behavioral Phenotype	77
Hybrids	79
Trait Profiles in Different Genotypes	83
Measures of Learning	85
Extreme Environments and Genotypes	87
Mice and Men	89
The Meaning of the Term "Race"	93
Conclusions and Summary	94
References	96

Discussion

LEONARD L. HESTON   99

Comment

NEWTON E. MORTON   103

CHAPTER 6   **HUMAN BEHAVIOR ADAPTATIONS: SPECULATIONS ON THEIR GENESIS**

I. I. GOTTESMAN AND L. I. HESTON

General Considerations: Evolutionary Outcomes and Kinds of Selection	106
A Brief Overview of Primate Phylogeny	107
Evolution of Brain Size and Tool Use	107
Within Species Behavioral Variability	110
Adaptability and Genotype–Environment Interaction	111
The Evolution of Milk Drinking	115
What Next?	120
References	120

Discussion

WILLIAM S. POLLITZER	123
References	127

CHAPTER 7   **BIOCHEMICAL GENETICS AND THE EVOLUTION OF HUMAN BEHAVIOR**

GILBERT S. OMENN AND ARNO G. MOTULSKY

Evolutionary Development of the Biological Substrate	131
Evolution of Allelic Gene Products	133

	Evolution by Gene Duplication	135
	Reductionistic Description of the Human Nervous System	137
	Protein Polymorphisms	142
	Current Studies of Enzyme Variation in Human Brain	144
	Approaches to Complex Behavioral Phenotypes	150
	The Central Role of Language in the Evolution of Man	157
	The Impact of Evolution of Man's Culture upon Man	163
	References	167

Discussion

	V. ELVING ANDERSON	172
	References	177

CHAPTER 8 GENE–ENVIRONMENT INTERACTIONS AND THE VARIABILITY OF BEHAVIOR

L. ERLENMEYER-KIMLING

	Introduction	181
	What Is Interaction?	182
	Just as the Twig Is Bent? (Illustrations from Early Experience Studies)	191
	Parameters of Interaction	197
	Concluding Remarks	199
	References	204

Discussion

	W. R. THOMPSON	209
	References	214

Comment

	ERNST CASPARI	215
	References	216

Comment

	NEWTON E. MORTON	217
	Reference	218

CHAPTER 9 THE MEANING OF THE CRYPTANTHROPARION

E. TOBACH

	Introduction	219
	The Formulation of the Scientific Problem	220
	The Logistics of Research in the Problem of Human Behavior and Genetics	229

The Relationship of the Scientific Problem to Society	231
Final Statement	232
References	233

Discussion

ARTHUR R. JENSEN	240

CHAPTER 10 HUMAN BEHAVIORAL GENETICS

N. E. MORTON

What Are the Effects of Single Genes on Behavior?	247
What Are the Effects of Chromosome Aberrations on Behavior?	249
How Can Behavior Whose Transmission Is Unknown Be Screened for Sensitivity to Genetic Differences?	250
How Can the Inheritance of Behavioral Attributes Be Studied?	257
Do Psychological Factors Have Genetic Significance?	259
To What Extent Are Group Differences in Behavior Genetic?	260
What Are the Effects of Behavior on Population Structure and Selection?	263
Summary	264
References	264

Discussion

P. L. WORKMAN	266
References	270
Editor's Comment	270
References	271

CHAPTER 11 THE FUTURE OF HUMAN BEHAVIOR GENETICS

S. G. VANDENBERG

Within and between Ethnic Group Comparisons	276
Integration of Behavior Genetics, Biochemistry and Physiology	277
Most Likely Future Research	278
Needed Ancillary Research	281
Need for "Basic" Theoretical Formation	283
References	284
Appendix 1: Suggestions for Ideal Body of "Core" Data to Be Collected in Cooperative Studies	285
Editor's Comment	287
References for Appendix	288

Discussion
BENSON E. GINSBURG 290
References 295

CHAPTER 12 COMMENTS ON SCHOOL EFFECTS, GENE–ENVIRONMENT COVARIANCE, AND THE HERITABILITY OF INTELLIGENCE
BRUCE K. ECKLAND

The Search for Explanation: Do Schools Make a Difference? 297
Prediction or Explanation: Gene–Environment Covariance 300
The Heritability of Intelligence and School Achievement 301
References 305

Epilogue
GILBERT S. OMENN, ERNST CASPARI, AND LEE EHRMAN 307

Author Index 311
Subject Index 320

List of Contributors

Numbers in parentheses indicate the pages on which the authors' contributors begin.

V. ELVING ANDERSON (172), Dight Institute for Human Genetics, University of Minnesota, Minneapolis, Minnesota

ERNST CASPARI (1, 215, 307), Department of Biology, University of Rochester, Rochester, New York

J. C. De FRIES (5), Institute for Behavioral Genetics, University of Colorado, Boulder, Colorado

THEODOSIUS DOBZHANSKY (72), Department of Genetics, University of California, Davis, California

BRUCE E. ECKLAND (297), Department of Sociology, University of North Carolina, Chapel Hill, North Carolina

LEE EHRMAN (307), Division of Natural Sciences, State University of New York, College at Purchase, Purchase, New York

JOHN L. FULLER (17), Department of Psychology, State University of New York, Binghampton, New York

BENSON E. GINSBURG (290), Department of Biobehavioral Sciences, University of Connecticut, Storrs, Connecticut

I. I. GOTTESMAN (105), Department of Psychology, University of Minnesota, Minneapolis, Minnesota

LEONARD L. HESTON (99, 105, 240), Department of Psychiatry, University of Minnesota, Minneapolis, Minnesota

ARTHUR R. JENSEN (23, 240), Institute for Human Learning, University of California, Berkeley, California

L. ERLENMEYER-KIMLING (181), New York State Psychiatric Institute, New York, New York

GERALD E. McCLEARN (55), Institute for Behavioral Genetics, University of Colorado, Boulder, Colorado

AUBREY MANNING (48), Department of Zoology, University of Edinburgh, Edinburgh, Scotland

NEWTON E. MORTON (217, 247), Population Genetics Laboratory, University of Hawaii, Honolulu, Hawaii

ARNO G. MOTULSKY (129), School of Medicine, University of Washington, Seattle, Washington.

GILBERT S. OMENN (129, 307), School of Medicine, University of Washington, Seattle, Washington

P. A. PARSONS (75), Department of Genetics and Human Variation, La Trobe University, Bundoora, Victoria, Australia

SATYA PRAKASH (68), Department of Biology, University of Rochester, Rochester, New York

CLAUDINE PETIT (27), Université Paris 7, Faculte des Sciences, Laboratorie de Biologie Animale, Paris, France

WILLIAM S. POLLITZER (123), Department of Anatomy and Anthropology, University of North Carolina, Chapel Hill, North Carolina

W. R. THOMPSON (209), Department of Psychology, Queens University, Kingston, Ontario

E. TOBACH (219), Department of Animal Behavior, American Museum of Natural History, New York, New York

S. G. VANDENBERG (273), Department of Psychology, University of Colorado, Boulder, Colorado

PETER L. WORKMAN (26, 266), Department of Anthropology, University of Massachusetts, Amherst, Massachusetts

Foreword

At the request of the U.S. Office of Education, the National Academy of Sciences (NAS), jointly with the National Academy of Education (NAE) established the Committee on Basic Research in Education (COBRE) in 1968 to support the conduct of research of a fundamental character in education.

This Committee is currently composed of a group of distinguished scientists, with Patrick Suppes (Stanford University) as Chairman, and James S. Coleman (The Johns Hopkins University) as Vice Chairman, and includes the following members: John B. Carroll (Educational Testing Service), Ernst W. Caspari (University of Rochester), Bruce K. Eckland (University of North Carolina), Robert M. Gagné (Florida State University), Wayne

H. Holtzman (The Hogg Foundation for Mental Health, University of Texas), H. Thomas James (The Spencer Foundation), Arthur W. Melton (University of Michigan), Julius B. Richmond, M. D. (Harvard Medical School), A. Kimball Romney (University of California at Irvine), Edgar H. Schein (Massachusetts Institute of Technology), and Theodore W. Schultz (University of Chicago). The program is administered by the Division of Behavioral Sciences of the National Research Council. Henry David, Executive Secretary of the Division, is the Project Director.

Also serving on the Committee in earlier years were: R. Taylor Cole (Duke University), Lawrence A. Cremin (Teachers College, Columbia University), John I. Goodlad (University of California at Los Angeles), Louis Hartz (Harvard University), and Fritz Machlup (Princeton University).

In its first two years of life, COBRE developed a project grant support program in the behavioral sciences in support of basic research on problems relevant to education. The selected projects were funded by the Office of Education. For its third year, a special small grant program directed toward recent doctoral recipients was established. The purposes of the grant program are to support research which will contribute to fundamental knowledge, and will deepen understanding of the critical problems in educational theory, policy and practice. In this effort, the behavioral sciences include anthropology, economics, geography, linguistics, political science, psychology, and sociology, and also the relevant areas of the biological sciences, engineering, history, philosophy, and other sciences.

A second phase of third-year activity was a series of eight research workshops. These workshops were invitational, informal, 15–20 participants and ran from five to ten days. Each workshop was directed by a member of COBRE, and their general goals were to identify significant researchable questions in the area, to define regions or groups of research efforts, and to identify individual scientific contributors.

This book is derived from such a workshop held at Wainwright House, Rye, New York in October 1971 under the direction of Ernst W. Caspari and the coordination of Lee Ehrman. The workshop was convened under the title "Genetic Endowment and Environment in the Determination of Behavior," and concentrated on the contributions of geneticists and psychologists. Specific, advance contributions were commissioned, and Dr. Ehrman prepared a summary and overview of the workshop. This book, with changes and additions, is a derivative of this process.

COBRE, the NAS and NAE, and the sponsors, the U. S. Office of Education, hope that this book will contribute to a richer and sharper development of genetic and biological research contributions to the problems of education. We thank the participants and the contributors, the director, the coordinator (and the editors) for their efforts.

SHERMAN ROSS
Executive Secretary
Committee on Basic Research In Education

Preface

From Sunday, October 3, 1971 through Friday, October 8, at Wainwright House, Rye, New York, with the State University of New York, College at Purchase, Purchase, New York, as host, a workshop was held on Genetic Endowment and Environment in the Determination of Behavior. This was the eighth and final workshop in a series concerned with the social, political, and biological aspects of educational policy.

A research workshop on the genetics of behavior and learning cuts across many disciplines. At the very least, those of animal behavior, anthropology, biochemical genetics, cytogenetics, demography, ecology, ethology, evolution, population genetics, psychiatry, psychology, and sociology are intimately involved—this is not to omit the new interdisciplinary field of behavior genetics itself. This hybrid subject has recently been graced with its own journal, *Behavior Genetics*. Its editors (Professors S. G. Vandenberg and J. C. DeFries of the University of Colorado) found it necessary to state the following in their introductory address: "It is most clear from recent events that the misunderstandings inherent in the old nature–nurture controversy are not dead and buried, but alive and well. In fact, this topic seems to generate today as much emotional reaction with as little information as in the past. Perhaps when appreciation of the substantive and methodological informations of behavior genetics becomes more widespread people will be able to cope more effectively with such issues."

It seemed wise, therefore, to make the theme of our workshop, *Genetic Endowment and Environment in the Determination of Behavior*.[1]

Our reason for planning this workshop was the recognition that *(1)* the question of the relationships between genetic characteristics and behavior is an interesting scientific question and an important one for understanding human behavior and learning, and *(2)* that this problem has been approached in different ways by geneticists and psychologists. There are important problems within each science. For example, geneticists, psychologists, and other scientists face the problem of conceptualizing, measuring, and manipulating the phenomena they study. Some of the problems could be approached by the application of expertise from related fields. Behavioral scientists are often ignorant of each others' research, or unable to interpret the findings of research because of a lack of familiarity with the methods, theories, and accomplishments of other disciplines of research into behavior.

Our intention was to bring together geneticists, psychologists, and other behavioral scientists, whose previous work was directly relevant to these problems, to interact with each other, and to learn from each other. We expected that discussion would isolate issues in these areas where the research of one science is applicable to the questions of another, and areas in which research using any approach or combination of approaches has not yet been done but could or should be. In format, the workshop consisted of discussion of papers prepared and circulated in advance by some of the participants, as well as more informal discussion.

LEE EHRMAN

---

[1] See Caspari, E. (1968). *American Educational Research Journal,* **5**, 43–55.

Acknowledgments

The Coordinator takes this opportunity to state her conviction of the value and the timeliness of such a workshop topic. Indeed, for anyone involved in any way in the science that is genetics, such efforts as coordination required were obligatory and gladly offered so that this meeting could take place. The Coordinator wishes to thank Drs. Sherman Ross and Barbara Meeker (and Mrs. S. Jobst) for both instructive and constructive assistance. I also wish to thank Dean Curtis A. Williams, Jr., Dr. and Mrs. Jack Leonard, and Mrs. J. Rucquoi for reading and rereading sections of the book.

One almost regrets the completion of this endeavor because of the friendships this entire project engendered over a period of more than a year. We hope that this written record proves as stimulating as were the actual exchanges in bridging the several disciplines that were represented.

LEE EHRMAN

I join the other participants in thanking Workshop Chairman Ernst Caspari, Coordinator Lee Ehrman, and the staff of the Committee on Basic Research in Education of the National Research Council for their remarkable efforts in organizing this conference. The intensive and constructive interaction of individuals from the several disciplines combined a searching examination of the state of knowledge in each field and a concern that interdisciplinary research be directed to provide basic understanding for some of the difficult social issues of our time. We are pleased that the proceedings of the workshop can be shared with all interested readers. Finally both Lee Ehrman and I wish to acknowledge the expert advice of Dr. L. Sandler. My thanks are also due to Miss S. Schneider, who assisted me with checking the proof.

GILBERT S. OMENN

Chapter 1 Introductory Remarks

ERNST W. CASPARI
University of Rochester
Rochester, New York

It may appear unexpected to some people that among the workshops organized by the Committee on Basic Research in Education, there should be one devoted to the genetic aspects of behavior. It may, therefore, be worthwhile to point out that the exclusive preoccupation of educational research with environmental variables is of relatively recent origin. Two of the fundamental papers in the field of behavior genetics, those by Burks (1928) and by Tryon (1940) actually first appeared in the Yearbook of the National Education Association. This indicates that both the authors and the editors of the Yearbook felt at that time that investigations in genetics were relevant to educational problems. In the meantime, educational research, following the lead of psychological learning theory, has become completely concerned with the environmental conditions of cognitive and emotional functioning. The reasons for this development are discussed in this book, particularly in the chapter by McClearn. Suffice it to state here that concentration on environmental factors alone neglects the dimension of heredity–environment interactions, which is expressed as different reactions of different genotypes to the same environmental conditions, that is, differences between individuals in their behavior under the same environments. Differences in genotype form the primary basis for the individuality of organisms, including human beings, and should therefore not be neglected in studies which have as their ultimate goal the understanding of individual behavior in society.

A few remarks on the present state of behavior genetics may be appropriate. People working in this field are divided into two camps, those who are applying the techniques of quantitative genetics to behavioral characters, and those who prefer to study the effects of individual genes affecting behavior. This difference, though more sharply expressed in behavior genetics, actually permeates all of genetics and may be regarded as a remnant of the controversy between Mendelian geneticists and biometricians in the beginning of this century. In an earlier publication (Caspari, 1967), I have pointed out that this difference may be merely a matter of technique. In animal experiments, waiting for mutants affecting a specific character to occur (or increasing their frequency by the use of mutagens) invariably results in single gene differences, while using a selection technique in a population usually results in polygenic differences. In human beings the study of rare diseases corresponds to the study of mutation—rare because natural mutation rates are low, and diseases, usually severe, because the genes are kept in the population by a mutation–selection equilibrium. Individual mutations lend themselves better to the study of mechanisms at the biochemical and molecular level, since each mutant gene is assumed to be responsible for a single primary effect. The study of the pleiotropic effects of the gene causing phenylketonuria in humans may serve as an example of the potentialities of this method. Polygenic effects are most important from the point of view of evolution, because selection is supposed to act on the natural variation found in a population and may lead to the establishment of coadapted gene complexes rather than individual genes. The difference between these two approaches can be well documented in the work of Benzer (1967) and of Hirsch and Boudreau (1958) on phototaxis in *Drosophila*. Benzer collected a large number of mutants interfering with phototactic behavior and attempted to identify the action of these individual genes, and thus analyzed the interaction of individual processes contributing to phototactic behavior. Hirsch subjected a population of *Drosophila* to selection for different degrees of phototaxis by giving them a choice of light and dark passages. He isolated strains that differed in an unspecified number of genes affecting phototaxis and localized the genes to specific chromosomes. Whether the behavior characters called "phototaxis" by Benzer and by Hirsch are identical is not certain, and doubt has been expressed by Dobzhansky and Spassky (1969).

The idea proposed in the preceding paragraph may be summarized by

1. INTRODUCTORY REMARKS

stating that the difference between the single gene and the polygenic approaches in behavior genetics is only methodological, and that the two methods lend themselves to the investigation of different problems. More specifically, the method of investigating individual mutants is more suitable for the investigation of mechanisms because of the presumed association of a single gene with a specific protein. Therefore, it should be added that in higher organisms, the regulation of many individual proteins may be itself considered to be under polygenic control. The cases of the control of catalase in the mouse (Schimke *et al.*, 1969) and of human hemoglobin may be mentioned as examples. Different mouse strains differ in their catalase levels, and biochemical investigation has shown that in this case two biochemically and probably genetically different mechanisms are involved: differences in specific activity, that is, in activity of individual enzyme molecules (presumably differences in the structure of the enzyme) and differences in the rate of degradation of the enzyme. Differences in rate of enzyme synthesis have not been described for this particular enzyme, but have been found in other cases. A system of this kind would therefore be expected to show, in natural populations, the characters of an additive polygenic system. Actually, in F_1 hybrids between different inbred strains, intermediacy, dominance, and overdominance for catalase level have been found.

Hemoglobin A consists of four polypeptide chains controlled by two nonlinked loci. Every molecule has, in normal individuals, two alpha and two beta chains per molecule. In addition, every hemoglobin molecule contains four heme molecules which are synthesized in the organism by a chain of reactions involving six separate enzymes. Therefore, in the production of a hemoglobin molecule, eight separate genes are presumably involved, and their activity must be regulated in such a way that all of them produce appropriate amounts of the final product. The mechanism of this regulation has been partly cleared up. From the present point of view, it should be considered that such a regulatory system constitutes a polygenic system, in this case with epistatic interaction.

The concepts of polygenic inheritance can therefore be applied to the regulation of proteins in higher organisms. This does not mean that the working of these systems could be analyzed by selection experiments. What selection experiments, on catalase at least, could show is the genetic constitution of a population with regard to genes affecting catalase levels, and in this way the developmental and evolutionary potentialities of the population, and possibly of the individuals composing the population, can be elucidated.

The general conclusion is that for understanding the genetic basis of a biological character, behavioral or otherwise, all available approaches are necessary. This consideration was reflected in the organization of this conference. The proposition was made to look at behavior, human and animal, at the molecular–biochemical, cellular, individual, population and evolutionary levels, to formulate the problems arising at each level, and to discuss the methods by which they have been attacked and by which they should be attacked in the future. It should be kept in mind that the different levels are not separate compartments, but that they interdigitate with each other and that methods conventionally used at one level may be applicable to other levels as well. The primary purpose of this conference was to review the material available and the methods of analysis which have been used in the past, and to discuss their validity, their limitations and their possible applications in research of interest for education.

References

BENZER, S. (1967). Behavioral mutants of Drosophila isolated by countercurrent distribution. *Proc. Nat. Acad. Sci. U. S. A.* **58**, 1112–1119.

BURKS, BARBARA (1928). The relative influence of nature and nurture upon mental development. *Yearb. Nat. Soc. St. Educ.* **27**, 219–321.

CASPARI, E. W. (1967). Behavioral consequences of genetic polymorphism in man: A summary. In: "Genetic Diversity and Human Behavior," (J. N. Spuhler, ed.), pp. 269–278. Wenner-Gren Foundation Inc.

DOBZHANSKY, T., and SPASSKY, B. (1969). Artificial and natural selection for two behavioral traits in Drosophila pseudoobscura. *Proc. Nat. Acad. Sci. U. S. A.* **62**, 75–80.

HIRSCH, J. and BOUDREAU, J. (1958). Studies in experimental behavior genetics I. The heritability of phototaxis in a population of Drosophila melanogaster. *J. Comp. Physiol. Psychol.* **51**, 647–651.

SCHIMKE, R. T., DOYLE, D., and GANSCHOW, R. E. (1969). Mutations in inbred mouse strains and the study of enzyme regulatory mechanisms. *In:* "Problems in Biology: RNA in Development." (E. W. Harnly, ed.), pp. 163–185. Univ. of Utah Press, Salt Lake City, Utah.

TRYON, R. C. (1940). Genetic differences in maze-learning ability in rats. *Yearb. Nat. Soc. Study Educ.* **39**(1), 111–119.

Chapter 2 Quantitative Aspects of Genetics
and Environment in the
Determination of Behavior

J. C. DeFRIES

University of Colorado
Boulder, Colorado

Quantitative genetic theory was developed by applied scientists faced with the practical problem of improving polygenic characters in domestic animals and plants. Although this theory has not been substantially altered within the last several decades, its utilitarian value remains. Thus, it seems likely that such concepts as heritability, genetic correlation, and selection index will be indispensable when the issue of qualitative population control is finally faced. However, since quantitative genetics was largely developed for application to animal and plant breeding, it is not surprising that its concepts and methods are not immediately applicable to important social issues such as the heritable nature of racial differences or the feasibility of modifying behavior by environmental means. The primary objective of this paper is to illustrate how the concepts of quantitative genetics might be extended to deal with such problems.

The Heritable Nature of Group Differences

Jensen (1969) has recently marshaled compelling evidence to demonstrate that intelligence, as measured by conventional IQ tests, is a highly heritable character within Caucasian populations. From this evidence, Jensen hypothesized that genetic factors are strongly implicated in the reported difference of 15 IQ points between the means of Caucasians and Afro-Americans. In a critique of Jensen's paper, Lewontin (1970) showed that the genetic basis of interracial differences is not a simple function of the within-group heritability; however, the formal relationship between these variables was not explored. In view of the obvious importance of this issue, an examination of the relationship between within-group heritability and the heritable nature of group differences is clearly in order.

Heritability. The concept of heritability has been discussed lucidly by both Jensen (1969) and Lewontin (1970); thus, only a relatively brief review of this concept will be presented here. In quantitative genetic theory (Falconer, 1960), the measured value of some character of an individual, that is, its phenotypic value, is assumed to be some function of its genotype and the environment in which it develops. For simplicity, we may assume the following linear mathematical model:

$$P = G + E, \tag{1}$$

where P is the phenotypic value, G is the genotypic value, that is, the value conferred upon the individual by its genotype, and E is a deviation caused by the environment. Thus, since the mean environmental deviation is zero, the mean phenotypic value would estimate G in a population of like genotypes.

If nonlinear interactions occur between G and E, another term should be included in Eq. (1). A method for assessing the importance of such genotype–environment interactions in human twin data has been suggested by Jinks and Fulker (1970).

The Jinks and Fulker method for assessing the presence of genotype–environment interactions in human twin data is analogous to those tests which have been widely applied in animal and plant genetics. The difference between the scores of two members of a set of monozygotic (MZ) twins provides a measure of the different environmental influences experienced by members of the same family. If all sets of twins are affected to the same extent by these within-family environmental influences, the absolute difference between members of the same set of MZ twins should be equal (within sampling errors) for all sets. However, if some twins react differently to the same environmental influences or if some sets of twins experience different environmental influences than others, the twin differences will be unequal. On the other hand, the sums of the twin scores (or the means) will be expected to differ due to both genetic and environmental differences among families. Thus, a correlation between the absolute twin difference and the sum (or mean) over sets would indicate that different genotypes are responding differentially to environmental influences. This method of assessing for genotype–environment interactions in human twin data has also recently been explicated and applied by Jensen (1970).

If such interactions are found to be important in a set of data, but are not of special interest to the investigator, the raw data may be subjected to a scalar transformation which may render the original model appropriate.

From Eq. (1), it may be seen that the phenotypic variance may be simply expressed as follows:

$$V_P = V_G + V_E, \qquad (2)$$

where V_P is the phenotypic variance, V_G is the genotypic variance, and V_E is the environmental variance. If a correlation exists between G and E, the assumptions underlying the simple model [Eq. (1)] are not violated; however, Eq. (2) should then contain a term corresponding to twice the covariance of G and E. Roberts (1967) has suggested an intriguing solution to this problem. He suggests that the environment should be defined as affecting the phenotype independently of the genotype. Thus, if the genotype of an individual influences its choice of environment, this effect should be considered to be genetic, even if it is mediated by such things as habitat selection. However, if certain genotypes are forced to live in inferior environments, such a procedure would unfairly ascribe any resulting effect to the genotype.

The genotypic variance may also be partitioned into components due to different causes. The gene, not the genotype, is the unit of transmission. Therefore, the resemblance of all except very close relatives is due chiefly to the average effects of genes. In principle, each allele has an average effect for a character measured on individuals in a population. When summed, these average effects result in an expected or additive genetic value. Dominance and epistasis, however, may cause the genotypic value to deviate from this value. Symbolically,

$$G = A + D + I, \qquad (3)$$

where G is the genotypic value, A is the additive genetic value (sum of the average effect of the alleles across all loci), D is the dominance deviation (nonlinear interaction between alleles at the same locus, summed across all loci), and I is the epistatic interaction (nonlinear interaction between alleles at different loci). A, D, and I are all independent; thus,

$$V_G = V_A + V_D + V_I, \qquad (4)$$

where V_A is the additive genetic variance, V_D is the dominance variance, and V_I is the epistatic variance. (See Lush, 1948, and Falconer, 1960, for a more detailed discussion of the principles underlying the partitioning of genotypic variance.)

The ratio of the additive genetic variance to the phenotypic variance is known as heritability in the narrow sense (Lush, 1949) or simply heritability (Falconer, 1960). The proportion of the phenotypic variance due to both additive and nonadditive genetic variance is referred to as heritability in the broad sense (Lush, 1949). Heritability (narrow sense) has both descriptive and predictive properties. In addition to indicating the proportion of the variance due to the average effects of genes in a population, it may also be shown that heritability is equivalent to the regression of the additive genetic value of an individual on its phenotypic value. Thus, heritability may be used to predict the additive genetic value of an individual and the change in a population due to various breeding systems (Falconer, 1960). For this reason, heritability in its narrow sense should be of particular importance to those purportedly interested in eugenic considerations.

Because of the predictive property of heritability, it is important to be clear about when and when not to adjust the estimate for lack of perfect test reliability. Such adjustment may be reasonable when one wishes to compare estimates obtained from data in which tests with different reliabilities have been used. However, in such a case, the resulting estimates no longer correspond directly to heritability based upon single records. Instead, the estimates correspond to the heritability of the average of N records on each individual, where N is equal to infinity. The heritability of the average of N records ($h_{\bar{P}}^2$) is as follows:

$$h_{\bar{P}}^2 = \frac{Nh^2}{1 + (N-1)m}, \qquad (5)$$

where N is the number of records on an individual, h^2 is the heritability based upon single records (not adjusted for test reliability), and m is the correlation between repeated records on the same individuals. It may be shown that $h_{\bar{P}}^2$ is equivalent to the regression of the additive genetic value of an individual on the mean of N records on that individual.

The various methods of estimating heritability will not be discussed here. These procedures, as well as some of the special problems encountered with human data, have been discussed previously by the author (DeFries, 1967).

Within-Group Heritability and the Heritability of the Group Average. When a population is composed of two or more groups, the genetic variance and phenotypic variance in the population may each be partitioned into two parts: that between groups and that within groups. The ratio of the additive genetic variance within groups to the phenotypic variance within groups yields the within-group heritability (h_w^2):

$$h_w^2 = h^2 \frac{(1-r)}{(1-t)}, \qquad (6)$$

where h^2 is the population heritability (narrow sense, not adjusted for test reliability), t is the phenotypic correlation (intraclass) among members of the same group, and r is the analogous genetic correlation, that is, the correlation of the additive genetic values of members of the same group. For groups composed of close relatives, r is equal to the coefficient of relationship. However, for groups which have been isolated for many generations, selection and/or genetic drift could change gene frequencies in the groups such that r may differ considerably from the coefficient of relationship. It may be shown that h_w^2 is equivalent to the regression of the additive genetic value of an individual on its observed phenotypic value, where the phenotypic value is expressed as a deviation from the group mean.

The ratio of the additive genetic variance between groups to the phenotypic variance between groups yields the heritability of the group average (h_f^2):

$$h_f^2 = h^2 \left[\frac{1 + (n-1)r}{1 + (n-1)t} \right], \qquad (7)$$

where n is the number of individuals measured within the group under consideration, and h^2, r, and t are defined as above. It may be shown that h_f^2 is equivalent to the regression of the mean additive genetic value of a group on its mean phenotypic value, expressed as a deviation from the grand mean; thus, h_f^2 may be used to estimate the mean additive genetic value of a group or to explore the heritable nature of group differences. (The symbols and expressions of h_w^2 and h_f^2 are those used by Falconer, 1960, in his discussion of the heritability of within-family deviations and family means, respectively.)

From the above expression it is obvious that h_f^2 is a function of h_w^2 as follows:

$$h_f^2 = h_w^2 \frac{(1-t)}{(1-r)} \left[\frac{1 + (n-1)r}{1 + (n-1)t} \right]. \qquad (8)$$

When the number of individuals measured within a group is large, h_f^2 reduces to the following approximation:

$$h_f^2 \cong h_w^2 \frac{(1-t)r}{(1-r)t}. \qquad (9)$$

Equation (9) clarifies the two troubling cases raised by Lewontin (1970), which suggested that the heritability of the group average (or the heritable nature of group differences) bore no logical relation to the within-group heritability. In his first case, two completely inbred lines were reared in similar environments. Although the difference between lines is thus entirely due to gene effects, h_w^2 in isogenic lines is zero. From Eq. (9) it may be seen that h_t^2 is not zero in this case; it is undefined. Thus, r will equal one with completely inbred lines.

In Lewontin's second case, two random samples from an open-pollinated variety (or genetically heterogeneous population) are reared in quite different environments. In this case, h_w^2 is nonzero, yet all the difference observed between groups should be environmental. If the random samples are sufficiently large that genetic equality between the two groups is ensured, r will approach zero, but t will be nonzero; thus, as seen from Eq. (9), h_t^2 will approach zero in this case.

Equation (9) may also be used to explore the heritable nature of racial differences in IQ. The value of h_w^2 suggested by Jensen (about .8) is almost certainly an overestimate of heritability in the narrow sense. Since it is largely based upon twin comparisons, it will include nonadditive genetic variance and possibly some variance due to common environmental effects. In addition, it is based upon correlations which have been adjusted for test reliability and thus is an overestimate of h_w^2 based upon single records. Of course, data from members of the Afro-American group are also necessary to obtain a valid estimate of h_w^2. Because of the uncertainty inherent in the estimate of h_w^2, three possible values will be considered: .4; .6 and .8.

From the reported difference in average IQ between the two groups (15 points) and the standard deviation within (also assumed to be 15 points), it is possible to obtain an estimate of t. Assuming that the group means are known with exactness so that two degrees of freedom are associated with the between-group sum of squares, an estimate of $t = .20$ is obtained.

Unfortunately, no valid estimate of r is available. In their genetic analysis of morbidity data obtained from the major racial groups of Hawaii, Morton et al. (1967) estimated that the inbreeding coefficient was .0009 for major races. With low levels of inbreeding, r is approximately twice the coefficient of inbreeding; thus, for morbidity data, r may be as low as .002. However, it seems likely that such data from the major races of Hawaii are not at all comparable to IQ data from mainland Afro-Americans and Caucasians.

Various possible values of h_t^2 are tabulated in Table 1 as a function of h_w^2 and r. In these calculations, it was assumed that $t = .20$. Dashes

TABLE 1. Values of h_t^2 as a Function of h_w^2 and r, Assuming $t = .20$

					r							
h_w^2	.001	.002	.004	.008	.01	.02	.04	.06	.08	.10	.20	.30
.4	.002	.003	.006	.013	.016	.03	.07	.10	.14	.18	.40	.69
.6	.002	.005	.010	.019	.024	.05	.10	.15	.21	.27	.60	—
.8	.003	.006	.013	.026	.032	.07	.13	.20	.28	.36	.80	—

in the second and third rows indicate that the maximum value of r must be less than .3 when $t = .20$ and $h_w^2 = .6$ or .8.

From Table 1, it may be seen that if r were as low as .002 (corresponding to that with morbidity data in Hawaii) and if h_w^2 were about .6, h_t^2 would be approximately equal to .005. If this were the case, of the reported 15-point IQ difference between Afro-Americans and Caucasians, less than .1 IQ point would be heritable. However, since no valid estimate of r exists for IQ data, it is impossible to choose a particular value of h_t^2 at this time. Nevertheless, it is abundantly clear from Table 1 that a high within-racial heritability by no means implies a highly heritable racial difference.

Quantitative Aspects of Environmental Determination

As indicated previously, in quantitative genetic theory the genotype is assumed to confer a certain value on an individual, whereas the environment causes a deviation from this value in one direction or the other. Environmental variance is thus a source of error which the experimenter attempts to minimize. Although the principles and techniques of quantitative genetics are directly applicable to the study of behavioral characters in laboratory and domestic animals, some modification of the usual quantitative genetic model may be useful for human behavioral genetics.

Unlike the researcher who studies behavior in laboratory animals, the human behavioral geneticist has little or no direct control over the environment in which his subjects develop. As a consequence, variance in human behavioral characters due to nongenetic causes is not simply a manifestation of random error. On the contrary, some portion of this variance is due to measurable environmental effects, which in principle are controllable. Of course, a portion of this environmental variance is caused by uncontrollable factors such as errors of measurement or other intangible effects.

The relative importance of controllable environmental factors or the proportion of the variance in human behavioral characters due to measured environmental effects is of both theoretical and practical interest. The objective of this section is to present an extended model and to consider some possible applications.

Theory. The following is a simple extension of the usual quantitative genetic model:

$$P = G + C + E, \tag{10}$$

where P is the phenotypic value of an individual, G is the genotypic value, C is the "environmental value," due to measured environmental effects, and E is a positive or negative deviation caused by unmeasured, nongenetic factors. In principle, if the system were completely understood, all environmental effects would contribute to C; thus, the distinction between C and E is a function of the state of knowledge which exists at any given time. When G, C, and E are uncorrelated and when no genotype–environmental interactions exist, the phenotypic variance (V_P) may thus be partitioned as follows:

$$V_P = V_G + V_C + V_E. \tag{11}$$

The extended model permits the formulation of a new population parameter, analogous to heritability, with both descriptive and predictive properties. Let

$$c^2 = V_C/V_P, \tag{12}$$

where c^2 is the "coefficient of environmental determination" and represents the proportion of the total variance due to measured environmental effects. As shown below, c^2 is predictive since it is equivalent to the regression of the environmental value on the phenotypic value. The covariance of the environmental value and the phenotypic value, $\text{Cov}(CP)$, is as follows:

$$\begin{aligned}\text{Cov}(CP) &= \text{Cov}(C)(G + C + E) \\ &= \text{Cov}(CG) + \text{Cov}(CC) + \text{Cov}(CE).\end{aligned} \tag{13}$$

When G, C, and E are uncorrelated, $\text{Cov}(CG) = \text{Cov}(CE) = 0$. Thus, $\text{Cov}(CP) = \text{Cov}(CC) = V_C$, that is, the covariance of C and P is equal to the variance due to C. The regression of C on P, b_{CP}, is as follows:

$$b_{CP} = \frac{\text{Cov}(CP)}{V_P} = \frac{V_C}{V_P} = c^2; \tag{14}$$

thus, the regression of the environmental value on the phenotypic value is equivalent to the coefficient of environmental determination.

In addition, the correlation between C and P, r_{CP}, is equal to the square root of the coefficient of environmental determination:

$$r_{CP} = b_{CP} \frac{\sigma_P}{\sigma_C} = c^2 \frac{1}{c} = c. \qquad (15)$$

It is important to note that the symbols C and c^2 have been used previously with different meanings (see Lerner, 1958, p. 54).

Application. Since $c^2 = b_{CP}$, the phenotypic value may be used as an index of the environment in which an individual developed. The expected environmental value ($\hat{C}$) may be estimated as follows:

$$\hat{C} = b_{CP}(P) = c^2(P), \qquad (16)$$

where P is the phenotypic value of an individual expressed as a deviation from the population mean.

The mean phenotypic value of individuals from an unmeasured population may be estimated from the properties of the normal distribution. The mean phenotypic value of individuals in a truncated portion of the normal curve should deviate from the population mean by $(z/p)\sigma_P$ units, where z is the height of the ordinate at the point of truncation of the normal curve, p is the proportion of the population in the truncated portion, and σ_P is the phenotypic standard deviation. Values of z for corresponding values of p may be found in various statistical tables (see Fisher & Yates, 1963, Tables II and II.I). For example, let us assume that an intelligence test is administered to a large, normally distributed population. The mean IQ score (phenotypic value) of individuals in the upper .01% of the population should be $(z/p)\sigma_P = (.0004/.0001)(15) = 60$ IQ points above the population mean. Three major factors are responsible for the scores of these individuals: (1) their heredity, (2) measured environmental effects, and (3) random environmental effects. The expected environmental value of individuals which rank in the upper .01% of the population is equal to $c^2(60)$.

The coefficient of environmental determination may also be used to predict the change that may occur in a population when offspring develop in a "selected" environment, that is, in an environment in which measured nongenetic effects are completely controlled. For example, let us assume that a random sample of children was reared in the same measured environment as individuals in the upper .01% of the population of the previous generation. Since these children were chosen at random, the expected phenotypic value would equal the expected environmental value of individuals in the upper .01% of the population. Therefore, the mean IQ score of these individuals should average $(c^2)(60)$ above the mean of the previous generation. Of course, unlike genetic selection, this new environment would have to be maintained in order to sustain this change. Although no estimate of c^2 is available, let us assume for illustrative purposes that the heritability (h^2) of performance on this test is .5 and that c^2 is .25 (the remaining 25% of the variance being due to both nonadditive gene effects and to unknown environmental causes). Therefore, children reared in the measured environment of individuals in the upper .01% of the population would be expected to score $(c^2)(60) = (.25)(60) = 15$ IQ points above the overall mean of the previous generation.

The effects of environmental selection on individuals which are not randomly selected from a population may also be estimated. In such prediction equations, the genotype, as well as the environment, must be considered. The expected phenotypic value ($\hat{P}$) is merely equal to the sum of the expected additive genetic value ($\hat{A}$) and the expected environmental value ($\hat{C}$), since $\hat{E} = 0$. If the population were subdivided into different racial groups, the estimate of A would be based upon the deviation of the phenotypic value of the individual from the mean of its group (P_w) and the deviation of the group mean from the population mean (P_f), each weighted according to its respective heritability. An analogous c_w^2 and c_f^2 could also be formulated. However, for the sake of simplicity, it shall be assumed that the population is not subdivided into such groups.

Let us, for example, consider the effect of environmental selection on the performance of children from "culturally disadvantaged" homes, where the average IQ test scores of the parents is 20 points below the mean. If the children were reared under the same measured environmental conditions as the parents, they would be expected to average $b_{AP}(P) + b_{CP}(P) = (h^2 + c^2)(P) = (.50 + .25)(-20) = -15$, or 15 IQ points below the mean of the population. If, however, these children were allowed to develop under average environmental conditions, the expected environmental value would be zero; hence, they would be expected to score only $(h^2)(-20) = (.5)(-20) = -10$, or 10 IQ points below the mean. But what would be the expected performance of these children if they were reared under an enriched environment, that is, the measured environment of individuals which scored in the upper .01% of the population? These children should average $(h^2)(-20) + (c^2)(60) = (.50)(-20) + (.25)(60) =$ IQ points above the mean of the population.

Discussion. A simple extension of the usual quantitative genetic model permits the formulation of a new population parameter, the coefficient of environmental determination. This parameter, symbolized c^2, has both descriptive and predictive roles: It indicates the relative importance of measured environmental effects as causes of individual differences in a population and also may be used to predict the change that will occur when a population develops in a selected environment. Such predictions may be of doubtful value, due to the impossible requisite of complete environmental control. However, such estimates may suggest the feasibility of changing the mean phenotypic value of a segment of the population by the control of existing environmental variation. If c^2 were large, much change could result from control of existing environmental variation. If c^2 were small, little change would result from such control.

However, it is important to note that a low value for c^2 would not necessarily imply that deficiencies could not be compensated by environmental factors. A low c^2 would merely indicate that measured environmental factors were not important causes of individual differences in the population. Thus, although control of measured environmental effects would not result in a substantial change in the mean phenotypic value when c^2 is low, special environmental regimes (for example, therapy, diets, special education, etc.) might still be effective. It is also important to recall that c^2 is a population parameter which, like h^2, may vary from character to character in the same population, from population to population for the same character, and from time to time for the same character in the same population.

No valid estimate of c^2 is currently available. In fact, even available estimates of h^2 for behavioral characters in human populations are of doubtful validity. Human relatives share a common environment. Therefore, the resemblance between relatives in the human population will almost certainly result in overestimates of h^2 unless the environmental contribution to the similarity is removed. However, it would seem that valid estimates of both h^2 and c^2 are obtainable for human behavioral characters. Such estimates could be obtained from large-scale family studies where behavioral scores on a large number of parents and their children are assessed and where the environment in which the children developed is indexed as accurately as possible.

Summary

Although quantitative genetic theory was primarily developed for application to animal and plant breeding, its concepts and methods are applicable to important social issues. The heritable nature of group differences may be expressed as a function of the within-group heritability. Application of IQ data demonstrates that a high within-group heritability does not imply that the observed difference between the means of Afro-Americans and Caucasians is also highly heritable.

The quantitative genetic model may also be extended to include measured environmental effects. This extended model facilitates the formulation of a new population parameter, the coefficient of environmental determination, which is defined as the proportion of the total variance for some character in a population due to measured environmental effects. This variance ratio, analogous to heritability, has both descriptive and predictive properties. It may be utilized as an index of the value of the environment in which an individual developed, and to predict the effects of controlling environmental variation in a population.

Acknowledgments

I thank C. S. Chung, J. F. Crow, D. W. Fulker, J. C. Loehlin, G. E. McClearn and N. E. Morton for their assistance, and E. R. Dempster for providing the germinal idea from which the derivation of Eqs. (8) and (9) developed. Supported in part by a Faculty Fellowship from the Council on Research and Creative Work, University of Colorado.

References

DeFRIES, J. C. (1967). Quantitative genetics and behavior: Overview and perspective. *In:* "Behavior-Genetic Analysis" (J. Hirsch ed.), pp. 322–339. McGraw-Hill, New York.
FALCONER, D. S. (1960). "Introduction to Quantitative Genetics." Oliver and Boyd, Edinburgh.
FISHER, R. A., and YATES, F. (1963). "Statistical Tables for Biological, Argicultural and Medical research." Hafner, New York.
JENSEN, A. R. (1969). "Environment, Heredity and Intelligence." Reprint Ser. No. 2, compiled from the *Harvard Educational Review*, Cambridge, pp. 1–123.
JENSEN, A. R. (1970). IQ's of identical twins reared apart. *Behav. Genet.* **1**, 133–148.
JINKS, J. L., and FULKER, D. W. (1970). Comparison of the biometrical genetical, MAVA, and classical approaches to the analysis of human behavior. *Psychol. Bull.* **73**, 311–349.
LERNER, I. M. (1958). "The Genetic Basis of Selection." Wiley, New York.
LEWONTIN, R. C. (1970). Race and intelligence. *Bull. At. Sci.* **26**, 2–8.
LUSH, J. L. (1948). *The genetics of populations.* Ames, Iowa. Mimeographed notes.
LUSH, J. L. (1949). Heritability of quantitative characters in farm animals. *Hereditas*, Suppl. Vol., 356–375.
MORTON, N. E., CHUNG, C. S., and MI, M. P. (1967). "Genetics of interracial crosses in Hawaii, Monographs in Human Genetics 3." S. Karger, Basel.
ROBERTS, R. C. (1967). Some concepts and methods in quantitative genetics. *In:* "Behavior-Genetic Analysis" (J. Hirsch ed.), pp. 214–257. McGraw-Hill, New York.

DISCUSSION

JOHN L. FULLER

State University of New York
Binghampton, New York

Dr. DeFries, in the introductory session of the symposium, raised many of the issues which concerned the group for the remainder of the sessions. Again and again we returned to the validity of the assumptions made in computation of heritabilities of IQ or other attributes, to the genetic implications of racial diversity, and to the consequences of gene-environment correlations and interactions. The following paragraphs represent a personal reaction to these matters.

Heritability: Broad or Narrow. Geneticists make a distinction between heritability in the "broad" and "narrow" senses. The former refers to the proportion of phenotypic variance attributable to variations in genotype. Narrow heritability is the proportion of phenotypic variance predictable from the phenotypes of parents (or other relatives) and may be substantially lower than broad heritability. Heritability in the narrow sense is the important one for evolution, since it expresses the effects of differential birth rates upon traits of interest in later generations. But educators, trying to fit educational procedures to a particular child or group of children, are concerned with all the biological variability in an individual. Special strengths or weaknesses arising from dominance and epistasis (and thus not contributing fully to narrow heritability) must be considered even though they will not be transmitted in predictable fashion to the child's descendants. Therefore, the argument runs, it is heritability in the broad sense which should concern educators. The flaw in this argument is that the genetic or environmental origin of a particular phenotype (for example, low IQ) is not usually the critical issue in designing an appropriate teaching procedure for an individual. The value of heritability determination is not for guidance regarding an individual, but as a clue to the way in which a population will change under selective pressures. Furthermore, narrow heritability has the greater potential for helping us understand an individual through knowledge of his relatives. Phenotypic correlations between relatives are attributable to communalities of both environments and genes. In partitioning global terms such as heredity and environment into specific components, narrow heritability should be used. The added genetic effects which are included in broad heritability are real enough, but they produce noise in the system that reduces the precision of predictions regarding transmission of characteristics.

Habitat Choice. There are other problems related to the allocation of sources of variance to the genetic or environmental categories. Quantitative geneticists are alert to the difficulties encountered when genotype and environment are correlated. In the laboratory, we can separate these influences experimentally; in human societies we do the best we can by studying foster children, separated twins and other individuals whose biological and experiential backgrounds are dissociated. But organisms sometimes choose their habitats, and the nature of these choices may be functions of their genotype. A choice, once made, has consequences for the further development of behavior patterns, which are functions of the nature of the selected habitat. Should these effects be classified as genetic or environmental?

It seems logical to call them genetic, since the original determination was dependent upon genotype, although without an opportunity for choice, that particular mode of development might have been closed. Furthermore, to complicate things, an environment exactly the same as that chosen might be imposed, and its effects would then turn up in the environmental section of a biometrical analysis. Or, the imposition of a particular habitat may be arranged by parents, who originally selected it on the basis of free choice. Suppose that the making of the choice has high heritability. The offspring will have the parental genes which could guide them to same choice, but since their surroundings are imposed they cannot demonstrate the trait. One can imagine additional complications and deduce their effects upon heritability estimates, but the main point is this: The distinction between environment and genetic sources of phenotypic variance is not as sharp as we would like, particularly when a developing phenotype such as behavior is considered.

Discussion

Genotype–Environment Interaction. That heritability is an attribute of a trait in a particular population living under specified conditions is a generally accepted precept. Nevertheless, one hopes that estimates of heritability will be sufficiently robust to have more general applicability. By and large, biometrical geneticists find this to be so. DeFries developed his equations without introducing a major term for genotype–environment interactions. In a comprehensive review, Jinks and Fulker (1970) found evidence for interactions, but for most measures their magnitude did not preclude the use of relatively simple biometrical models.

This situation is surprising, for experimental behavior geneticists who look for genotype–environment interactions generally find them. For example, Henderson (1967) observed the effects of three levels of stimulation in early life upon open field behavior of all possible crosses between four inbred strains of mice. Results of his treatment were highly variable among the 16 different genotypes, no one of which conformed to the group average. The implication for basing psychological laws applicable to all individuals upon observations of a single genotype or upon means from genetically heterogeneous subjects is obvious. Fuller (1967) reported that the effects of experiential deprivation in two breeds of dogs upon locomotor activity and intensity of social interaction were opposite in direction. If these examples are typical (see Chapter 8 by Erlenmeyer-Kimling, Table 11), as they seem to be, one wonders why quantitative genetic analysis can come out fairly well so often without requiring much correction for interaction.

There are two possible explanations. DeFries suggested that the interactions are complex and nonlinear and that they simply turn up as error variance. I should like to propose another possibility. The variance term which experimental psychologists label as genotype–environment interaction is really a phenotype–environment interaction. In Henderson's and Fuller's experiments, controlled breeding of subjects is a method for obtaining groups differing in nervous, endocrine, and other characteristics. In inbred and selected lines, these may correspond to particular genotypes so that either of the labels is appropriate. But in natural populations, there may be many independent genetic ways of specifying the same phenotype, all with unique interactions which on the average cancel out. This concept is compatible with DeFries's view, but centers on the physiological processes involved rather than the mathematical model.

Coefficient of Environmental Determination (c^2). DeFries's suggestion that we begin to be more explicit concerning the sources of environmental variance is excellent. There is considerable folklore dealing with the truly essential requirements for converting a neonate into an effective adult participant in the life of its species, but we still find it difficult to explain the extremely intelligent child from a culturally deprived background, or the dullard among the offspring of a pair of geniuses. I believe we shall have to develop new scales for measuring environment if DeFries's c^2 is to become a useful device for behavior genetic analysis. Such scales must pay attention not only to physical features of the environment, but also to the temporal characteristics of their availability to an individual. The experimental literature is rich in examples of sensitive developmental periods for the maximal effect of certain kinds of stimulation. Imprinting of the following response in birds is a classic instance. The spacing and duration of experiences may also be critical.

Fuller (1967) found that a few minutes of contact with other puppies and with humans counteracted most of the effects of enforced isolation of dogs aged four to fifteen weeks. Social experience was necessary to socialization, but apparently ordinary rearing practice in his laboratory provided more stimulation than was really needed for adequate learning (or elicitation) of common social responses. It will not be easy to develop scales which give proper weight to sequential features of life histories, but they would have practical application if DeFries's hypothetical illustration of the effect of rearing children in a "selected" environment is to be realized. Simply as a matter of economics, society will have to allocate its resources for environmental improvement to those features which can be demonstrated to be most effective in promoting healthy development. We need to explore a field which might be called developmental psychogenetics.

Heritability between and within Groups. Finally, we come to the central issue of the DeFries paper, the relationship between heritabilities within groups to the heritability of group differences. His exposition is clear, and, given his assumptions, his conclusions are justifiable. But I have reservations concerning the applicability of a model based on heritabilities within and between families of a single population to a comparison involving populations whose genetic structures differ to an unknown degree. His illustration of the use of inbreeding coefficient to estimate r, the coefficient of relationship, very likely underestimates the actual correlation between gene pools of Caucasian and Afro-Americans, which has been placed at approximately .30 in several studies (Glass and Li, 1953). Other estimates vary rather widely, depending upon the genes chosen for study. Naturally we should like to know what r is for those genes that affect intelligence and similar adaptive traits, but it is hopeless to make direct measurements with such polygenic systems.

Despite these reservations, I regard Dr. DeFries's contribution as an important step toward dispelling some of the fog which obscures meaningful discussion of the genetics of intelligence. We need clear thinking in this area, for what we believe concerning the relationship between genes and behavior affects decisions regarding the degree to which education should be individualized. If children are blank slates whose talents and skills are determined solely by the experiences to which they are exposed, no injustice is done by assigning children at random to the various kinds of schooling (and other experiences) which will give a well-proportioned mix of laborers, scientists, politicians, etc., to staff our highly organized technological society. Parental pride might be damaged if one's child were assigned by lot to a low-status position, but if there are no innate strengths to be encouraged, neither are there any to be frustrated, so that no injustice is done the child. Conversely, if genes do make a difference, an attempt should be made to allow them expression by permitting individual choice of environments with only the constraints imposed by granting similar opportunities to others. The results should be beneficial to society and to individuals.

My point in bringing up these issues, admittedly overly simplified, is that an emphasis on the importance of heredity in psychological variation has been equated with a conservative attitude toward social and political issues. Depreciation, or even denial, of significant genetic effects has been considered a liberal attitude. It is possible to argue exactly the opposite, as I have done. I suspect, however, that the correlation between sociopolitical attitudes and beliefs regarding the heritability of psychological traits is spurious. The important point is not whether ethnic or social groups differ in genes that affect behavior; it is how we propose to help individual children develop maximally. Wisely used, knowledge of behavior genetics can assist in this objective.

References

FULLER, J. L. (1967). Experiential deprivation and later behavior. *Science,* **158,** 1645–1652.
GLASS, B., and LI, C. C. (1953). The dynamics of racial intermixture and analysis based on the American Negro. *Amer. J. Hum. Genet.,* **5,** 1–20.
HENDERSON, N. D. (1967). Prior treatment effects on open field behaviour of mice—a genetic analysis. *Anim. Behav.* **15,** 364–376.
JINKS, J. L., and FULKER, D. W. (1970). Comparison of the biometrical genetical, MAVA, and classical approaches to the analysis of human behavior. *Psychol. Bull.* **73,** 311–349.

Reply to Professor Fuller: J. C. DeFries

In his thoughtful discussion of my paper, Professor Fuller raises two issues which deserve further comment. The first concerns the assumption of the absence of a significant genotype–environment correlation in my derivations. Such simplifying assumptions were made for purposes of clarity of presentation, not because quantitative genetic theory is incapable of handling more general cases.

As indicated in my paper, it may be shown that the between-group heritability (h_t^2) is equivalent to the regression of the mean additive genetic value of a group (A_t) on its mean phenotypic value (P_t) in the absence of a genotype–environment correlation (r_{AE}). When r_{AE} is nonzero, this regression ($b_{A_t P_t}$) is as follows;

$$b_{A_t P_t} = \frac{\text{Cov } A_t P_t}{V_{P_t}} = \frac{\text{Cov } A_t(A_t + D_t + I_t + E_t)}{V_{P_t}}$$

$$= \frac{V_{A_t}}{V_{P_t}} + \frac{\text{Cov } A_t E_t}{V_{P_t}} = h_t^2 + \frac{(V_{A_t} V_{E_t})^{1/2}}{V_{P_t}} r_{A_t E_t}$$

$$= h_t^2 + h_t e_t r_{A_t E_t}.$$

Thus, when r_{AE} is nonzero, prediction is still possible; however, more information is required. If r_{AE} is positive, prediction using only h_t^2 would yield an *underestimate* of the extent to which the observed group difference is heritable (h_t^2 is less than $b_{A_t P_t}$ when r_{AE} is positive). If r_{AE} is negative, h_t^2 will overestimate the heritable nature of the group difference. If r_{AE} were negative and large, the expected mean additive genetic value of the group with the lower mean phenotypic value might actually exceed that of the group with the higher mean phenotypic value.

Professor Fuller also raises the point that 20–30% of Afro-American genes are of Caucasian origin. The additive genetic correlation of members of the same group (r) is a function of the group difference, that is, the smaller the genetic difference between groups, the lower the corresponding value of r. Thus, the sharing of many of the same genes by Afro-Americans and Caucasians should result in a *lower* value of r, not a higher value.

COMMENT

ARTHUR R. JENSEN

*University of California
Berkeley, California*

DeFries's formulation of the logical relationship between heritability within groups (h_w^2) and heritability of between-group means (h_t^2) is a valuable conceptual contribution. It puts an end to the mistaken notion that there is no connection whatever between within-groups and between-groups heritability.

While the basic formulas [Eqs. (8) and (9)] in DeFries's paper appear to be correct, I believe some of the things he says about the formulas are in error, either partially or wholly.

First, I regard the formula as theoretically important, because it shows precisely the relationship between h_t^2 and h_w^2, given the values of t and r. But it is empirically useless, at the present state of our knowledge, because we have no estimates of r for any traits of interest, least of all for intelligence. In fact, if we knew r (the genetic intraclass correlation), the question of h_t^2 would be trivial. If we knew r, we would already know what we really wanted to know in the first place. If $r > 0$, there is a genetic mean difference between the groups. But we have no estimate of r for intelligence for any pair of populations we might want to compare, and so the formula is empirically inapplicable. The figure for r used by DeFries for comparing populations in IQ is most surely wrong, being based on the inbreeding coefficient estimated from morbidity data. Thus, r is not independent of the particular trait in question, nor of the trait variance within populations or the mean difference between the two populations being compared. The inbreeding coefficient seems to me irrelevent here. Assortative mating is different for various traits, and some overall estimate of degree of inbreeding in the population simply will not work. It is possible to have a high degree of genetic correlation among individuals for a particular trait, even though a coefficient of inbreeding would be close to zero. Imagine a group of Tibetans and a group of Germans in which pairs of individuals, one from each group, were perfectly matched for height, for example. The genetic correlation among these Tibetans and Germans would be very high, but the coefficient of inbreeding would be extremely low. So the deductions DeFries makes about the heritability of the Negro–white mean IQ difference is quite meaningless, based, as it is, on an inappropriate estimate of r.

Second, I do not see why DeFries insists upon h^2 in the narrow sense. The difference between group means, in so far as it is genetic, does not exclude nonadditive gene effects; dominance and epistasis can constitute a part of the between-groups variance. If so, a better estimate of h^2 for the formula is something lying between the limits of narrow and broad heritability. Since the formula is only theoretical at present, it doesn't really matter, except that we might as well try to be as conceptually correct as possible.

Third, I am not sure about DeFries's stipulation that the estimate of h^2 as used in his formula should not be corrected for attenuation (that is, unreliability of measurement). I say this because we are estimating the heritability of a difference between two *means*, and we know that if the mean is based on a sizable N it is unaffected by errors of measurement. (The within-group variance of course is affected by unreliability.) So I would suggest that h_f^2 be corrected for attenuation. At what stage of the game this should be done algebraically makes no difference. I think it best to do it as a final step, by taking h_f^2/r_{tt}, where r_{tt} is the reliability of the test or measurement, and then report both the corrected and uncorrected values of h_f^2.

Reply to Professor Jensen: J. C. DeFries

The primary objective of the first section of my paper, to which all of Professor Jensen's comments have been directed, was to demonstrate that the heritability of between-group means (h_f^2) may be very low, in spite of a high within-group heritability (h_w^2). It appears that Professor Jensen now agrees with this conclusion.

In my paper it was shown that h_t^2 is a function of three variables: h_w^2, r and t, where r and t are the additive genetic and phenotypic correlations (intraclass) of members of the same group. (For the purposes of this derivation, it was assumed that genotype–environment correlations were absent.) Thus, I must agree with Professor Jensen that if r were known for IQ, we would have the answer to our problem, since we already know something about h_w^2 and t. In general, however, knowing r does not make the question of h_t^2 trivial; r by itself is no more informative about the heritable nature of group differences than h_w^2 by itself.

The inbreeding coefficient to which I referred was that used by Morton et al. (1967, p. 130). This inbreeding coefficient may be estimated for different characters, is a function of r, and thus may be important in the empirical application of Eqs. (8) and (9).

I do not follow Professor Jensen's argument about height in Tibetans and Germans. If their mean genotypic values for height are equal, the corresponding r will be zero. Nevertheless, as indicated in my paper, application of the r obtained from morbidity data was solely for illustrative purposes.

I have insisted upon the use of heritability in the narrow sense for three reasons: First, h^2 could be used to predict the performance of children, based upon their parents' performance. Heritability in its narrow sense would be appropriate for such prediction, since the covariance of parents and offspring contains one-half of the additive genetic variance, none of the dominance variance, and only a relatively small fraction of the epistatic variance. Second, reports of a high heritability for IQ in Caucasian populations have led some people to advocate eugenic programs. However, the response so selective breeding is a function of heritability in the narrow sense. Thus, it is important not to advocate eugenic measures based upon inappropriate estimates of h^2. Finally, heritability in the narrow sense is used in Eqs. (8) and (9), due to the definition of r (the *additive* genetic correlation among members of the same group). To be consistent, the heritability employed in these equations should reflect only that part of the genetic variance which is additive. It is certainly possible to formulate h_t^2 as a function of heritability in the broad sense and r_G, where r_G is the correlation among members of the same group due to additive genetic values, dominance deviations and epistatic interactions. However, it would be empirically impossible to obtain estimates of r_G, even for close relatives, since the gene frequency and type of gene action would have to be known for every locus which influenced the character under investigation.

Heritability is both descriptive and predictive. I have urged that h^2 not be corrected for attenuation because of the predictive validity of the uncorrected estimates.

COMMENT

PETER L. WORKMAN

University of Massachusetts
Amherst, Massachusetts

It is important to remember that estimates of heritability are defined for a specific population in a specific environment and therefore, they represent an average over the individuals (and their particular environments) who make up that population. Thus, there is no way we can use estimates of heritability to predict how much any given individual will be affected by a change in his environment. That is, some individuals may have a phenotypic value highly representative or highly unrepresentative of their genotypic potential. Moreover, there may be a few relatively rare environments which permit extensive modifications of an individual's performance. If most individuals in a population were able to develop in such environments then, possibly, the population heritability would become quite low. Thus, if the heritability of a trait is quite high, since we have no inference about the possible range of environments, we cannot assume that a search for an optimal environment will fail. In particular, the observation that the heritability of IQ seems to be quite high should not preclude our attempt to find an educational environment in which the heritability of IQ would be low and in which we could make substantial improvements in the achievement levels of children with reportedly low IQ scores.

Chapter 3   Qualitative Aspects of Genetics and Environment in the Determination of Behavior

CLAUDINE PETIT
Université Paris 7
Paris, France

The quarrel between "innate" and "acquired" has been going on for years. The demonstration that gene action is variable, depending upon the conditions of gene regulation, has allowed better understanding in the fields of morphology and physiology. However, the problem must be viewed from a different perspective when it comes to behavior, for here the importance of learning makes genetic analyses particularly difficult.

From a geneticist's viewpoint, this is a false quarrel. Any trait, fundamentally, is genetic: Whatever it may be, either the weight of a cow, the performance of a racehorse, or the sexual advantage of a *Drosophila,* the range of expression of the character is genetically determined. Genetic variability alone reaches the minimum value of 40% of loci in a population, as shown by the works of Dobzhansky and his collaborators on genetic load (Dobzhansky, 1957) and that of Lewontin on enzymatic polymorphism (Hubby and Lewontin, 1966; Lewontin and Hubby, 1966). However, nongenetic factors such as the internal, physical or external biological environment may interact with the genetic variability. When applied at the right time in the life of an individual, including embryonic development, these factors can significantly influence phenotypic expression. The problem is not only to evaluate the *genotype–environment* interaction, but the *genotypes–environment* interaction, when unknown numbers of genotypes are involved. Since many unknown genotypes interact with environment, mathematical or experimental analyses of the "nature–nurture" relation become extremely difficult.

Mankind, whose behavior is subjected to all types of internal, physical and external biological environments, might appear to be the best species for this kind of study, especially since sound knowledge of these interactions would be important in education. However, culture must be added to the three other environments, and is so tightly intertwined with them that they are practically inseparable without experimentation. Furthermore, when man is involved, the most serious scientist finds his reason hampered by all sorts of emotional factors. Last of all, man has so many genes that the problem reaches staggering proportions.

Since the topic, "Qualitative Aspects of Genetics and Environment in the Determination of Behavior," encompasses nearly all behavioral science, I will only give its broad outlines and base my arguments on some well-analyzed examples. Three points will be considered:

1. Genetic determination of some types of behavior and the influence of physical environment on genotype;
2. The influence of physiological factors and internal environment on gene action;
3. The influence of external biological and social environment: imprinting, conditioning, learning, and a newly-recognized phenomenon, which may play an important part in evolution, the advantage of the rare type.

Genetic Determinism and the Influence of Physical Environment on Some Types of Behavior

Behavior is sometimes determined by responses of the organism to external stimuli. These signals are sent out by alien or conspecific individuals, by objects or physical phenomena; they generally release simple behavior such as attraction or flight. They are perceived by sensory receptors and integrated by the central nervous system.

The easiest cases to analyze are those that involve Mendelian gene-dependent types of behavior. Such behaviors, generally abnormal, are determined by metabolic dysfunctions. Mayer *et al.* (1951), and Fuller and Jacoby (1955), for example, have observed that the recessive gene responsible for obesity in some strains of mice causes the selection of fattening food. Schizophrenic behavior, possibly the result of metabolic dysfunction that involves perception (Huxley *et al.*, 1964), is probably determined in many cases by a single dominant gene of low penetrance (Slater, 1958). Circling and choreic behavior, well known in mammals and birds, results from nervous system injuries caused by lethal Mendelian genes. In mice, some audiogenic seizures are determined by a recessive gene (Lehman and Boesiger, 1964). In *Drosophila*, the *Hyperkinetic* genes (Hk^{1p} or Hk^{2p}) produce a leg-shaking action in response to ether vapor (Kaplan *et al.*, 1971).

It is interesting to note that, on the other hand, apparently simple types of behavior, such as taxis, are determined polygenically. The difference may involve the fact that abnormal behaviors are pathological, while taxis have a serious adaptive value for the species and require precise sensillae and central nervous system integration. Such complex and different elements are unlikely to be controlled by a single gene.

Phototaxis was studied in *Drosophila melanogaster* by Hirsch and Tryon (1956) and in *Drosophila pseudoobscura* by Spassky and Dobzhansky (1967). A multiple-choice apparatus was used to allow the selection of positive or negative phototaxis in inbred or outbred strains. The responses indicated polygenic determination. Geotaxis was studied in the same species using the same technique. Results were identical, with the selection of strains having positive or negative geotaxis (Hirsch and Boudreau, 1958; Ehrman *et al.* (1965). Medioni (1961) showed different orientations in *Drosophila melanogaster* strains of different geographic origins. Erlenmeyer-Kimling and Hirsch (1961), working with marked chromosomes, made strains homozygous for specific chromosomes; they demonstrated that genes of the X chromosome control positive geotaxis and genes of the third chromosome, negative geotaxis. In another strain, the second chromosome was shown to bear factors which control positive geotaxis. Thus, all three chromosome pairs contain genes controlling such behavior.

However, these taxis are not independent of environment. Environment may alter phototaxis. The beetle *Blastophagus pinniperda* L. is positively phototatic in spring, at temperatures between 10° and 35°C, but negatively phototactic below and above this temperature (Perttunen, 1958, 1959). In the fall, the temperature range is reduced to 20°–30°C. Thus, a temperature of 15° induces positive phototaxis in spring, and negative phototaxis in autumn, when the animal starts looking for winter shelter. In the weevil, *Calandra granaria,* where genetic changes in phototaxis were achieved by selection, Richards (1951) and Perttunen (1963) found phototaxis more and more positive as dryness increased. There is a good chance that this kind of change in the behavior through physical environmental factors is mediated by hormonal changes, the action of which will be considered later.

The best examples of detailed analysis of the interaction of genetics and environment are provided by courtship and sex behavior. The importance of sexual selection, discovered a century ago by Darwin (1871), is now generally accepted: Sexual selection contributes to speciation and plays a prominent part in the maintenance of genetic variability among populations. Complex sexual behaviors have been studied in animals as different as mice, guinea pigs, *Phatypoecilus,* and *Drosophila,* whose different species provide a tremendous amount of genetic, behavioral and sensory information.

The description of courtship of *Drosophila melanogaster* by Bastock (1956) is classical and may be used as a basis to study any courtship of *Drosophila.* Sexual selection in *Drosophila* was first demonstrated between geographical origins, as examples of incipient isolation. Thus, Dobzhansky and Streisinger (1944) showed a north–south gradient in the vigor of *Drosophila prosaltans* males. Mayr and Dobzhansky (1945) described selective matings between strains of different geographical origins in *Drosophila pseudoobscura* and *Drosophila persimilis.* Incipient isolation occurred in some cases, for example, between *Drosophila arizonensis* and *mojavensis* (Baker, 1947), or among *Drosophila sturtevanti* individuals (Dobzhansky, 1944). Sexual selection was also demonstrated between strains differing by inversions. Brncic and Koref Santibanez (1964) studied *Drosophila pavani* and *Drosophila gaucho,* Spiess and his collaborators *Drosophila persimilis* (Spiess and Langer, 1964a,b; Spiess *et al.,* 1966; Spiess and Spiess, 1967), Dobzhansky and his collaborators a wide variety of species, including *Drosophila pseudoobscura* and *D. paulistorum* (see Petit and Ehrman, 1970).

All of these results seem to favor polygenic determination of mechanisms of sexual selection. However, some authors (see Spiess, 1970) describe sexual selection between strains that differ theoretically by only one gene. Petit (1951, 1954) demonstrated sexual selection between strains that differed only by the *Bar* or *white* gene in *Drosophila melanogaster.* Bastock (1956) studied competition between *yellow* and the wild type and demonstrated that the *yellow* male was at a disadvantage with wild-type females. Elens (1957) found the same results with *ebony,* which, in some cases, seemed to be caused by a cytoplasmic factor. But other investigations showed that sexual selection was in fact the result of interaction between the mutant gene and the residual karyotype (Petit, 1955, 1958).

Polygenic determination seems reasonable when one describes the complexity of courtship in three phases:

1. *orientation,* during which the male stays behind the female, or follows her if she starts walking;

2. *vibration,* which corresponds to a period of great agitation in the male; the male vibrates and circles the female with one wing kept horizontal and his head always turned toward her;

3. *licking,* which immediately precedes mating; during this phase, the male proceeds to lick the abdomen of the female with his proboscis, just prior to the attempt at copulation.

It seems likely that the first phase involves olfactory or visual stimulation, the second, auditory and tactile stimulation, and the third, tactile and chemical stimulation. Ablation of effectors or receptors made it possible to define the importance of the different kinds of stimuli. The importance of the antennae as stimulus receptors during courtship was first shown by Mayr (1950): antennaeless females of *Drosophila pseudoobscura* or *Drosophila persimilis* discriminated less against males of a foreign species than did normal females. In later studies, the role played by different parts of the antennae was defined by Petit (1958, 1959) and Manning (1967).

The antenna includes three segments (Fig. 1): the *scape,* the *pedicel* including Johnston's organ which receives vibrations, the *funiculus* including the olfactory pit, and the *arista.* It was therefore thought that ablation of different parts might provide some information; although many different sensillae cover the body of a fly, experimental results seem sufficiently clear. Ablations were done during narcosis necessary for sexing; virgins were left to grow older in a unisexual group and two days later, males and females were put together, without anesthesia, for 24 hours. The percentages of mating were compared with those of normal flies for the aristaectomized flies, and with those of injured flies for the antennectomized ones. It was concluded (Petit, 1958) that ablation is more harmful for the female than for the male: 54% of the antennaeless males fertilized females, while only 8% of

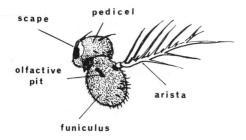

FIG. 1. The antenna of *Drosophila melanogaster.*

antennaeless females were inseminated. This implies that the receptors of the females are essentially limited to the antennae, while those of the male are found all over the body. This conclusion is in agreement with direct observations of courtship, in which the male touches the female with his legs and proboscis. It leads to the assumption that stimuli received by the male are essentially chemical and tactile, and those received by the female are auditory and perhaps olfactory. The aristaectomy, only slightly harmful for the males (74% of inseminations as compared to 88% in the controls), was significant for females where the percentage of mating fell to 43%. An explanation of the role of aristae in mating was given by Manning (1967), who proved that aristae and funiculus have to act as a unit to stimulate Johnston's organ and make the female receptive. Olfactory stimulation of the female may exist in this species, since experimental destruction of the funiculus, which includes the olfactory pit, lowers the percentage of mating to 30%. However, since this injury is quite different and the statistical comparison of test percentages only slightly significant, interpretation of these experimental results is less certain. In *Drosophila melanogaster,* vibration thus appears to be an important stimulus for the female, while the male responds to tactile and olfactory stimuli.

In a precise acoustic study of *Drosophila* courtship songs, Ewing (1969, 1970) demonstrated that the genes controlling the song patterns are located on the X chromosome, while quantitative features are controlled by autosomal genes. Tan (1946) found similar results in *Drosophila persimilis,* where sexual isolation differs when the X and second chromosomes are modified. Songs are one of the main stimuli of courtship, so studies of their patterns open the way to a more precise genetic analysis of courtship behavior.

It is conceivable that all the components of physical environment acting on vibration can change the intensity of sexual selection. Reed, *et al.* (1942) demonstrated a positive correlation between the mean frequency of vibration and temperature on four different species of *Drosophila*. In addition, vibration frequency was found to be proportional to the volume of flight muscles and size. Although temperature does influence sexual selection and isolation, one cannot definitely conclude that temperature exerts its influence via its relationship with vibration frequency; temperature changes the rate of development and, consequently, delays maturation and changes body size. Mayr and Dobzhansky (1945) demonstrated that isolation between *Drosophila pseudoobscura* and *Drosophila persimilis* is lower at 16° than at 25°C, the *D. persimilis* males being the most sensitive (Table 1). The intensity of the selection between white and wild type varies with the temperature in *Drosophila melanogaster* (Petit, 1958). The differences in mating speed found by Parsons and Kaul (1966) and Spiess *et al.* (1966)

TABLE 1. Mate Discrimination between *Drosophila pseudoobscura* and *Drosophila persimilis* at Different Temperatures[a]

$t°$	Females	Males	Homogamic		Heterogamic		x^2	Isolation index
			n	%	n	%		
$24^1/_2°$	pseudoobscura, persimilis	pseudoobscura	30	83.3	28	3.6	20.4	0.92
18°	pseudoobscura, persimilis	pseudoobscura	21	85.7	18	0.0	15.4	1.00
$16^1/_2°$	pseudoobscura, persimilis	pseudoobscura	42	92.9	40	12.5	24.2	0.76
$24^1/_2°$	pseudoobscura, persimilis	persimilis	65	93.8	64	39.1	14.6	0.41
21°	pseudoobscura, persimilis	persimilis	56	53.6	63	12.7	15.4	0.62
18°	pseudoobscura, persimilis	persimilis	21	4.8	20	55.0	8.7	−0.84
$16^1/_2°$	pseudoobscura, persimilis	persimilis	86	32.6	90	52.2	4.0	−0.23

[a] From Mayr and Dobzhansky, 1945. Reproduced by permission.

with AR and PP karyotypes[1] of *Drosophila pseudoobscura* are probably due to this factor. The vigor of the two karyotypes is the same when they are kept at 15°C. But, when they are kept at 25°C, the mating speed of PP suddenly increases. Similar results were found for *Drosophila persimilis* (Spiess, 1970). WT and KL[1] do not have the same optimal temperatures; WT mates more quickly when the temperature is low, and KL when it is high. In any case, heterosis is greater as one approaches these limits and not at the optimal temperature. This is true for all components of fitness as well as for sexual selection (Dobzhansky and Levene, 1955); it appears to be one of the aspects of genetic homeostasis.

Moisture probably influences sexual selection, as I have often observed, but to my knowledge, this question has not been the object of any systematic research. It is likely that the optimal level of humidity varies from one species to another and that moisture may change the taxis in some species.

Light is important too. Species able to mate in the dark are inhibited by complete darkness (Spieth, 1952; Spieth and Hsu, 1950). Some species, such as *Drosophila sudobscura* are unable to mate in the dark. The total amount of copulation varies with light in *Drosophila prosaltans* (Mayr and Dobzhansky, 1945). In *Drosophila pseudoobscura,* there is a negative correlation between mating ability and light intensity (Elens and Wattiaux, 1970). In *Drosophila melanogaster* some mutants, such as ebony, are sensitive to light (Rendel, 1951). The ebony males, at a disadvantage in competition with wild males in the light, are, on the contrary, at an advantage in

---

[1] AR and PP karyotypes differ by the inversion of a part of the chromosome; so do WT and KL. These are abbreviations of the name of the area where the inversions were first found. The inversions are seen very easily in the polytenic chromosomes of the salivary glands of Diptera.

the dark. A complete study of the relative importance of vibration and light was made by Grossfield (1966, 1968).

It is evident that the influence of environment on sexual selection or sexual isolation may imply important evolutionary consequences. Certain species, slightly isolated when the temperature is low, are almost completely isolated when it is high. Therefore, the territorial enlargement of two populations to warmer countries or a change in climate would produce sexual isolation between strains that formerly showed none. Moreover, the alternating advantage incurred by either of two forms, depending upon their environment, may be a way of maintaining polymorphism, the adaptive advantage of which need not be demonstrated here.

This detailed analysis of the consequences of the interaction of genotype and environment in sexual behavior on evolution does not mean that sexual behavior is the only part of behavior liable to evolutionary implications. These results are but examples, and different aspects of behavior that certainly have a primordial adaptive role need to be looked for carefully.

Influence of Physiological Factors, Including Hormones, on Genes

Environmental factors may change behavior via physiological mechanisms. Physiological factors that are dependent upon growth conditions, age, or composition of blood are well known in the determination of behavior, especially that of sexual behavior. In *Drosophila,* the components of sexual behavior (Faugères *et al.* 1971) include the "athletic ability" of Maynard Smith (1956), and male vigor, evaluated as the greater ability of heterozygous males to inseminate more females when competing with homozygous males (Boesiger, 1958, 1962). Moreover, factors of learning can influence sexual behavior; but I shall devote my attention to those later. It is not surprising that growth conditions have an important influence on sexual activity and that flies reared in overcrowded conditions are at a disadvantage when competing with well-fed flies (Petit, 1958; Robertson, 1963; Kaul and Parsons, 1965; Spiess and Spiess, 1969a). Lack of yeast during larval growth and adult maturation delays mating and lowers receptivity in females (Manning, 1967; Spiess and Spiess, 1969a).

Age is another factor that influences sexual activity. All drosophilists know that sexual maturity does not appear at the same age in the different species, even when the growth conditions are the same. In addition, sexual activity can change during the course of life: in *Drosophila melanogaster,* one can see that wild type males mature very quickly, and their activity remains constant; in contrast, the activity of *white*[2] males develops more

---

[2] The *white* mutation blocks the synthesis of eye pigments; *white* mutants have white eyes.

slowly, but becomes equal and even superior to that of the wild type (Petit, 1958). Sexual receptivity in *Drosophila melanogaster* appears on the second day after hatching; simultaneously, there is an increase in hormonal secretion from the *corpora allata* (Doane, 1960). Hormonal influence was demonstrated by injections of extracts made from the *corpora allata*. Pupa were injected with the extract 17 or 19 hours before hatching; the controls received a small amount only of the aorta. Injected flies were receptive on the first day; control and normal flies were receptive on the second day (Manning, 1966).

However, invertebrates are not good material for this kind of research; more precise information can be obtained from vertebrates. The effect of hormones on behavior—especially that of steroids, estrogens, and testosterone on the sexual behavior of mammals and birds—has been known for a long time. Still, the nature of interactions that induce hormones to change genetically determined behavior patterns remains undetermined. The problem is even more difficult because social experience can interfere with genetically-determined behavior patterns.

A comprehensive study of sexual behavior in the guinea pig was made, taking into consideration the different components of sexual behavior. Valenstein *et al*. (1954) looked at the behavior of males in two inbred strains and in one heterogeneous strain. In one of the inbred lines, the amount of preliminary courtship behavior was greater than in the other, whereas the other line had higher frequencies of behavior in the activities related to actual impregnation. However, both had a sexual drive lower than that of the heterogeneous stock. Similar, genetically determined differences were demonstrated in female guinea pigs (Goy and Jakway, 1959). Grünt and Young (1952, 1953) distinguished three levels of sexual behavior in their male guinea pigs (high, medium, and low). After castration, all animals had a low sexual drive. Sixteen weeks later, when the castrated guinea pigs were injected with testosterone propionate, their sexual behavior reappeared. However, although the three stocks received the same amount of hormone, the levels of regained sexual activity were different and identical to those of the three categories before castration. The same results were found in female guinea pigs by Goy and Young (1957), suggesting that the differences in sexual behavior do not result from differential release of hormones, but from differential response to sexual hormones in tissues responsible for sexual behavior.

On the molecular level, the problem is even more puzzling. Hormones have long been known to stimulate enzyme activity, but the molecular mechanisms underlying hormone action in the genetic regulatory machine are quite unknown. In any case, it is not my purpose to develop this point that concerns typically molecular genetics.

Influence of Biological and Social Environment

To the action of physical and internal environment, the influence of biological and social environment must be added. Such phenomena as imprinting, learning and the advantage of the rare type are superimposed on the genotype. The action of environment is then even more difficult to predict, since the physiological and biochemical mechanisms of these phenomena are unknown. The term "learning" is especially ambiguous, since it includes the ability to gather information and the ability to exploit it, such as the bird's song or the use of a stick to take a banana.

Thus, we return to the quarrel of "innate" and "acquired" within the formulation of genotypic limits. The problem in the study of man is to evaluate the influence of education on the development of intellectual faculties. Man is a bad species for this kind of study, because of the importance of cultural influences on his development. I shall devote my attention to animal experimentation, in hopes that some models may be applied to the human species.

Imprinting, the influence of social environment during the first hours or days of life, and learning are often considered completely different phenomena. But one can question, with Bateson (1966), the legitimacy of the distinction. Imprinting is concerned with the first social ties in young animals and their influence on the social behavior of the adult. The influence of social environment, as studied by Valenstein *et al.* (1955), is limited to the consequences on sexual behavior of the adult of intraspecific associations during the first period of life. So it might be the same kind of phenomenon as imprinting, but spanning a longer period of the developmental process.

Lorenz (1935) was the first to pay attention to imprinting, with his work on the graylag goose. Goslings follow the first moving thing that they catch sight of when they are born; if this object is not their mother, they stay as tightly bound to it as they would to their mother. As adults, their social behavior can then be disturbed; they may court this object, instead of animals of their own species. In 1909, Craig observed that two species of wild pigeons could mate when the young of one species were brought up by the parents of the other species. These young, as adults, preferred the birds of their foster parents' species. Heinroth (1910; Heinroth and Heinroth 1924–1933) ascertained the same kind of behavior in several species of European birds. These phenomena only occur during a precise stage of development; some experiments demonstrate sensitive periods. For example, in the mallard duckling *Anas platyrhynchos,* Ramsay and Hess (1954) found that the sensitive period for imprinting is between 13 and 16 hours after hatching.

Whether this imprinting phenomenon is fundamentally different from

3. Qualitative Aspects of Genetics and Environment

sexual integration into the group as a consequence of socialization during the first days of life is unclear. The study by Valenstein et al. (1955) on the sexual behavior of the guinea pig provided interesting results. Besides the genetic and hormonal factors, they looked at social components in the determination of this particular behavior. The males of two inbred strains and one heterogeneous strain were reared either together, or isolated. They were all separated from their mother at the age of 25 days. When 77 days old, males were presented to females in estrus and different parameters were recorded, the most typical being the number of ejaculations. The results are given in the following tabulation.

Frequency of Ejaculations

	Isolated	Controls
Line 2 (inbred)	6%	84%
Line 13 (inbred)	0%	57%
Heterogeneous line	71%	100%

In the two inbred lines, the sexual performances of the animals reared together were better than those of the animals reared separately. Since there was only a slight difference in performance in the heterogeneous line, the authors thought that the 25-day-old animals might have been socialized earlier due to a faster rate of development. To test this hypothesis, they separated the heterozygous guinea pigs from their mother at 10 days of age. Under these conditions, the sexual behavior of isolated males was not as competent as that of the males reared together. This shows an acceleration of the developmental process in the heterozygous animals and, as a consequence, a shortening of the sensitive period.

A similar phenomenon was described in an animal capable only of more limited types of behavior, Drosophila. Mayr and Dobzhansky (1945) reared males of *Drosophila pseudoobscura* and *Drosophila persimilis* with either females of the same species or with alien females. Although no difference appeared in *Drosophila pseudoobscura,* in *D. persimilis,* males reared with females of the same species discriminated more against alien females than males reared with *D. pseudoobscura* females.

Too little is known about developmental genetics and the physiological and genetic determination of these phenomena to allow a genetic interpretation. However, these experiments call to mind some well-known facts in morphological genetics. For example, the Bar mutation in *Drosophila melanogaster* reduces the number of ommatidia, and this reduction increases in magnitude as the temperature rises; but this action is only possible during a short period of development (Chevais, 1943).

Persistent behavioral patterns may result from earlier life experiences, as suggested by the experiments of Rosenblatt and Aronson (1958a, b) on sexual behavior of the cat. Male cats with sexual experience (from 32 to 81 copulations) and those with none were castrated. They were then tested weekly with receptive females. Fifteen weeks later, sexually experienced males, whose sexual activity slowly declined, were still greatly superior to inexperienced males. When the animals were castrated before sexual maturity, sexual behavior never developed at all. The genotypic sexual behavior, despite its hormonal dependency, is so much improved by experience that it is able to survive elimination of the responsible hormones.

The maternal behavior of rabbits may be of the same kind. Differences exist from one strain to another in nest building, nesting time, plucking hair and aggressive protection of the young, indicating that these activities are genetically determined (Sawin and Curran, 1952). Nest building is not an all-or-nothing process, and many degrees exist between the absence of nest and the perfect nest, a hole in the straw covered with plucked hair. Observation of the quality of nests after the first three litters from 84 females demonstrated an improvement during the first three litters; after that, no more progress was registered. Unfortunately, since physiological changes cannot be excluded, some uncertainty remains as to the importance of learning in this improvement.

A more elaborate behavior, with both genetic and learning aspects, is the song of the chaffinches. To an inherited, basic pattern, learning adds all of the finer details and much of the pitch and rhythm (Thorpe, 1954, 1958a, b). An analysis proves that the normal song, territorial proclamation, and stimulation for females consists of three phases: The first phase has from one to four notes, usually somewhat crescendo, and normally with a gradual and stepwise decrease of mean frequency; the second phase, generally distinct, but not always so, is made by a series of two to eight notes, of constant frequency, lower than that of phase one; the song concludes with phase three, consisting of one to five notes, with a more or less complex terminal flourish.

In a first series of experiments, birds normally reared by their parents were separated from them in September, in order to study their song the following spring. Their song was normal, even if the young birds were exposed not only to adult chaffinches, but to all bird songs. In a second series of experiments, the chaffinches, separated from their parents at the same age as before, were kept far from any bird song except that of their companions from September to May. The results were different: phase one and two were normal, but phase three, specific to each community of young birds, was slightly abnormal. In a last series of experiments, the birds were hand-

3. QUALITATIVE ASPECTS OF GENETICS AND ENVIRONMENT

reared and never allowed to hear any adult song; phases one and two were correct, but phase three was partly or completely lacking. Each community had an entirely different, but extremely uniform, community pattern.

So it seems that some elements, modulated by environment, can be added to innate song, which is a fixed expression of the genes given in phases one and two. The modulated elements of phase three are learned during the socialization that follows birth; they can be considered as a phenotype developed perhaps by sexual selection. More generally, a look at the learning ability of genetically controlled strains would be necessary to predict the genetic limits of learning. This may be a technically difficult problem with primates, but is probably possible with rats or any other clever laboratory animal.

Let us now consider a quite different aspect of the influence of external biological environment on behavior, the advantage of the rare type in sexual selection. When male and female *Drosophila* of different geographical origin, or reared at different temperatures, or marked by different visible muations, are put together, they do not mate at random. One of the types is usually at an advantage in either sex, generally the male, because it mates repetitively. This advantage, calculated in experimental populations of 200 to 2000 flies, is constant as long as the frequency of the two genotypes is constant. However, it varies as a function of the frequency of the two competing genotypes. In some cases, the genotype that is at a disadvantage when it is abundant in the population is at an advantage when it becomes rare.

Frequency-dependence was discovered in *Drosophila melanogaster* (Petit, 1951) when competition between *Bar* and its normal allele was studied. The advantage of the rare type was clearly demonstrated with the *white* mutant (Petit, 1954). In these two cases, selection occurred between the males, and the female genotype had no influence. The relative selective value K was calculated; K is the ratio of the probabilities for a female to be inseminated by one type of male or the other. For *Bar* and *white,* it was proved to be frequency-dependent (Figs. 2 and 3). The disadvantage of *Bar* appears to be more important when its frequency in the population is above 50%. The results are more striking for *white,* which is at a disadvantage when its frequency lies between 40–80%, and at an advantage when its frequency falls below 40%.

This advantage of the rare type was found in various species of *Drosophila*. Ehrman demonstrated it between strains of different geographic origin, between lines selected for geotaxis and phototaxis, and between lines reared at different temperatures in *Drosophila pseudoobscura* (Ehrman *et al.,* 1965; Ehrman, 1966) (see Table 2). Spiess demonstrated its existence in *Drosophila persimilis* (Spiess in Ehrman, 1966; Spiess, 1968; Spiess and

TABLE 2. The Influence of Frequency on the Numbers of Matings Recorded in Observation Chambers Containing Two Kinds of Drosophila pseudoobscura[a]

No.	Pair per chamber A	Pair per chamber B	Runs	Matings $A_♀ \times A_♂$	Matings $A_♀ \times B_♂$	Matings $B_♀ \times A_♂$	Matings $B_♀ \times A_♂$	Have mated $A_♀$	Have mated $B_♀$	Have mated $A_♂$	Have mated $B_♂$	$x^2 \, \sigma$
1	12 Cal.[b]	12 Texas	7	29	21	26	28	50	54	55	49	0.35
2	20 Cal.	5 Texas	6	57	27	13	12	84	25	70	39	16.96
3	5 Cal.	20 Texas	7	13	17	26	48	30	74	39	65	19.91
4	23 Cal.	2 Texas	5	73	20	4	4	93	8	77	24	34.75
5	2 Cal.	23 Texas	10	4	8	26	62	12	88	30	70	65.76
6	10 Cal.	15 Texas	11	16	44	23	46	60	69	39	90	5.13
AR Mather, 16° v 25°												
7	12–16°	12–25°	8	44	18	28	28	62	56	72	46	5.72
8	20–16°	5–25°	6	67	18	15	1	85	16	82	19	0.09
9	5–16°	20–25°	6	11	12	21	57	23	78	32	69	8.61
10	23–16°	2–25°	10	67	29	13	3	96	16	80	32	64.26
11	2–16°	23–25°	9	3	11	20	72	14	92	23	83	27.02

[a] From Ehrman, 1966. Reproduced by permission.
[b] Cal = California; both California and Texas flies are homozygous for the AR gene arrangement in their chromosomes. The AR Mather individuals were raised at the different temperatures noted. Flies raised at 16° are larger.

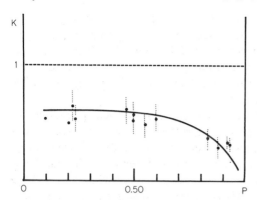

FIG. 2. The relative selective value in sexual competition between *Bar* and wild in *Drosophila melanogaster*. Abscissa: *Bar* males frequency; Ordinate: Relative selective value. (From Petit, 1958. Reproduced by permission.)

Spiess, 1969b), Ehrman and Petit (1968) in the *D. willistoni* group, and Borisov (1969, 1970a,b) in *Drosophila funebris*.

Unfortunately, systematic investigations have been made only of *Drosophila*. Nevertheless, it is known that a black ewe in a white herd is mated first (J.P. Signoret, personal communication), and there are some spotty indications of an advantage of the rare type in the human species: The charm of

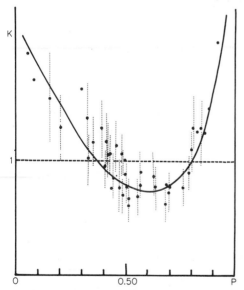

FIG. 3. The relative selective value in sexual competition between *white* and wild in *Drosophila melanogaster*. Abscissa: *white* males frequency; Ordinate: Relative selective value. (From Petit, 1954. Reproduced by permission.)

the exotic may be viewed as one of its manifestations. Following another trend of thought, it seems difficult to believe that the oral tradition conveyed by fairy tales is absolutely gratuitous. In the fairy tales of both Perrault and Grimm, the beloved hero or heroine is always an exceptional individual either in social status or in physical aspect. Serious anthropological studies should be undertaken on this subject.

It would be interesting to know the reasons for this curious phenomenon, but we only have a few indications. Mating is the result of an interaction between male and female; so the level of receptivity of females is important as well as the activity of males. Female receptivity depends on the courtship that she personally receives, and on the general amount of stimuli emitted by the males of the population. Male activity depends on metabolic and physiological factors, and it can be limited by competition for space. Ehrman (1966, 1967, 1969) emphasizes the influence of olfactary factors, and Petit (1970), the ecological competition between males. In fact, this problem is not yet clear and a simultaneous study of male and female behavior in different species is necessary. Since, when there is no sexual isolation, the advantage of the rare type has been found every time it has been looked for, it may well be a general phenomenon. If so, it is an unexpected example of the importance of the interaction between genotype and biological environment in evolution, and it might be responsible for the maintenance of a great number of balanced polymorphisms.

Conclusion

Most behavior exhibits wide genetic variability, which is predictable when one considers the amount of polymorphism discovered during the last 20 years. All the consequences of the interaction between genotype and environment are to be added to this genotypic variability. Such interaction is one of the aspects of genetic homeostasis.

From an evolutionary point of view, the ability of a genotype to react to environment is an advantage important enough to be selected during the course of evolution. Thus, evolutionarily more advanced species may have developed greater potentiality for adaptive behavior through enhanced influence of the environment on their genotypes.

When it comes to man, a large environmental variability is superimposed on the considerable genetic polymorphism in the biological substrate of behavior and intellectual abilities. The range of phenotypic variability is all the greater as the extent of evolution and length of time of development of each individual have made the action of social and cultural factors more important.

References

BAKER, W. K. (1947). A study of the isolating mechanism found in *D. arizonensis* and *D. mojavensis*. *Univ. Tex. Publ.* **4720**, 126–136.
BASTOCK, M. (1956). A gene mutation which changes a behavior pattern. *Evolution*, **10**, 421–439.
BATESON, P. P. G. (1966). The characteristics and context of imprinting. *Biol. Rev.* **41**, 177–220.
BOESIGER, E. (1958). Influence de l'hétérosis sur la vigueur des mâles de *D. melanogaster*. *C. R. Acad. Sci.* **246**, 489–491.
BOESIGER, E. (1962). Sur le degré d'hétérozygotie des populations naturelles de *D. melanogaster* et son maintien par la sélection sexuelle. *Bull. Biol. Fr. et Belg.* **96**, 3–122.
BORISOV, A. I. (1969). Interaction of *D. funebris* chromosomes from urbanian and rural races in experimental populations. *Genetika* **5** (5), 119–122.
BORISOV, A. I. (1970). Interaction of *Drosophila funebris* chromosomes from urbanian and rural races in experimental populations. *Genetika* **6** (2), 81–90. (a)
BORISOV, A. I. (1970). Valeur adaptive du polymorphisme chromosomique. IV. Observations prolongées sur une population de *Drosophila funebris*. *Genetika*, **6** (3), 115–122. (b)
BRNCIC, D., and KOREF SANTIBANEZ, S. (1964). Mating activity of homo and heterokaryotypes in *D. pavani*. *Genetics* **49**, 585–591.
CHEVAIS, S. (1943). Déterminisme de la taille de l'oeil chez le mutant *Bar* de *Drosophila melanogaster:* Intervention d'une substance diffusible spécifique. *Bull. Biol. Fr. Belg.* **77**, (4), 109.
CRAIG, W. (1909). The voices of pigeons regarded as a mean of social control. *Amer. J. Sociol.* **14**, 86–100.
DARWIN, C. (1871). "The Descent of Man and Selection in Relation to Sex." John Murray, London.
DOANE, W. W. (1960). Developmental physiology of the mutant female sterile (2) Adipose of *Drosophila melanogaster*. I. Adult morphology, longevity, egg production and egg lethality. II. Effects of altered environment and residual genome on its expression. *J. Exp. Zool.* **145**, 1–23, 23–42.
DOBZHANSKY, T. (1944). Experiments on sexual isolation in Drosophila. III. Geographic strains of *D. sturtevanti*. *Proc. Nat. Acad. Sci.* **30**, 335–339.
DOBZHANSKY, T. (1957). Genetic loads in natural populations. *Science*, **126**, 191–194.
DOBZHANSKY, T., and LEVENE, H. (1955). Genetics of natural populations. 24: Developmental homeostasis in natural populations of *D. pseudoobscura Genetics*, **40**, 797–808.
DOBZHANSKY, T., and STREISINGER, G. (1944). Experiments on sexual isolation. II. Geographic strains of *D. prosaltans*. *Proc. Nat. Acad. Sci.* **30**, 340–345.
EHRMAN, L., SPASSKY, B., PAVLOVSKY, O., and DOBZHANSKY, T. (1965). Sexual selection, geotaxis and chromosomal polymorphism in experimental populations of *D. pseudoobscura*. *Evolution* **19**, 337–346.
EHRMAN, L. (1966). Mating success and genotype frequency in Drosophila. *Anim. Behav.* **14**, 332–339.
EHRMAN, L. (1967). Further studies on genotype frequency and mating success in Drosophila. *Amer. Natur.* **101**, 415–424.

EHRMAN, L. (1969). The sensory basis of mate selection in Drosophila. *Evolution,* **23,** 59–64.

EHRMAN, L., and PETIT, C. (1968). Genotype frequency and mating success in the *willistoni* species group of Drosophila. *Evolution* **22,** 649–658.

ELENS, A. A. (1957). Importance sélective des différences d'activité entre mâles ekoni et sauvage dans les populations artificielles de *D. melanogaster. Experientia,* **7,** 293–294.

ELENS, A. A., and WATTIAUX, J. M. (1966). Direct observation of sexual isolation. *D.I.S.* **39,** 118–119.

ELENS, A. A., and WATTIAUX, J. M. (1970). Influence of light intensity on mating prospensity of *D. ambigua* and *D. subobscura. D.I.S.* **45,** 110.

ERLENMEYER-KIMLING, L., and HIRSCH, J. (1961). Measurements of the relation between chromosomes and behavior. *Science* **134,** 835–836.

EWING, A. W. (1969). The genetic basis of sound production in *D. pseudoobscura* and *D. persimilis. Anim. Behav.* **17,** 555–560.

EWING, A. W. (1970). The evolution of courtship songs in Drosophila. *Rev. Compar. Anat.* **4,** 3–8.

FAUGÈRES, A., PETIT, C., and THIBOUT, E. (1971). The components of sexual selection. *Evolution* **25,** 265–275.

FULLER, J. L., and JACOBY, G. A. (1955). Central and sensory control of food intake in genetically obese mice. *Amer. J. Physiol.* **183,** 279–283.

GOY, R. W., and JAKWAY, J. S. (1959). The inheritance of patterns of sexual behavior in female guinea-pigs. *Anim. Behav.* **7,** 142–149.

GOY, R. W., and YOUNG, W. C. (1957). Strain differences in the behavioral responses of female guinea-pig to alpha-estradiol benzoate and progesterone. *Behaviour* **10,** 340–354.

GROSSFIELD, J. (1966). The influence of light in the mating behavior of Drosophila. Studies in Genetics:III.*Univ. Tex. Publ.*6615, 147–176.

GROSSFIELD, J. (1968). The relative importance of wing utilization in light dependent courtship in Drosophila. Studies in Genetics: IV. *Univ. Tex. Publ.* **6818,** 147–156.

GRÜNT, J. A., and YOUNG, W. C. (1952). Differential reactivity of individuals and the response of the male guinea-pig to testosterone propionate. *Endocrinology* **51,** 237–249.

GRÜNT, J. A., and YOUNG, W. C. (1953). Consistency of sexual behavior patterns in individual male guinea-pigs following castration and androgen therapy. *Comp. Physiol. Psychol.* **46,** 138–144.

HEINROTH, O. (1910). Beitrage zur Biologie, namelicht Ethologie und Physiologie des Anatidien. *Verh. Fünfte Int. Ornithol. Kongr.* 589–702.

HEINROTH, O., and HEINROTH, M. (1924–1933). "Die Vögel Mitteleuropas." Berlin Lichterfelde.

HIRSCH, J., and BOUDREAU, J. (1958). The heritability of phototaxis in a population of Drosophila melanogaster. *J. Comp. Physiol. Psychol.* **51,** 647–651.

HIRSCH, J., and TRYON, R. C. (1956). Mass screening and reliable individual measurement in the experimental behavior genetics of lower organisms. *Psychol. Bull.* **53,** 402–410.

HUBBY, J. L., and LEWONTIN, R. C. (1966). A molecular approach to the study of genic heterozygosity in natural populations. I. The number of alleles at different loci in *D. pseudoobscura. Genetics* **54,** 577–594.

3. QUALITATIVE ASPECTS OF GENETICS AND ENVIRONMENT 45

HUXLEY, J., MAYR, E., OSMOND, H., and HOFFER, A. (1964). Schizophrenia as a genetic morphism. *Nature* **204**, 220–221.

KAPLAN, W. D., IKEDA, K., TROUT, W. E., and WONG, P. (1971). Studies in the genetics and behavior of neurological mutants of *Drosophila melanogaster*. Genetical Society of Great Britain, 166th meeting. *Heredity* (in press).

KAUL, D., and PARSONS, P. A. (1965). The genotypic control of mating speed and duration of copulation in *D. pseudoobscura*. *Heredity* **20**, 381–392.

LEHMAN, A., and BOESIGER, E. (1964). Sur le déterminisme génétique de l'épilepsie acoustique de Mus musculus domesticus. *C. R. Acad. Sci.* **258**, 4858–4861.

LEWONTIN, R. C., and HUBBY, J. L. (1966). A molecular approach to the study of genic heterozygosity in natural populations. II. Amount of variation and degree of heterozygosity in natural populations of *D. pseudoobscura*. *Genetics* **54**, 595–609.

LORENZ, K. (1935). Der Kumpan in der Umvelt des Vogels. *J. Ornithol.* **83**, 289–413.

MANNING, A. (1966). Corpus allatum and sexual receptivity in *D. melanogaster* female. *Nature* **211**, 1321–1322.

MANNING, A. (1967). Antennae and sexual receptivity in female *D. melanogaster*. *Science* **158**, 136–137.

MAYER, J., DICKIE, M. M., BATES, M. W., and VITALE, J. J. (1951). Free selection of nutrient by hereditary obese mice. *Science*, **113**, 745–746.

MAYNARD SMITH, J. (1956). Fertility, mating behavior and sexual selection in *Drosophila subobscura*. *J. Genet.* **54**, 261–279.

MAYR, E. (1950). The role of the antennae in the mating behavior of female Drosophila. *Evolution* **4**, 149–154.

MAYR, E., and DOBZHANSKY, T. (1945). Experiments on sexual isolation in Drosophila. IV. Modification of the degree of isolation between *D. persimilis, D. pseudoobscura* and of sexual preferences in *D. prosaltans*. *Proc. Nat. Acad. Sci.* **31**, 75–82.

MEDIONI, J. (1961). Contribution à l'étude psycho-physiologique et génétique du phototropisme d'un insecte: *Drosophila melanogaster*. *Thèse Fac. Sci., Univ. Strasbourg*. **203**, (Ser. E.), 11–157.

PARSONS, P. A., and KAUL, D. (1966). Mating speed and duration of copulation in *D. pseudoobscura*. *Heredity*, **21**, 219–225.

PERTTUNEN, V. (1958). The reversal of positive phototaxis by low temperatures in *Blastophagus pinniperda* L. *Ann. Ent. Fenn.* **24**, 12–18.

PERTTUNEN, V. (1959). Effect of temperature on the light reactions of Blastophagus pinniperda L. *Ann. Ent. Fenn.* **25**, 65–71.

PERTTUNEN, V. (1963). Effect of dessication on the light reactions of some terrestrial arthropods. *Ergeb. Biol.* **26**, 90–97.

PETIT, C. (1951). Le rôle de l'isolement sexuel dans l'évolution des populations de *D. melanogaster*. *Bull. Biol. Fr. et Belg.* **85**, 392–418.

PETIT, C. (1954). L'isolement sexuel chez *D. melanogaster*. Etude du mutant white et de son allélomorphe sauvage. *Bull. Biol.* **88**, 435–443.

PETIT, C. (1955). Le déterminisme génétique de l'isolement sexuel. *C. R. Acad. Sci.* **241**, 521–522.

PETIT, C. (1958). Le déterminisme génétique et psychophysiologique de la compétition sexuelle chez *D. melanogaster*. *Bull. Biol.* **92**, 248–329.

PETIT, C. (1959). De la nature des stimuli responsables de la sélection sexuelle chez *D. melanogaster*. *C. R. Acad. Sci.* **248**, 3484–3485.

PETIT, C. (1970). Quelques facteurs responsables de l'avantage du type rare chez *D. melanogaster*. *C. R. Acad. Sci.* **270**, 1719–1722.
PETIT, C., and EHRMAN, L. (1970). Sexual selection in Drosophila. *In:* "Evolutionary Biology," (Dobzhansky, Hecht, and Steere, eds.), Vol. 3, pp. 177–223. Appleton, New York.
RAMSAY, A. O., and HESS, E. H. (1954). A laboratory approach to the study of imprinting. *Wilson Bull.* **66**, 196–206.
REED, S. C., WILLIAMS, C. M., and CHADWICK, L. E. (1942). Frequency of wing beat as a character for separating species, races and geographic varieties of Drosophila. *Genetics,* **27**, 349–361.
RENDEL, J. M. (1951). Mating of ebony, vestigial and wild type *D. melanogaster* in light and dark. *Evolution* **5**, 226–230.
RICHARDS, O. W. (1951). The reactions to light and its inheritance in grain-weevils *Calandra granaria*. *Proc. Zool. Soc. London* **121**, 311–314.
ROBERTSON, F. W. (1963). The ecological genetics of growth in Drosophila. 6: The genetic correlation between the duration of the larval period and body size in relation to larval diet. *Genet. Res.* **4**, 74–92.
ROSENBLATT, J. S., and ARONSON, L. R. (1958). The decline of sexual behavior in male cats after castration with special reference to the role of prior sexual experience. *Behaviour* **12**, 285–338. (a)
ROSENBLATT, J. S., and ARONSON, L. R. (1958). The influence of experience on the behavioral effects of androgen in prepuberally castrated male cats. *Anim. Behav.* **6**, 171–182. (b)
SAWIN, P. B., and CURRAN, R. H. (1952). Genetic and physiological background of reproduction in the rabbit. I. The problem of its biological significance. *J. Exp. Zool.* **120**, 165–201.
SLATER, E. (1958). The monogenic theory of schizophrenia. *Acta Genet.* **8**, 50–56.
SPASSKY, B., and DOBZHANSKY, T. (1967). Responses of various strains of *Drosophila pseudoobscura* and *Drosophila persimilis* to light and to gravity. *Amer. Natur.* **101**, 59–63.
SPIESS, E. B. (1968). Low frequency advantage in mating of *D. pseudoobscura* karyotypes. *Amer. Natur.* **102**, 363–379.
SPIESS, E. B. (1970). Mating propensity and its genetic basis in Drosophila. *In* "Essays in Evolution and Genetics in Honor of Theodosius Dobzhansky," (Hecht and Steere, eds.), pp. 315–379. Appleton, New York.
SPIESS, E. B., and LANGER, B. (1964). Mating speed control by gene arrangements in *D. pseudoobscura* homokaryotypes. *Proc. Nat. Acad. Sci.* **51**, 1015–1019. (a)
SPIESS, E. B., and LANGER, B. (1964). Mating speed control by gene arrangements carriers in *D. persimilis*. *Evolution,* **18**, 430–444. (b)
SPIESS, E. B., LANGER, B., and SPIESS, L. D. (1966). Mating control by gene arrangement in *D. pseudoobscura*. *Genetics,* **54**, 1139–1149.
SPIESS, E. B., and SPIESS, L. D. (1967). Mating propensity, chromosomal polymorphism and dependent conditions in *D. persimilis*. *Evolution* **21**, 672–678.
SPIESS, E. B., and SPIESS, L. D. (1969). Mating propensity, chromosomal polymorphism and dependent conditions in *D. persimilis*. II: Factors between larvae and adults. *Evolution,* **23**, 225–236. (a)
SPIESS, E. B., and SPIESS, L. D. (1969). Minority advantage in interpopulational matings of *D. persimilis.* *Amer. Natur.* **103**, 155–172. (b)
SPIETH, H. T. (1952). Mating behavior within the genus Drosophila. *Bull. Amer. Mus. Natur. Hist.* **99**, 401–474.

SPIETH, H. T., and HSU, T. C. (1950). The influence of light on the mating behavior of 7 species of the *D. melanogaster* species group. *Evolution* **4**, 316–325.
TAN, C. C. (1946). Genetics of sexual isolation between *Drosophila pseudoobscura* and *Drosophila persimilis*. *Genetics* **31**, 558–573.
THORPE, W. H. (1954). The process of song learning in the chaffinch as studied by means of the sound spectrograph. *Nature* **173**, 465.
THORPE, W. H. (1958). The learning of song patterns by birds with special reference to the song of the chaffinch (*Fringilla coelebs*). *Nature*, **101**, 535–570. (a)
THORPE, W. H. (1958). Further studies on the process of song learning in the chaffinch (*Fringilla coelebs gengleri*). *Nature* **182**, 554–557. (b)
VALENSTEIN, E. S., RISS, W., and YOUNG, W. C. (1954). Sex drive in genetically heterogeneous and highly inbred strains of male guinea-pigs. *J. Comp. Physiol. Psychol.* **47**, 162–165.
VALENSTEIN, E. S., RISS, W., and YOUNG, W. C. (1955). Experiential and genetic factors in the organization of sexual behavior in male guinea pigs. *J. Comp. Physiol. Psychol.* **48**, 397–403.

DISCUSSION

AUBREY MANNING
University of Edinburgh
Edinburgh, Scotland

Perhaps the most quoted reference, considering all the papers submitted for this workshop, is Hirsch (1967), which was itself the result of similar meetings in 1961 and 1962. So far as the qualitative nature of gene action on behavior is concerned, our concepts have advanced little in the past decade. We have more examples, but they offer little scope for generalization.

It is often difficult for an ethologist to appreciate the struggle that behavior geneticists have had to convince some psychologists of the genetic component in behavioral development. Not that the ethologists were free from their own brand of naïveté over what was implied by the genetic control of behavior, from which state they are now emerging. Petit is absolutely right in emphasizing that all behavior presents us with problems concerning the manner in which genes and environment interact during development. A great deal of modern ethological work is concerned with ontogeny, and it continues to concentrate on studies of animals in their natural environment as well as in the laboratory. This emphasis has particular advantages for the study of genetics and behavioral evolution. It is interesting to note that in addition to Petit, McClearn and Erlenmeyer-Kimling in their papers for this workshop both emphasize the need to study animals in the environments in which they have evolved.

Petit provides a lucid review of the literature on the genetic determination of behavior. Our ignorance of mechanisms remains profound and because of the ontogenetic "distance" between the genes and behavioral performance, it is difficult to scratch the surface of this problem at present. The work of Ikeda and Kaplan (1970a,b) referred to by Petit, is a landmark in that it represents the only case where gene action can be located at the single motoneuron level. Many of the "fixed action patterns" described by ethologists —the stereotyped postures or movements which are species-specific and often function as communicating displays—must have a strong genotypic component in their development. Knowledge of how the genes operate is very desirable, but the ontogeny of fixed action patterns is, in most respects, complete before the student of behavior gets at them. Most of their extreme stereotypy must result from the selective growth of neural connections in the motor areas of the brain. The classic work of Sperry and that of his successors (see Gaze, 1970) indicates what kinds of factors may be involved, but we do not have a suitable system in which to study the effects of genetic changes on the development of neural networks.

A further continuing problem for qualitative behavior genetics concerns the selection of appropriate behavioral units. Several contributors to Hirsch's volume discussed this, and it still remains a matter for pragmatic treatment, although modern methods of factor analysis have been of some help. The remarkable work of Rothenbuhler (1964) remains the only good example of a classification into behavioral units that correspond with genetic units. The fixed-action pattern whereby "hygienic" honeybees uncap the cells which contain dead larvae and then remove the corpses would be naturally broken down into two subunits by most ethologists observing it for the first time, namely uncapping cells and removing larvae. Rothenbuhler found that one pair of alleles controls the performance threshold of each of these units in virtually an all-or-nothing fashion.

All the evidence suggests that whereas single genes commonly exert a quantitative effect on performance thresholds, the development of fixed-action patterns must involve numerous loci. Occasionally we have tantalizing glimpses as to how these loci may be organized. Studies of interspecies hybrids reveal that in both *Drosophila* (Ewing, 1969) and crickets (Bentley and Hoy, 1972) genes controlling the performance of courtship songs are more concentrated on the X chromosome than would be expected by chance. Perhaps selection favors the evolution of linkage and we have some other evidence that fixed-action patterns are inherited as unbroken units if at all (see Manning, 1972, Chapter 7). However, suitable material for the necessary interspecific hybridization studies is hard to come by and it is unprofitable to speculate with the thin material available to date.

Petit's review covers some of the recent work on imprinting and bird song development. This is of great importance for our understanding of the principles of how genes must operate in development, although it must be admitted there is no formal genetics of any kind accessible to us at present. Nevertheless, it is worth drawing attention to some details of the work of Marler and Tamura (1964) and Konishi (1965) on the song development of the white-crowned sparrow *(Zonotrichia leucophrys)* (see Manning, 1972, for a summary). By a series of elegant experiments, it has been possible to define the limits of the song potential which is inherited (subject to an environment which supports healthy growth). The sparrows inherit (a) a tendency to sing when testosterone is circulating in the bloodstream, (b) a tendency to sing in bursts of about 2.0 sec (the normal song length) at the species-specific pitch (probably largely determined by the structure of the syrinx), and (c) some kind of neural template of the species-specific song against which it can compare and modify its own utterances and which can itself be modified by hearing the song of adults.

Note that very little in the way of motor performance potential is inherited. Unlike the songs of insects, which are perfectly preprogrammed, that of the sparrow is not, and the song of deafened isolates is nothing more than an irregular jangle of the correct length and pitch. The template can only function if the young bird can hear itself, when it rapidly adjusts its motor output to match as soon as it begins singing. During the first few weeks of its singing life the bird can modify its song further as a result of hearing the song of neighboring males, but beyond this period no further modification is possible. (Some other species have a more protracted and open-ended development; see reviews in Hinde, 1969.)

The most fascinating experiments from our point of view concern the demonstration that the young sparrow can modify the inherited template as a result of hearing adult songs, *before it has ever sung itself*. The results of this modification are revealed only some months later when it first begins to sing. Further, the template is sensitive to modification only by the song of white-crowned sparrows; the songs of other species have no effect, that is, there is an inherited predisposition to respond to particular inputs.

We see a similar phenomenon with imprinting. Bateson and Reese (1969) have shown that conspicuous objects are reinforcing to ducklings in the first few hours of life, before any imprinting has taken place. Further evidence comes from the work of Schutz (1965) and Immelman (1969) on sexual imprinting in ducks and Estrildine finches. The sexual choice of males depends very largely on the characteristics of the mothers who rear them. While it is simple to imprint males upon foreign foster species, they are most easily imprinted upon females of their own species, as revealed by the reduced length of exposure required to develop a firm attachment. Again we may suspect an inherited predisposition.

This ethological work is paralleled now by the work of several psychologists who have recently been reexamining the generality of the so-called "laws of learning." Seligman (1970) reviews a wide range of studies which show that rats and other mammals come to a learning situation with a good deal of built-in bias to learn particular things. It is simply not the case that any stimulus can be associated equally well with any reinforcement.

The ingenious experiments of Garcia and Koelling (1966) show that rats have a predisposition to associate gustatory stimuli with subsequent sickness, and will do so even if the sickness reinforcement is delayed for an hour or more.

The conclusion that animals may inherit the tendency to learn particular things—that the genes, so to speak, program for an expectancy—is of the greatest significance for the general theme of this workshop. It leads us to speculate on the degree to which social animals may have inherited expectancies which affect the development of their social contacts with conspecifics. At the conclusion of this discussion, Ginsburg described his observations that young wolves, reared outside their normal social context, nevertheless formed a typical wolf social group when put together.

Kummer (1968) described "cultural transplants" between the two baboons *Papio hamadryas* and *P. cynocephalus*. The former has one-male bands in which a male herds together a harem of four or five females. *Cynocephalus* has a multimale group in which females form temporary consort relationships with a dominant male as they come into estrus. Kummer found that female *P. cynocephalus* apparently learn their social role and can rapidly adjust when transplanted into a *P. hamadryas* culture. The herding behaviour of males appears more resistant to change, although some of the crucial experiments have yet to be carried out.

Dobzhansky and Tobach raised some objections to Manning's use of the term "culture" when applied to animal social behavior. Dobzhansky defined culture as "the sum of behaviors which the individual learns as a member of a society" and was very dubious about applying the term to the baboon situation. Manning avoided defining culture, but preferred to use the term *cultural transmission* to apply to a particular mode of behavioral development. He was prepared to use the term for the development of bird song dialects such as are found in the white-crowned sparrow.

Tobach disliked the lack of precision in Manning's definition and thought it lacked useful meaning if it could be applied so generally. Should one apply the term cultural transmission to the phenomenon whereby prairie-dogs or even some fish groups detect and discriminate against strangers who enter the stable social group? Manning suggested that these groups represented social structures rather than cultures, but one might usefully contrast a cultural transmission of intersociety *differences* with, for example, inherited differences. The argument was not fully resolved.

The remarkable phenomenon of rare-male advantage in *Drosophila*, leading to frequency dependent selection when there are two genotypes of males present, was described in detail by Petit, its discoverer, and Ehrman. Some aspects of the situation remain puzzling.

1. The interaction of the rare male advantage with the need for the evolution of sexual isolation between two populations that meet after a long period of separate evolution. If sexual isolation is to evolve, then a rare-male advantage might hamper its progress. Ehrman made it clear that the rare-male phenomenon was only seen when there was no sexual isolation between the two genotypes. Thus, she had never found it within the *D. paulistorum* group of incipient species, where sexual isolation is well developed.

2. The degree of difference required between males in order to produce a rare-male advantage. Ehrman and Petit quoted a number of cases showing that, for example, males from the same stock reared at different temperatures also showed the phenomenon. Differences in temperature might be enough to produce chemical changes that led to changed smell. Ehrman (1969) has convincing experiments showing that the phenomenon of rare-male advantage depends on the females' use of their sense of smell. However, Petit quoted experiments showing that removing the antennae (the chief, although probably not the only receptors for airborne chemicals in *Drosophila*) did not abolish rare male advantage in a different *Drosophila* species.

Ehrman thought that the rare-male advantage may not occur often in nature, because the phenomenon probably depends on having a high density of courting males. She was reluctant to ascribe to it an important role in the maintenance of natural polymorphisms.

There was some speculation on the possible occurrence of rare-allele advantages in man. Probably the most significant comment was made by Dobzhansky, who pointed out that blondes were highly sought after by South American males (as well as other males). The group was dubious as to how far this could be said to demonstrate their increased biological fitness!

References

BATESON, P. P. G., and REESE, E. P. (1969). The reinforcing properties of conspicuous stimuli in the imprinting situation. *Anim. Behav.* **17**, 692–699.

BENTLEY, D. R., and HOY, R. R. (1972). Genetic control of the neuronal networks generating cricket song patterns. *Anim. Behav. in press.*

EHRMAN, L. (1969). The sensory basis of mate selection in *Drosophila*. *Evolution* **23**, 59–64.
EWING, A. E. (1969). The genetic basis of sound production in *Drosophila pseudoobscura* and *D. persimilis*. *Anim. Behav.* **17**, 555–560.
GARCIA, J. and KOELLING, R. (1966). Relation of cue to consequence in avoidance learning. *Psychonom. Sci.* **4**, 123–124.
GAZE, R. M. (1970). "The Formation of Nerve Connections." Academic Press, New York.
HINDE, R. A. (ed.), (1969). "Bird Vocalizations." Cambridge Univ. Press, London and New York.
HIRSCH, J. (ed.), (1967). "Behavior-Genetic Analysis." McGraw-Hill, New York.
IKEDA, K., and KAPLAN, W. D. (1970). Patterned neural activity of a mutant *Drosophila melanogaster*. *Proc. Nat. Acad. Sci. U. S. A.* **66**, 765–772. (a)
IKEDA, K., and KAPLAN, W. D. (1970). Unilaterally patterned neural activity of gynandromorphs, mosaic for a neurological mutant of Drosophila melanogaster. *Proc. Nat. Acad. Sci. U.S.A.* **67**, 1480–1487. (b)
IMMELMAN, K. (1969). Über den Einfluss frühkindlicher Erfahrungen auf die geschlechtliche Objektfixierung bei Estrildiden. *Z. Tierpsychol.* **26**, 667–691.
KONISHI, M. (1965). The rôle of auditory feedback in the control of vocalization in the white-crowned sparrow. *Z. Tierpsychol.* **22**, 770–783.
KUMMER, H. (1968). Two variations in the social organization of baboons. *In* "Primates: Studies in Adaptation and Variability," P. Jay (ed.), pp. 293–312. Holt, New York.
MANNING, A. (1972). "An Introduction to Animal Behaviour," 2nd ed. Arnold, London.
MARLER, P., and TAMURA, M. (1964). Culturally transmitted patterns of vocal behavior in sparrows. *Science* **146**, 1483–1486.
ROTHENBUHLER, W. C. (1964). Behavior genetics of nest cleaning in honey-bees. IV Responses of F_1 and backcross generations to disease-killed brood. *Amer. Zool.* **4**, 111–123.
SCHUTZ, F. (1965) Sexuelle Prägung bei Anatiden. *Z. Tierpsychol.* **22**, 50–103.
SELIGMAN, M. E. P. (1970). On the generality of the laws of learning. *Psychol. Rev.* **77**, 406–418.

Chapter 4   Genetic Determination of Behavior (Animal)

GERALD E. McCLEARN[1]
University of Colorado
Boulder, Colorado

The Comparative Method

It is not often the case that research conducted on some nonhuman animal species is motivated solely by interest in that species per se. To some extent, explicitly or implicitly, the results are expected to have some degree of phyletic generality. Evolutionary theory provides the basis for expecting some generality. If each species were separately created, there would be no particular reason for expecting common principles from one to another.

No biological discipline has had greater success in the comparative approach than has genetics, where the spectacular advances in understanding of the nature of inheritance of the whole spectrum of living forms have

[1] This work was sponsored by the National Academy of Education and the National Academy of Sciences.

come from, among others, peas, *Drosophila, Neurospora,* and bacteriophage.

For our present topic we need to inquire also about the comparative method in behavioral research. In some branches of psychology the results of animal research have been readily accepted, but in other branches of the field, and in some other social and behavioral science disciplines, the relevance of animal data to man has been challenged on the ground that, once man developed culture, he became something apart from the rest of the biological world, and exempt from the rules applicable to that world. In the simple-minded expression of the nature–nurture controversy, many felt that there were two opposing teams and that one had to choose sides. Because culture has an undeniable influence, many social scientists therefore came to reject "nature's" influence entirely. This attitude was strongly reinforced by the ascendancy of behaviorism within psychology. John B. Watson, the promoter of this movement, was eager to exorcise the circular "instinctive" explanations of behavior from psychology's lexicon. In so doing, he confused, as many others have since done, the instinctive and the inherited, and evidently felt that heredity had to be cast aside entirely as a class of behavioral determinants. He did so, not in ignorance of the then-current understanding of genetics, with which he was, in fact, quite well acquainted, but out of the conviction that a given behavior could be put together in such a variety of ways that genetic constraints on one system could be compensated for by other systems. In this view, the possibilities were so complicated that knowledge of genetics would not add predictive or explanatory power with respect to behavior. This attitude was opposed by a substantial amount of data available at the time, some rather weak and some quite sound, implicating heredity as having influence on behavioral properties of animals and man.

Watson's influence on psychology was very great indeed; his hyperbolic promise to take any "normal" child and turn it into anything he wished by environmental manipulations—doctor, lawyer, merchant, chief—was widely accepted as a demonstrated fact, and psychology came to have an almost exclusively environmentalistic orientation. Similarly, in sociology, Durkheim's admonition that explanations for sociological phenomena should be sought only at the sociological level had an extremely influential impact on other social sciences, directing them away from any examination of biological determinants.

Added to this purely theoretical rejection of heredity as a causal force in behavior was the revulsion at the genocide practiced by Nazi Germany. Since the Nazis justified and rationalized their extermination programs by a perverted and distorted brand of eugenics, and since eugenics was a program concerned with social consequences of genetic determination of many human traits, including behavioral ones, a moral rejection of the concentration camps seemed to many to demand denial of the possibility that genes might influence behavior. That this conclusion was a *non sequitur* did not prevent it from becoming widespread, contributing to the fact that the social and behavioral sciences effectively cut themselves off from the biological discipline that made such astonishing progress in the past 20 years. Nowadays, the dichotomous view of nature versus nurture is no longer supportable, and the interaction and mutual action of genotype and environment, both in generating variability within populations and in the evolutionary process, must provide the conceptual framework. A renewed interest in behavioral genetics with this perspective made explicit appeared in the 1950s, and a large body of data has since been generated (see below, p. 58). New perspectives in physical anthropology have also clarified the gradual nature of the development of culture and of the evolution of the human brain to cope with a given level of culture and to generate more. There has been, in effect, a mutual boot-strapping operation. First steps toward culture provided a new environment in which some individuals were more fit, in the Darwinian sense, than others; their offspring were better adapted to culture and capable of further innovations; and so on. The argument can be made that, far from removing mankind from the process of evolution, culture has provided the most salient natural selection pressure to which man has been subject in his recent evolutionary past.

However, a simple ladder conception of evolution, with species arranged in a unidimensional array, will not suffice in evaluating the comparative approach. The branching and subbranching of the phylogenetic tree leads us to expect that some characters will be quite general and that others will be quite restricted. Lacking advance notice, the test will be a pragmatic one. It would seem, therefore, that data from a nonhuman source should be viewed as suggestive with respect to man; one should be neither too eager to generalize to man nor to deny potential relevance.

Another related objection sometimes raised with respect to the comparative method in behavior is that the animal model may be an incomplete representation of the human situation. For example, the possibility of relating animal research on alcohol preference to human alcoholism is rejected out of hand by some on the grounds that the measures of alcohol ingestion of the mice or rats did not at the same time assess all other aspects of addiction, particularly tissue tolerance and physical dependence. It is difficult to account for this requirement of complete isomorphism of the animal model to the human situation in the case of behavioral traits. In other scientific contexts it seems to be agreed generally that simplification is often a useful precondition for understanding of complex phenomena. A complete model would be desirable, without doubt, but it is not obvious that partial models will not shed important light. Again, it would seem to be an empirical question for any particular trait. The proof of the model will be in its application, and it is likely that we will discover that some models are extremely useful and that others are worthless.

Alexander Pope may have been correct in asserting that the proper study of mankind is man. In some cases, however, we may advance this study most rapidly by an apparent detour through research on his phyletic relatives.

Animal Behavioral Genetics

A major practical reason for using infrahuman animals in genetics research is that mating can be controlled. Species of choice tend to be those that have large numbers of progeny and short generation intervals. An additional requirement for behavioral genetics research is that the animal display some behavior of interest. "Interest" is, of course, largely in the eye of the beholder, but there has been a strong tendency to deal with behavior related to central issues within psychology.

The compromises over these sometimes conflicting desiderata have given rise to research that has concentrated on a few species, with most of the work involving *Drosophila,* mouse, or rat. The breeding procedures have variously involved selective breeding, crossing of inbred strains, and to a lesser extent random mating, with study of correlations among relatives and techniques appropriate to single locus analyses. The behaviors have included geotaxis, phototaxis, activity, hoarding, sexual behavior, social dominance and aggression, emotionality, alcohol preference and audiogenic seizures. The basic fact that there is some genetic influence upon the trait has been clearly demonstrated for all of these. A few years ago, this simple demonstration was regarded as noteworthy, because a long tradition of exclusive environmentalism was being challenged within psychology. The success of efforts to demonstrate a genetic component has been so consistent that it is now a foregone conclusion, and efforts have been largely directed to quantitative analysis or to analysis of the physiological mechanisms. The different phenotypes have lent themselves differentially to these enterprises. The behavioral domain of activity has been particularly amenable to quantitative genetic analysis, for example, and a large number of papers have been published in this area. Audiogenic seizures, as another example, have been particularly useful in the search for neurochemical bases of the influence of the genes. Overall, the results of these studies lead to the conclusion that the domain of behavioral phenotypes is not particularly unique, and that no rules of inheritance other than those described for nonbehavioral characters need be invoked to account for their transmission.

4. Genetic Determination of Behavior (Animal)

The growth of the field of animal behavioral genetics has been very robust in recent years, and the total literature now is too extensive for review here. Recent reviews elsewhere may be consulted for overviews (Lindzey et al. 1971; McClearn, 1970). For present purposes, the methodologies employed and the type of evidence adduced may be summarized by the work in one behavioral area. Because learning has often been placed in an antithetical position with respect to "native" traits, it seems particularly appropriate to examine the data on the inheritance of the learning process itself.

Genetics of Learning

Learning was one of the earliest foci of interest of behavioral genetics. Because the rat had early become established as the "standard" psychological research animal, it is natural that the earliest work made use of this animal. Tolman's (1924) pioneering selective breeding program for rat maze-learning served as a pilot study for Tryon's (1940) classical work on "maze-brightness" and "maze-dullness." This work was paralleled by that of Heron (1941), who was also successful in breeding selectively for rat learning performance in a different type of maze. Other more recent selection studies have included Thompson's (1954) work with the Hebb–Williams maze, which may be superior as a model of human "intelligence" because of its graded difficulty, and Bignami's (1965) study which dealt with avoidance learning rather than appetitive learning.

The mouse had low popularity as a behavioral research animal, but was studied thoroughly genetically. The growing interest in the genetics of mouse learning has most often been expressed in strain comparison work, as contrasted with selective breeding in rats, perhaps because of the availability of a large number of highly inbred mouse strains. Rather surprisingly, in view of the quantitative distribution of learning in these studies, there has been relatively little effort expended upon classical quantitative genetic analysis. Most of the research can be subsumed under the rubric of a search for "correlated characters." Sometimes these researches have been oriented towards other characters also at a behavioral level of analysis; sometimes they have reflected a reductionist orientation, and have sought to relate differences in learning performance to physiological properties.

One of the earliest concerns was to determine the generality of the difference between selected lines. Searle (1949), for example, administered a series of learning tasks to a sample of Tryon maze-bright and maze-dull rats. One finding was that on some learning tasks, specifically escape from water, maze-dull animals were brighter than maze-brights. Tryon himself had been very explicit about the fact that his selection was for a particular phenotype, operationally defined as the number of errors in his particular maze. This point was not always understood, however, and the failure of the maze-brights and maze-dulls to be universally bright and dull was interpreted by some, who did not really understand or particularly like the idea anyway, as weakening the argument that genes could influence learning ability at all.

In a similar vein, the generality of differences in learning performance among inbred strains of mice has been explored. McClearn (1958, 1961) found C3H mice to be poorer performers than C57BL or BALB/c mice in an elevated maze, a visual discrimination apparatus, and a tactual discrimination apparatus. Bovet *et al.* (1969) found striking strain differences in both shuttle box avoidance learning and Lashley III maze learning. A general consistency was found across situations, with those strains performing well in one apparatus also performing relatively well in the other. Among the strains tested was the C3H strain, and animals of this group proved to be inferior to C57BL, BALB/c, DBA/2 and several others. These investigators were also successful in selectively breeding for the shuttle box behavior, beginning with a foundation population of genetically heterogeneous Swiss mice. Lindzey and Winston (1962), using a six-unit multiple T maze, also found C3H animals to be relatively inferior to C57BL, DBA, and A mice. C3H's performed less well than C57BL and BALB/c in a wheel-turn shock-avoidance apparatus (Zerbolio, 1967), and were only mediocre in a jump box shock-avoidance situation (Schlesinger and Wimer, 1967). Winston (1963) again found C3H's to be inferior to A and DBA animals in an enclosed maze, but superior to them in a water-escape situation. C3H mice have also been found to perform relatively well in another water-escape study (Winston and Lindzey, 1964) and in shuttle box avoidance situations (Bovet and Oliverio, 1967; Carran *et al.* 1964; Collins, 1964; Royce and Covington, 1960).

These results testify to the complexity of the phenotypic category of learning performance. One way of exploring this complexity has been the examination of what might be regarded to be components of performance. Krechevsky (1933), for example, tested some of Tryon's strains in an apparatus that permitted analysis of an animal's performance in terms of responsiveness to visual and spatial cues. He found that the maze-bright rats employed more spatial hypotheses (responded more to spatial stimuli) and the maze-dull used more visual hypotheses (responded more to visual stimuli). This result is, of course, consistent with the fact that the maze employed in the selection study was enclosed and therefore offered minimal visual stimuli pertinent to the correct choice.

Heron and Skinner (1940) reasoned that since error reduction is the elimination of incorrect responses, animals differing in maze learning ability should differ also in rates of extinction in a bar-press situation. When tested on Heron maze-bright and maze-dull rats, however, this expectation was not confirmed. Another exploration of the nature of the difference between Heron's bright and dull animals was undertaken by Harris (1940). The maze used by Heron permitted scoring of two types of error: first errors and repeat errors at each of the successive choice points. In examining the error scores over trials, typical learning curves were found for both strains, with the error curve of the dulls being, of course, higher than that of the brights. Closer examination revealed that both types of error were reduced in the learning performance of the brights, but only the repeat errors were reduced by the dulls. That is to say, the dull rats' learning consisted solely of learning not to repeat a mistake once made; they learned essentially nothing about correct initial responses at the choice points.

The matter of different error types was explored in detail by Wherry (1941), who analyzed some of Tryon's original data in terms of a forward-going error producing factor, a food-pointing factor, and a goal gradient factor. The relative importance of the forward-going factor declines over trials in a similar manner in both strains. The goal gradient factor rises in both, but more rapidly and to a higher relative level in the maze-bright strain. The food-pointing factor, which begins at a moderate level and subsequently declines in the brights, begins at a low level and rises rapidly to become the predominant factor in the latter trial performance of dull animals.

More recently, McGaugh and colleagues (McGaugh et al., 1962; McGaugh and Cole, 1965) have studied the influence of distribution of practice on the behavioral differences between descendents of the Tryon maze-bright rats (now called S-1's) and maze-dull rats (S-3's). This particular parameter of the learning situation is a central one, because it is related to the consolidation of memory traces. Briefly stated, strain differences in the expected direction were found in performance in a Lashley III maze when a 30-sec interval was provided between trials, but no differences were found with intervals of 5 min, 30 min, or 24 hours. Age of the animals has also been found to affect the strain difference in response to distribution of practice. These results clearly imply genetically influenced differences in rates of neural consolidation.

Genetic differences in response to intertrial interval have also been found in mouse research. In one study (Wimer et al., 1968) both the active shock escape learning and passive shock avoidance learning of C57BL mice were better under a long (24-hour) intertrial interval condition than with brief (5–40 sec) intervals; for DBA/2 mice, the converse was true. Another mouse study on distribution of practice has yielded strain differences (Bovet et al. 1968). In shuttle box avoidance learning, 500 trials were presented either in one continuous 250-min session or in five 50-min sessions at daily intervals. The distribution of practice over 5 days resulted in a dramatic enhancement of learning compared to the continuous session performance in DBA/2 mice, but resulted in poorer performance in C3H and BALB/c mice. Similar strain differences were found in continuous sessions when the intertrial interval was either 30 or 120 sec. These results for DBA/2 mice appear inconsistent with those described in the Wimer *et al.* (1968) study (although differences did exist in apparatus and tasks), and further study obviously is required to sort out the matter. Nevertheless, these demonstrations of strain differences have amply shown a genetic influence on memory and consolidation mechanisms.

An early, rather different approach to the inheritance of mouse learning was taken by Vicari (1929). In time scores on a maze learning task, she found several inbred strains to be characterized by one of three types of learning curve: a flat curve, a classical descending curve, and a descending–ascending curve. Results from F_1's and F_2's suggested dominance for the alleles influencing faster response time, and there was even some evidence that only a single locus might be involved in the difference between the flat curve and the classical one.

In relating strain differences in learning to physiological systems and events, it has been natural to look to the nervous system. Rosenzweig, *et al.* (1960) have described results of a major program seeking to discover neurochemical bases of the behavioral difference between the S-1 and S-3 Tryon strain descendents. They hypothesized that the differences in learning performance are related to neural efficiency and that neural efficiency is related to the biochemistry of the neurotransmitter, acetylcholine. The first investigations dealt with the enzyme acetylcholinesterase, and results were in accord with a straightforward hypothesis: S-1 rats had more acetylcholinesterase than did S-3 rats. As part of the same overall program, Roderick (1960) began selective breeding from a heterogeneous foundation population of rats for levels of the enzyme; after lines were satisfactorily separated for this measure, they were tested for learning ability. The results did not confirm the earlier ones; the high enzyme strains were generally inferior to the low enzyme strains in performance. Subsequent work has led to the position that the ratio of the substrate, acetylcholine, to the enzyme acetylcholinesterase is consistently related to learning in all of the rat strains tested. Other recent work by Schlesinger and his colleagues (see Schlesinger and Griek, 1970) has explored the genetics of the neurotransmitters serotonin, norepinephrine and gamma-aminobutyric acid in the context of seizure susceptibility.

The work just briefly reviewed has followed the basic tactic of observing trait B in two or more strains already discovered (in the case of inbred strains) or bred (in the case of selected strains) to differ with respect to trait A. Useful as this procedure is in generating hypotheses or in their initial testing, it suffers from shortcomings, as well. Given a difference in trait A between two strains, there are three possibilities with respect to any other trait B: the high-A strain may also be significantly higher on B than is the low-A strain; it may be significantly lower; or it may not differ significantly. The latter outcome is quite positive information, allowing the rejection of an hypothesis that traits A and B are related. A significant mean difference in B contrary to the hypothesized direction would also permit rejection of the initial hypothesis and would probably prompt some intellectual gymnastics. Unfortunately, an outcome confirming an hypothesis is extremely weak in these circumstances. The differences in trait B might be entirely fortuitous, reflecting no causal connection at all between A and B, but only chance fixation of alleles.

A solution to this difficulty is very straightforward. One need only examine the correlation between A and B in a segregating population. If A and B share no loci, then segregation should yield a phenotypic correlation of zero. If there are shared genes, a correlation will be expected. One appropriate segregating population for tests of this sort is an F_2 or subsequent generation derived from an F_1 between the two parent strains. The possibility of linkage can complicate interpretations of F_2 data somewhat, although subsequent generations should clarify the issue with respect to all but very closely linked loci.

Populations with greater genetic heterogeneity than F_2's also provide useful animals for examining associations between traits. An example pertinent to the topic of learning is provided by Tyler and McClearn (1970), who studied straight runway learning in a parent and offspring generation of HS mice. This stock was established by crossing of eight inbred strains, and is maintained by systematic matings which minimize inbreeding. A polynomial of the simple form $Y = a + bX + CX^2$, where Y is in terms of running time and X is number of trials, was fitted to each animal's acquisition record. Separate estimates of heritability were then determined for a, b, and c. These ranged from .19 to .40. Examination of the genetic correlations of various indices of acquisition and extinction led to the conclusion that the genes influencing the initial part of acquisition performance contribute less to performance as learning progresses, but come into play once again during the later parts of the extinction process.

Another way of dealing with association between traits presents itself when one of the traits is already known to be inherited in a simple fashion. In learning phenomena, this situation arose with respect to the albino locus. A simple observation that animals of an albino inbred strain perform more poorly than do those from a pigmented strain is subject to the limitations cited above: The behavioral difference may be due to thousands of loci other than the albino locus. Indeed, the a priori odds would seem to be quite long against an association with any particular locus such as this, which is singled out largely because its phenotypic effect is obvious. Clearly, the association can be put to the test in a segregating population. Winston and Lindzey (1964) found that albino segregants in an F_2 between the albino A strain and the pigmented DBA strain were poorer in water-escape learning than their pigmented littermates. Further work (Winston *et al.* 1967) showed differences in response style also to be associated with the albino locus in that albino segregants employed passive avoidance almost exclusively, whereas pigmented animals used both passive and active avoidance about equally.

It is true, of course, that the F_2 data cannot rule out a locus linked to the albino locus as the responsible agent. The data of Tyler (1970) on HS animals in which a number of segregating generations had occurred is confirmatory, however. His albino segregants were inferior in straight runway learning to their pigmented controls. Even more persuasive, however, is the evidence from animals in which a mutation to the recessive allele for albinism has occurred on a pigmented inbred strain background. In this case, the albino animals are presumably like their pigmented strain mates at all loci other than the albino locus, and any differences in behavior can be clearly ascribed to that locus alone. Fuller (1967) and Henry and Schlesinger (1967) employed a stock of C57BL mice in which such a mutation had occurred. They were able to show inferiority in performance of the albinos in a water escape and in a shock avoidance situation, respectively.

In concluding this brief review of the animal learning literature, it may be said that these studies have clearly demonstrated an hereditary basis for learning performance of several kinds in mice and rats, and have also shown the utility of genetic techniques in analyzing traits correlated with, and mechanisms underlying, learning behavior.

Summary and Conclusions

Overall, we might ask, what has the animal behavioral genetics literature contributed to knowledge? It seems to me that a general contribution has been in demonstrating, in company with human data, the plausibility of the modern genetic perspective in application to the realm of behavior, and helping thereby to lay to rest the old nature–nurture formulation. In itself, this would represent a reasonable accomplishment, since this dichotomous view still is a formidable barrier between parts of the social sciences and the biological sciences.

As a corollary, the animal data have helped to provide a new perspective on individuality. Lamentably, many social science formulations would have it that the only source of variability is environmental, and that we all start life essentially as uniform, interchangeable biological units, devoid of individuality. The genetic view of variability as a biological necessity, and an appreciation of the mechanisms that assure it, add a whole dimension of explanatory power to a simple environmental model, and permit the analysis to consider interactive effects between environmental factors and the biological uniqueness of the individual. Such perspective should be particularly valuable in respect to the educational process.

Finally, the animal behavioral genetics literature has strong implications of a pragmatic sort for the conduct of animal behavioral research in general. The genotype dependence of so many effects, even, as we have seen, such "robust" effects as those of distribution of practice, is a clear warning about the generalizability of results obtained on genetically unspecified animal subjects. Replicability, that *sine qua non* of a science, suffers when research is conducted on the nondescript groups used by so many contemporary researchers, and thus the cumulative build up of knowledge that is supposed to characterize a science is severely impaired. The use of genes as variables, to be held constant by choice of a single strain for investigation; to be manipulated as fixed effects by making strain comparisons; to be manipulated by selective breeding; or to "randomize" by use of a deliberately genetically heterogeneous stock, can increase research efficiency greatly.

References

BIGNAMI, G. (1965). Selection for high rates and low rates of avoidance conditioning in the rat. *Anim. Behav.* **13**, 2–3.
BOVET, D., and OLIVERIO, A. (1967). Decrement of avoidance conditioning performance in inbred mice subjected to prolonged sessions: Performance recovery after rest and amphetamine. *J. Psychol.* **65**, 45–55.
BOVET, D., BOVET-NITTI, F., and OLIVERIO, A. (1968). Memory and consolidation mechanisms in avoidance learning of inbred mice. *Brain Res.* **10**, 168–182.
BOVET, D., BOVET-NITTI, F., and OLIVERIO, A. (1969). Genetic aspects of learning and memory in mice. *Science* **163**, 139–149.
CARRAN, A. B., YEUDALL, L. T., and ROYCE, J. R. (1964). Voltage level and skin resistance in avoidance conditioning of inbred strains of mice. *J. Comp. Physiol. Psychol.* **58**, 427–430.
COLLINS, R. L. (1964). Inheritance of avoidance conditioning in mice: A diallel study. *Science,* **143**, 1188–1190. (Abstract)
FULLER, J. L. (1967). Effects of the albino gene upon behaviour of mice. *Anim. Behav.* **15**, 467–470.
HARRIS, R. E. (1940). An analysis of the maze-learning scores of bright and dull rats with reference to motivational factors. *Psychol. Rec.* **4**, 130–136.
HENRY, K. R., and SCHLESINGER, K. (1967). Effects of the albino and dilute loci on mouse behavior. *J. Comp. Physiol. Psychol.* **63**, 320–323.

HERON, W. T. (1941). The inheritance of brightness and dullness in maze learning ability in the rat. *J. Genet. Psychol.* **59**, 41–49.

HERON, W. T., and SKINNER, B. F. (1940). The rate of extinction in maze-bright and maze-dull rats. *Psychol. Rec.* **4**, 11–18.

KRECHEVSKY, I. (1933). Hereditary nature of "hypotheses." *J. Comp. Psychol.* **16**, 99–116.

LINDZEY, G., and WINSTON, H. (1962). Maze learning and effect of pretraining in inbred strains of mice. *J. Comp. Physiol. Psychol.* **55**, 748–752.

LINDZEY, G., LOEHLIN, J., MANOSEVITZ, M., and THIESSEN, D. (1971). Behavioral genetics. *Ann. Rev. Psychol.* **22**, 39–94.

McCLEARN, G. E. (1958). Performance differences among mouse strains in a learning situation. *Amer. Psychol.* **13**, 405. (Abstract)

McCLEARN, G. E. (1961). The generality of mouse strain differences in learning. Paper presented at the meeting of the Western Psychological Association.

McCLEARN, G. E. (1970). Behavioral genetics. *Ann. Rev. Genet.* **4**, 437–468.

McGAUGH, J. L., and COLE, J. M. (1965). Age and strain differences in the effect of distribution of practice on maze learning. *Psychonom. Sci.* **2**, 253–254.

McGAUGH, J. L., JENNINGS, R. D., and THOMSON, C. W. (1962). Effect of distribution of practice on the maze learning of descendants of the Tryon maze bright and maze dull strains. *Psychol. Rep.* **10**, 147–150.

RODERICK, T. H. (1960). Selection for cholinesterase activity in the cerebral cortex of the rat. *Genetics,* **45**, 1123–1140.

ROSENZWEIG, M. R., KRECH, D., and BENNETT, E. L. (1960). A search for relations between brain chemistry and behavior. *Psychol. Bull.* **57**, 6.

ROYCE, J. R., and COVINGTON, M. (1960). Genetic differences in the avoidance conditioning of mice. *J. Comp. Physiol. Psychol.* **53**, 197–200.

SCHLESINGER, K., and GRIEK, B. J. (1970). The genetics and biochemistry of audiogenic seizures. *In* "Contributions to behavior-genetic analysis: The mouse as a prototype." (G. Lindzey and D. D. Thiessen eds.). Appleton, New York.

SCHLESINGER, K., and WIMER, R. (1967). Genotype and conditioned avoidance learning in the mouse. *J. Comp. Physiol. Psychol.* **63**, 139–141.

SEARLE, L. V. (1949). The organization of hereditary maze-brightness and maze-dullness. *Genet. Psychol. Mono.* **39**, 279–325.

THOMPSON, W. R. (1954). The inheritance and development of intelligence. *Proc. Ass. Res. Nerv. Ment. Dis.* **33**, 209–231.

TOLMAN, E. C. (1924). The inheritance of maze-learning ability in rats. *J. Comp. Psychol.* **4**, 1–18.

TRYON, R. C. (1940). Genetic differences in maze-learning ability in rats. *Yearb. Nat. Soc. Study Educ.* **39**(1), 111–119.

TYLER, P. A. (1970). Coat color differences and runway learning in mice. *Behav. Genet.* **1**, 149–155.

TYLER, P. A., and McCLEARN, G. E. (1970). A quantitative genetic analysis of runway learning in mice. *Behav. Genet.* **1**, 57–69.

VICARI, E. M. (1929). Mode of inheritance of reaction time and degrees of learning in mice. *J. Exp. Zool.* **54**, 31–88.

WHERRY, R. J. (1941). Determination of the specific components of maze ability for Tryon's bright and dull rats by means of factorial analysis. *J. Comp. Physiol. Psychol.* **32**, 237–252.

WIMER, R., SYMINGTON, L., FARMER, H., and SCHWARTZBROIN, T. (1968). Differences in memory processes between inbred mouse strains C57BL/6J and DBA/2J. *J. Comp. Physiol. Psychol.* **65**, 126–131.

WINSTON, H. D. (1963). Influence of genotype and infantile trauma on adult learning in the mouse. *J. Comp. Physiol. Psychol.* **56**, 630–635.
WINSTON, H. D., and LINDZEY, G. (1964). Albinism and water escape performance in the mouse. *Science,* **144**, 189–191.
WINSTON, H. D., LINDZEY, G., and CONNOR, J. (1967). Albinism and avoidance learning in mice. *J. Comp. Physiol. Psychol.* **63**, 77–81.
ZERBOLIO, D. J., Jr. (1967). Differences between three inbred mouse strains on a wheel-turn avoidance task. *Psychonom. Sci.* **7**, 201–202.

DISCUSSION

SATYA PRAKASH

*University of Rochester
Rochester, New York*

This discussion deals with the following questions in the genetic analysis of behavior: First, how many genes control a given behavior trait and what is the mode of action of such genes? A most fruitful approach to this problem is being used by Benzer and colleagues in *Drosophila;* this involves the study of genetic mutants of behavioral traits. *Drosophila melanogaster* shows circadian rhythms for eclosion of adults from pupal cases, locomotor activity, mating activity and fecundity. Konopka and Benzer (1971) obtained three independent mutants of eclosion and locomotion rhythm by ethyl methanesulfonate treatment. All three mutants map at the same site on the X chromosome; this fact indicates that circadian rhythm of eclosion may be controlled by very few genes. All three mutations affect both the eclosion rhythm and the locomotion activity rhythm; this indicates that the same clock system probably controls both rhythms. It is hoped that such work will be done with other characters. While it will be comparatively easy to get an idea of the number of gene loci controlling a behavior trait, it will be more difficult to determine the physiological mode of action of these loci.

Second, how much genetic variation exists for a given behavior trait in

DISCUSSION

natural populations? What is the optimum for a behavior trait and what mechanisms are responsible for the maintenance of the genetic variation and the optimum of the trait? While most characters are probably controlled by a large number of loci, it may be that only a small proportion of genetic loci affecting a behavior trait are present in polymorphic state in natural populations. Practically all of the work on behavior characters done for this purpose in *Drosophila,* rat, and mouse has employed the methods of quantitative genetic analysis—estimations of heritabilities from analysis of variance in inbred lines, diallel crosses, or the estimation of realized heritabilities from selection experiments. Selection experiments have been carried out in *Drosophila* for mating speed (Manning, 1961; Kessler, 1969), geotaxis and phototaxis (Hirsch and Erlenmeyer-Kimling, 1962; Dobzhansky and Spassky, 1967), and amount of red eye pigment (Barthelmess and Robertson, 1970). Certain interesting facts, summarized below, emerge from these experiments. In selection experiments for mating speed in *D. melanogaster* and *D. pseudoobscura,* it is observed that most or all of the selection response occurs in the first six to seven generations. Selection response is asymmetrical; not much progress is obtained in selection for fast mating, while selection for slow mating is far more effective. This shows that natural selection keeps the mating speed close to a maximum in laboratory populations. Studies on F_1 populations obtained from crosses between flies from lines selected for slow and fast mating show that genes for fast mating speed are dominant over slow mating speed (Kessler, 1969). Directional selection for positive and negative phototaxis and geotaxis in *D. pseudoobscura* leads to about equal response in both directions (Dobzhansky and Spassky, 1967). Mating speed is then maintained in populations at a nearly maximum value while geotaxis and phototaxis are maintained at an intermediate value. There is then directional selection for high mating speed and stabilizing selection for geotaxis and phototaxis in *D. pseudoobscura.*

Drosophila males with different amounts of eye pigment differ in their mating behavior. A correlation is observed between mating success and eye pigment density in *D. melanogaster* males (Connolly *et al.,* 1969). Low amounts or absence of pigment from the compound eye affects the visual acuity of the fly, which in turn affects the mating success. Selection experiments for high and low red pigment content in the eyes have been carried out in *D. melanogaster* (Barthelmess and Robertson, 1970). Selection was effective in both directions. Relaxation of selection in low lines showed no increase in the amount of eye pigment, which suggests that partial fixation of genes may have occurred. High lines, however, do respond to relaxation of selection; there is a distinct decrease in the amount of pigment. It may be concluded that high lines were still polymorphic.

Mating speed as well as the quantity of eye pigment are both characters associated with fitness. Both characters display similarities in selection experiments; low lines become fixed for genes but high lines keep segregating. This means that polymorphisms are responsible for higher fitness. But we have no idea of the kind of gene frequency changes which occur at different loci during selection experiments. None of these experiments provide any information about the frequencies of alleles at various loci affecting a certain character. How might we begin to get such an answer?

Enzyme polymorphisms have been described in *Drosophila* (Prakash *et al.* 1969), in mice (Selander and Yang, 1969) and man (Harris, 1969). Moreover, instances of differential activity of allozymes—the enzymes formed by different alleles of a locus—are accumulating in *Drosophila* (Doane, 1969) and in man (Harris, 1970). It is very likely that most allozymes will have different specific activities. Selection for simple traits, such as amount of eye pigment in *Drosophila,* can then be combined with a study of polymorphisms at loci such as Xanthine dehydrogenase and tryptophane pyrrolase involved in the pathway of eye pigment formation. Gene frequency analysis by electrophoretic procedures of the initial population and of several replicate selection lines may provide some interesting information.

Next, I will consider learning, since most of Dr. McClearn's discussion focusses on learning in mice and rats and it is an important trait in human context. From selection experiments in rats for "maze dullness" and "maze brightness," it can be concluded that genetic variation controlling learning behavior existed in the rat strains used in these experiments and that "maze brights" selected for a particular maze are not universally bright. In the human context, this information might not have any direct applicability, except for the possibility that genetic variation for learning ability may also exist in human populations. But this fact has to be independently demonstrated in human populations themselves. This has been done for several traits, including IQ, in some human populations. The methods used are those of quantitative genetics, and estimation of heritabilities is the usual procedure. Unfortunately, heritability estimates do not tell us much about the genetic make up of the trait—how many genes affect the trait, and what proportion of the genes affecting the trait are polymorphic, what are the frequencies of alleles at polymorphic loci and what mechanisms are responsible for the maintenance of these polymorphisms. Moreover, the information about a character's heritability in a certain population cannot be used in any way that might benefit the individual. It seems to me that genetic, physiological and behavioral studies of major gene mutations and of chromosome abnormalities in humans will prove to be most useful both for purposes of curing the ill as well as in genetic counselling. But how does one deal with continuous polygenic variations of behavior traits. There appears to be no easy solution; but here again we will obtain deeper understanding and valuable information from genetic biochemical studies. It may be possible to correlate some of the allozymes of brain enzymes with certain behavior traits as is being done by Omenn and Motulsky. I also find their suggestion of psychometric study of heterozygoes of rare recessive defects very attractive.

References

BARTHELMESS, I. B., and ROBERTSON, F. W. (1970). The quantitative relations between variation in red eye pigment and related pteridine compounds in *Drosophila melanogaster*. *Genet. Res. Camb.* **15**, 65–86.
CONNOLLY, K., BURNET, B. and SEWALL, D. (1969). Selective mating and eye pigmentation: An analysis of the visual component in the courtship behavior of *Drosophila melanogaster*. *Evolution* **23**, 548–559.
DOANE, W. W. (1969). Amylase variants in *Drosophila melanogaster*: Linkage studies and characterization of enzyme extracts. *J. Exp. Zool.* **171**, 321–342.
DOBZHANSKY, T., and SPASSKY, B. (1967). Effects of selection and migration on geotactic and phototactic behavior in Drosophila. I. *Proc. Roy. Soc. London BB* **168**, 27–47.
HARRIS, H. (1969). Enzyme and protein polymorphism in human populations. *Brit. Med. Bull.* **25**, 5–13.
HARRIS, H. (1970). "The Principles of Human Biochemical Genetics." American Elsevier, New York.
HIRSCH, J., and ERLENMEYER-KIMLING, L. (1962). Studies in experimental behavior genetics. IV. Chromosome analysis for geotaxis. *J. Comp. Physiol. Psychol.* **55**, 732–739.
KESSLER, S. (1969). The genetics of Drosophila mating behavior. II. The genetic architecture of mating speed in *Drosophila pseudoobscura*. *Genetics*, **62**, 421–433.
KONOPKA, R. J., and BENZER, S. (1971). Clock mutants of *Drosophila melanogaster*. *Proc. Nat. Acad. Sci. U. S. A.* **68**, 2112–2116.
MANNING, A. (1961). The effects of artificial selection for mating speed in *Drosophila melanogaster*. *Anim. Behav.* **9**, 82–92.
PRAKASH, S., LEWONTIN, R. C., and HUBBY, J. L. (1969). A molecular approach to the study of gene heterozygosity in natural populations: IV. Patterns of genic variation in central, marginal and isolated populations of *Drosophila pseudoobscura*. *Genetics* **61**, 841–858.
SELANDER, R. K., and YANG, S. Y. (1969). Protein polymorphism and genic heterozygosity in a wild population of the house mouse (*Mus musculus*) *Genetics* **63**, 653–667.

COMMENT

THEODOSIUS DOBZHANSKY

University of California
Davis, California

Let me call your attention to some of my work on geotaxis and phototaxis, published in a provincial journal called *Proceedings of Royal Society, Section B*. Now, it is true that *Drosophila* is a rather stupid animal, in the sense that it has not shown any ability to learn. Yet *Drosophila* is rather remarkable in having numerous behavior traits which are subject to rather precise measurement, thanks to the ingenious apparatus, mazes invented by Jerry Hirsch, and by his student N. Hadler. It is possible to do experiments on selection for these behavior traits, rather different from the experiments of S. Benzer which have been mentioned here. Benzer worked with mutants, with major genes. The selection experiments in mazes are experiments on polygenic traits. Incidentally, polygenic traits need not be connected with heterochromatin; that has been an odd noise which K. Mather produced some time ago. As far as I know, he has not mentioned it for many years. Polygenic traits may be due to the presence of numerous regulatory genes, operators, suppressors, and so on. Selection for positive and negative reaction to light and to gravity gives rather interesting results, which at least I would not have predicted or expected when these experiments were started. Most of the wild strains are, on the average, neutral to light and to gravity.

Selection results in strains decidedly positive and decidedly negative. Computed by D. Falconer's methods of realized heritability, these traits have a remarkably low heritability: about 9% for phototaxis, only 4% for geotaxis. Compared with the famous (or infamous) IQ in man, that is a low heritability indeed. Nonetheless, selection is able to shift the characteristics of the population in a very decisive manner: One obtains sharply positive and sharply negative populations. A very remarkable fact is that these traits are subject to what I. Lerner has termed genetic homeostasis. When the artificial selection is relaxed, there is a trend to go back to photo- and geotactic neutrality. In an unpublished paper, Howard Levene has analyzed this mathematically, and quantified what he calls "homeostatic strength." The pressure to return back to the original state is almost equal to half of the intensity of the selection which we have practiced and which caused divergence. The behavior traits have the remarkable property of genetic homeostasis or buffering. This shows that what natural selection has accomplished in nature, making *Drosophila* populations on the average neutral, was not accomplished by fixing major genes for neutrality. It was done in a far more interesting, and I believe biologically meaningful, way. Natural populations are genetically variable. They contain both positive and negative genetic variants and are balanced in such a way that the population as a whole is neutral. What is the biological meaning of this phenomenon? Assuming that the average reaction norm of the population is most favorable when it is neutral, this structure, this genetic architecture, of the behavior trait permits the species, or the population, to rather quickly change its adaptive level if the environment changes so that positivity or negativity becomes adaptive. Moreover, when this selection process operates, because of this genetic architecture of the character, the species does not burn the bridges for possible retreat. If the change of the environment is only temporary, the species is able as quickly to retrace its steps. If the species became temporarily positive, it can quickly go back to neutrality or to negativity, if necessity so demands. Consider another possible genetic architecture. Selection could establish major genes or polygenes which would fix the population at the point of neutrality, and thus make the population genetically uniform for the "ideal" neutrality genes. The population then would be commited to that situation; it could not easily change in either direction. I think the genetic architecture of the character actually observed is a very interesting one. It combines the virtues of plasticity with the ease of responding to environment; the species does not commit itself to a fixed specialized state. This type of genetic architecture of behavior traits at least I would not have predicted on theoretical grounds before doing the actual experiment. Now, I do not contend that all behavior traits are identically constructed. However, phototaxis and geotaxis are interesting kinds of behavior traits, and eminently worth studying.

Chapter 5 Genetic Determination of
Behavior (Mice and Men)

P. A. PARSONS

*La Trobe University
Bundoora, Victoria, Australia*

Introduction

Behavioral traits influenced primarily by single genes and chromosomes can be studied with few problems by segregating families in laboratory animals, as well as in human pedigrees. However, in dealing with quantitative traits that are continuously distributed, greater difficulties are apparent. The methods and techniques of biometrical genetics must be employed in their analysis, one of the main aims being to assess the relative importance of genotype and environment. In laboratory animals such as *Drosophila,* mice, rats, and guinea pigs, rather sophisticated experiments can be carried out to separate genotype and environment, and assess the importance of interactions between them. The techniques of plant breeding are also being applied to the study of behavior in animals (Parsons, 1967a), since, as pointed out by Caspari (1968), animal behavior and plant morphology are analogous in that the effects of environment are comparatively larger than for animal morphology. At the behavioral level, animals are much more environment-sensitive than at the morphological level, hence the usefulness of techniques aimed at detecting and estimating environmental effects.

One important technique is the study of inbred strains, which provides an estimate of the heritability in the broad sense. An extension of this, taken from agricultural practice, is to study a trait, behavioral or otherwise, in a number of inbred strains in a number of environments. To take a behavioral trait as an example, we could have a strains at b temperatures and c light-intensities, with r replicates at each temperature and light intensity. A simple analysis of variance enables the estimation of variance components of strains Va and various strain × environment interactions, Vab, Vac, and $Vabc$, and these can be compared with the environmental variance. Interactions of this sort might be of importance in behavioral work, especially if extreme environments are used. Another technique consists of taking inbred strains and studying the F_1, F_2, and backcross generations. This provides estimates of the heritability in the narrow sense. Probably the best general technique for studying a trait about which we have little information is the diallel cross, which is a powerful technique for a general survey of a series of strains, perhaps in several environments, whatever the aim or method used. It is an extensive analytical method rather than intensive, as a number of inbred strains and hybrids can be surveyed at once, and it permits an estimate of the heritability in the narrow sense. It has been used a number of times for behavioral traits, but usually in only one environment (for references see Parsons, 1967a; Broadhurst, 1967). The other major technique, somewhat less commonly used in behavioral work, is that of correlation between relatives.

The problem of obtaining estimates of genotype, environment, and of genotype × environment interaction is stressed because immense difficulties arise when we turn to man. To begin with, environments generally cannot be defined, and therefore the techniques of studying a series of individuals in a series of known and defined environments is not possible. However, some information can be derived from twin studies, correlations between relatives, and the comparison of adopted and natural children. In some cases, reasonably reliable results have been obtained (see Jinks and Fulker, 1970).

The object of this chapter is to discuss some data on three inbred strains of mice, C57 (C57Bl), C3H, and Ba (BALB/c) and their hybrids for various behavioral measures, weight, and skeletal divergence. The results will be discussed especially from the point of view of making possible inferences about human behavior, in the hope that they will complement the more conventional biometric approach. Some of the experimental data on behavior are reported in Rose and Parsons (1970), and on skeletal divergence in Howe and Parsons (1967).

The Behavioral Phenotype

The observations to be described were based on mice 59 days old on the day of first observation, and the following behavioral measures were made.

1. *Open field activity* was measured as the number of squares entered in an arena in exactly 2 min. The arena consisted of an open perspex box, the floor of which was marked off into 16 4-inch squares. A square was defined as being entered if all four feet were within it.

2. *Open field emotionality* was evaluated by the sum of the number of urinations and fecal boluses deposited in the arena in 2 min.

3. *Exploratory activity* was measured as the number of crossings of the central barrier in a shock apparatus in 1 min. The shock apparatus consisted of a perspex box with a grid floor divided into two equal parts by a low central barrier.

4. *Initial reaction to shock.* After 1 min of exploration, a light source was switched on above the apparatus and this was followed by a shock to the feet through the floor 2 sec later. The shock, consisting of a 60-volt source which supplied a 250 mA current, could be applied to either side and the central barrier, the latter being shocked to prevent the mouse from "sitting on the fence." The times recorded for the first jump were used as a measure of "initial reaction to shock."

5. *Learning in the conditioned avoidance apparatus.* The mouse was rested outside the apparatus, and then given further trials. In total, the mouse received ten shock trials in the following sequence: (a) four trials 1 min apart, (b) after a rest of 1 hour, three further trials, and (c) after a rest of 24 hours, three further trials. The ten trials were used to determine the ability of mice from different strains to learn to avoid shock and so provide an estimate of learning ability.

A summary of the results obtained is presented in Table 1. In all cases, C3H was intermediate, and C57 and Ba extreme, and the same results were

TABLE 1 Order of Inbred Strains for Behavioral Traits, Weight, and Skeletal Differences

Open field activity	C57	>	C3H	>	Ba
Open field emotionality	Ba	>	C3H	>	C57
Exploratory activity	C57	>	C3H	>	Ba
Initial reaction to shock (order of superiority)	C57	>	C3H	>	Ba
Average time for all jumps (order of superiority)	C57	>	C3H	>	Ba
Weight	Ba	>	C3H	>	C57
Skeletal divergence	Ba	>	C3H	>	C57

obtained for body weight. This shows that C57, the lightest strain, was the most active, with the highest exploratory ability, responded to shock the most rapidly, and learned best. It was also the least emotional. The Ba strain was the complete opposite for all traits, and C3H was intermediate. Therefore, evidence is emerging for behavioral phenotypes corresponding to particular genotypes.

It may be significant that there is an apparent association of weight with the behavioral phenotype, since in man associations between the behavioral phenotype and morphology have been postulated. However, before considering this, we will look at morphology as assessed by the incidence of minor skeletal variants in mice. By the use of inbred strains and mutant stocks, it can be shown that much of the variation in skeletal morphology between strains is genetic (Grüneberg, 1952; Searle, 1954). Deol and Truslove (1957) and Grüneberg (1963) have suggested that many, if not most, minor skeletal variants are expressions of generalized or localized size variations. Therefore, Howe and Parsons (1967) classified skeletons of mice in the three strains for the presence or absence of 25 minor skeletal variants, consisting of 15 of the skull, 8 of the vertebral column, and 2 of the appendicular skeleton.

From the percentage incidences of each variant in the strains, a mean measure of divergence between strains can be obtained. The incidence of each variant p is transformed to an angular value ϕ, such that $\phi = \sin^{-1}(1 - 2p)$. A measure of difference or divergence between the two populations is given by

$$X = (\phi_1 - \phi_2)^2 - \left(\frac{1}{N_1} + \frac{1}{N_2}\right),$$

where ϕ_1, ϕ_2 are angular values corresponding to p_1 and p_2, and N_1 and N_2 are the sizes of the two populations. If a number of variants are taken, a mean measure of divergence can be computed by dividing the sum of the individual measures of divergence for each variant by the number of variants, so in this case the mean measure of divergence will be $\Sigma X/25$ (for further details, including expressions for variances, see Berry, 1963). The means measures of divergence provide a quantitative expression of the separation of populations. The method assumes that all variants have an equal effect on fitness. This is almost certainly an incorrect assumption, but it provides a reasonable assessment of relative divergences between populations, especially since Truslove (1961) found that the occurrences of nearly all the variants she studied were uncorrelated, indicating that the sensitivity of detection of differences between populations increases with the number of variants studied.

TABLE 2 Mean Measures of Divergence and Their Standard Deviations between Inbred Strains[a]

	Ba	C3H
C57	1.326 ± .221	1.012 ± .275
Ba	—	.348 ± .156

[a] After Howe, W. L. and Parsons, P. A. (1967). *J. Embryol. Exp. Morphol.* **17,** 283–292,

Divergences between strains indicate again that the difference between C57 and Ba is the greatest (Table 2), followed by that between C57 and C3H. The difference between Ba and C3H is considerably smaller than the other two comparisons, which is reasonable, since the Ba strain is derived from the Bagg albino strain, and the C3H from a cross between the Bagg albino and Little's DBA strain.

The similarity between weight and pattern of skeletal variation supports an association between the incidence of many skeletal variants and the size of structures associated with body weight. Admitting that the number of strains is limited naively allows one to argue for an association between genotype, skeletal morphology, weight, and various behavioral parameters. This may be reasonable, since skeletal variants are presumably associated with variants of the muscular, nervous and vascular systems, and such variants would presumably have consequences at the behavioral level.

This leads us again to the question of a possible relationship between behavior and morphology in man, as was put forward by Sheldon (1940, 1942) in his classification of individuals according to their degree of endomorphy, mesomorphy, and ectomorphy, with ratings for each dimension derived from a standardized set of photographs. Based on 200 male college students, he assigned ratings for somatotype and for temperamental variables. A considerable association between temperament and physique was found, perhaps overestimated by Sheldon, but even taking this into account, an association still occurred. It may seem to an experimentalist rather poor to search in this way for correlations, knowing as we do that correlation does not imply causation. However, it seems one step better to look at such situations with the knowledge of experimental animals in mind.

Hybrids

We have so far considered the three inbred strains of mice, but not the hybrids between them. For most traits, one or the other inbred strain showed dominance in the F_1. However, for measures of learning in the conditioned avoidance response apparatus, heterosis was quite marked (Fig.

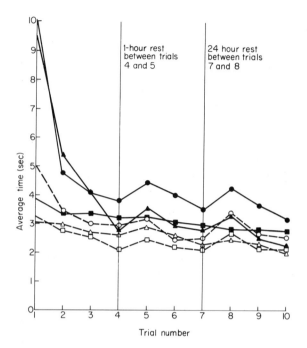

FIG. 1. Average time (in seconds) for each of the 10 successive trials for the three inbred strains and hybrids. Male data only are given; the female data were similar.
●:BALB/c ▲:C3H; ■:C57BL; ○:Ba × C3H; △:Ba × C57; □:C57 × C3H.
[After Rose and Parsons (1970).]

1). Heterosis was greatest for crosses involving one C57 parent. Hybrids from crosses between Ba and C3H tended to show a rather lower level of heterosis, probably because these two strains are more closely related to each other, than to C57. As well as showing heterosis, hybrids between inbred strains showed less variability compared with the inbred strains. The hybrids thus show homeostasis as compared to inbred strains, presumably because the processes leading to the observable phenotype are better buffered against environmental variation. This has been well documented in the literature for many morphological and fitness traits, and the same would be expected for behavioral traits.

If the observation of an association between heterosis and homeostasis for learning is correct, can an explanation be offered? In outbreeding species, heterosis for quantitative traits tends to be fairly marked for traits directly related to fitness such as viability, fecundity, etc. Such traits, which are direct components of fitness, show considerable inbreeding depression when artificial inbreeding is carried out, because of reduced relational balance compared with the outbred situation. Under natural conditions, such traits are subject to directional selection for higher fitness. In comparison, morphological traits such as weight of mice, which must be subjected more to stabilizing selection, are less prone to inbreeding depression, and hence crosses between inbred strains tend to yield less heterosis. Breese and Mather (1960) and Mather (1966) and others have discussed genetic architectures and their consequences under various modes of selection in some detail, and the above comments are compatible with these discussions.

Of the discussed behavioral traits in mice, it seems reasonable to assume that learning would be subject to fairly intense directional selection, while traits such as activity and emotionality would be more subject to stabilizing selection. The observation that learning in the conditioned avoidance apparatus does in fact show considerable heterosis associated with homeostasis is in agreement with the argument presented.

The reduced variability of hybrids compared with the inbred strains also represents a special form of genotype × environment interaction. The difficulty of dealing with the problem of genotype × environment interaction for behavioral traits has already been pointed out. In particular, if such interactions are most marked for those traits related to fitness, that is, subjected to directional selection, which in our example presumably consist of traits with a learning component, there are real problems in extrapolation to man. It is precisely these sorts of traits which are studied most in man, especially by psychologists, and if these show the greatest influence of genotype × environment interactions as suggested, we face problems of acute difficulty in man, where neither genotype nor environment can be controlled. On the other hand, simpler traits, say of a sensory perception nature, might well be subjected to less intense directional selection if not stabilizing selection, and are probably more amenable to accurate study, both in mice and men.

This is not to say that heterosis associated with homeostasis does not occur for other traits; Bruell (1964a,b) reported on data in mice for 31 hybrids derived from 11 inbred strains for wheelrunning and exploratory behavior, both traits probably having a lower component of learning than conditioned avoidance learning. Of the hybrids, 18 were derived by crossing unrelated parents, nine by crossing related inbreds, and four by crossing inbreds belonging to sublines of C57 mice. It is clear that heterosis is general in both sexes for wheelrunning when unrelated strains were crossed (Table 3), and less general when related strains and sublines were crossed, presumably because crosses between related strains and sublines led to rather homozygous individuals, showing less relational balance than for crosses between unrelated strains.

As well as showing heterosis, the hybrids often showed less variability as determined by coefficients of variation compared with the inbred strains;

TABLE 3 Wheelrunning in Mice[a]

a. Number of hybrids showing heterosis according to the degree of relatedness of the inbred parents

	Relatedness of inbred parents			
	Unrelated	Related	Sublines	Total
Heterosis	35	9	2	46
No heterosis	1	9	6	16

b. Number of hybrids showing less variability than both parents (positive homeostasis, +), variability between both parents (o), and more variability than both parents (negative homeostasis, −), according to the degree of relatedness of the inbred parents

	Unrelated	Related	Sublines	Total
+	21	15	—	36
0	9	4	5	18
−	3	1	4	8

c. The degree of homeostasis plotted against the occurrence or lack of heterosis in the hybrids noted above

	Heterosis	No heterosis	Total
+	30	3	33
0	15	5	20
−	1	8	9

[a] Date of Bruell, 1964a, as analyzed by Parsons, 1967b.

thus they showed homeostasis. Homeostasis is most common in crosses between unrelated individuals and least frequent in those between sublines (Parsons, 1967b). In other words, there is an association between heterosis and homeostasis (Table 3c). Certain other published data on other traits also show an association between heterosis and homeostasis (Parsons, 1967b). The same strains were tested for exploratory behavior (Bruell, 1964b), as assessed by placing mice individually in a four-compartment maze. As a mouse moved from one compartment of the maze to another, it interrupted a light beam and activated a photorelay and counter. The exploration score of an animal consisted of the total count registered in 10 minutes. Heterosis and homeostasis were found more often than not, but less often than for wheelrunning, therefore it is not surprising that no real association between homeostasis and heterosis was found.

While these data show, especially for wheelrunning, an association between heterosis and homeostasis, it is difficult to make comparisons with the data of Rose and Parsons (1970), since in the latter data (1) a gradation of traits with increasing learning components was employed, and (2) different genotypes and behavioral tests were used. Therefore, at this stage, the suggestions made about the association of heterosis with homeostasis for traits measuring learning should remain a working hypothesis only. It must be stressed that inferences about the genetic architecture of traits may depend on the sample of strains used, a principle frequently put to one side in quantitative genetic theory, but one which should lead us to treat with equal caution results that agree and disagree with a given hypothesis. The dependence of results on the degree of relatedness of strains shows this to some extent. Thus, we can conclude that for a given series of inbred strains and hybrids, a result, showing heterosis associated with homeostasis, was found for traits with learning components, but not for traits with little or no learning component. This result based on the given series is probably meaningful, although it needs to be extended.

Trait Profiles in Different Genotypes

Guttman (1967) compared the correlations between finger print ridge counts of an English sample with those of the Parsis of India. She found that in both cases the adjacent fingers (except the thumb) are more highly correlated than those further apart. In other words, correlation matrices from both populations show the same pattern, indicating developmental relationships of the measures. A general relationship was also found, for example, for bone lengths in several animal species. Even where the cause of the relationship is unknown, the presence of such constancies is highly suggestive of an underlying similarity of the population samples with respect to the variables forming the pattern. For mental tests, crosscultural stability was found for American college students and for Chinese students studying at American universities. The same was found for some sensory variables (hand-preference, arm-folding, and hand-clasping) for five Israeli subpopulations, even though the actual incidences of these variables differed between groups. Therefore correlational patterns are frequently similar in different groups.

The mouse data provide us with five measures where simple observations were made on all individuals, namely weight, open field activity, open field emotionality, exploratory activity, and initial reaction to shock. Correlation matrices are given in Table 4 for the inbred strains and hybrids. There is a general positive association between open field activity and exploratory activity for all strains, as would be expected. For these two traits, which are essentially activity measures, and for initial reaction to shock, negative correlations within inbred strains were found, in particular for Ba. Thus, within the Ba strain, the most active mice in both the arena and shock apparatus are significantly better at escaping from shock than are the less active mice.

As far as similarity of correlation matrices between inbred strains and hybrids are concerned, since most coefficients do not differ significantly from 0, there is less pattern than in most of Guttman's examples. However, it can be said that there tends to be a similarity for those contrasts where significant results were found. Another point is that there are only two correlation coefficients significant at $p < .05$ for hybrids out of a total of 30 calculated. In contrast, strain Ba shows three significant results, two at $p < .01$ and one at $p < .05$, and in general the deviation from zero of values for inbreds exceeds that of the hybrids (even disregarding C3H, where rather few mice were classified). Viewed in another way, in the inbreds there are seven and in the hybrids two correlation coefficients $> |.24|$. Thus it seems that the hybrids show less extreme associations between traits than the inbred strains. Since within inbred strains and hybrids, we are presumably dealing with identical or similar genotypes, this provides further evidence for greater stability of hybrids or homeostasis. It represents a form of genotype × environment interaction analogous to the lower variability of hybrids for learning traits, as discussed in the previous section. From the genetic architecture

TABLE 4 Correlation Matrices between Open-Field Activity (A), Open-Field Emotionality (B), Exploratory Activity (C), Initial Reaction to Shock (D), and Weight (E)[a]

	\multicolumn{4}{c}{Strain Ba}				\multicolumn{4}{c}{Strain C3H}			
	B	C	D	E	B	C	D	E
A	.012	.373**	−.269**	.009	−.170	.359*	−.181	.085
B		.085	.160	.044		−.303	.117	.035
C			−.272**	−.076			−.099	.176
D				.112				−.331
	\multicolumn{4}{c}{Strain C57}				\multicolumn{4}{c}{Ba × C3H}			
	B	C	D	E	B	C	D	E
A	−.069	.289***	.084	−.007	.124	.172	.003	−.142
B		−.046	.006	.093		.063	−.128	−.006
C			−.176	−.112			−.082	.246*
D				−.105				−.057
	\multicolumn{4}{c}{Ba × C57}				\multicolumn{4}{c}{C57 × C3H}			
	B	C	D	E	B	C	D	E
A	.010	.270*	.007	.008	.032	.200	.143	−.216
B		.078	−.118	.122		.081	.039	−.095
C			.036	.131			−.074	.125
D				−.086				−.121

[a] Adapted from Rose and Parsons, 1970.
* $p<.05$.
** $p<.01$.
*** $p<.001$.

point of view, the hybrids would approximate more to the situation in man, which is an outbreeding species. This leads us to another difficulty in studying human behavior, for, unlike animal work, it is not possible to study extreme genotypes from which inferences may be made, which may throw light on less extreme genotypes. The fact that one cannot carry out selection experiments in man is another manifestation of this problem.

Measures of Learning

In the conditioned avoidance apparatus at trial two (T2), C57 > Ba, C3H in learning ability, and Ba and C3H were almost equivalent (Fig. 1). At T4, however, C3H > C57 >> Ba. Thereafter, Ba was always the poorest at learning, but C3H tended to drop in learning ability after a rest (at T5 and T8), whereas C57 did not. All of these observations represent genotype × environment interactions, since the ordering of genotypes varies according to trial number, although overall C57 was just superior to C3H, which were both definitely superior to Ba, as already pointed out. This result also suggests that the learning component of the behavioral phenotype may not fit in quite as well as previously indicated with simpler forms of behavior and with morphology (as in Table 1). This became first evident with the heterosis found for learning, which was not shown for the other behavioral traits under study.

A measure of learning different from that used so far is the percentage of no-shock jumps, or the percentage of trials where the mouse jumped to the safe side of the apparatus after the light signal was switched on but before the shock was applied. Trials 2–10 were used to assess this, trial 1 being omitted because any crossing of the barrier before experiencing the shock cannot be regarded as a conditioned avoidance response. The highest proportion of no-shock jumps occurred for trials 4, 7, and 10, that is, at the end of each set of trials, thus showing learning during each set of trials. This was followed by a lower percentage following the first trial after resting, as might be expected. The overall order of the genotypes was (Table 5)

$$C3H > C57 > Ba,$$

in contrast with the measure of learning previously discussed, measured as the average time for all jumps. Thus C3H and C57 are reversed for the two measures, and show that different rankings may occur according to the mode of assessing learning.

Therefore we face two problems about learning in mice.

1. genotype × environment interactions between trials, and
2. variable results according to the mode of assessing learning.

TABLE 5 Percentage of "No-Shock Jumps" for Trials 2–10 (Male Data Only)[a]

Trial no.	Ba	C3H	C57	Ba × C3H	Ba × C57	C3H × C57
2	—	—	—	—	—	1.3
3	.7	14.8	—	3.1	2.2	9.3
4	2.8	22.2	2.0	10.2	9.6	22.7
5	1.4	16.0	1.0	2.0	7.4	12.0
6	.7	20.0	3.2	19.4	14.8	17.3
7	3.6	20.0	7.4	20.4	29.6	32.0
8	1.6	4.5	6.9	12.7	18.9	12.0
9	4.0	13.6	9.7	11.4	23.6	24.0
10	9.5	31.8	12.5	16.5	37.8	29.3
Total	2.6	15.8	4.3	10.4	15.8	17.8

[a] After Rose and Parsons (1970).

A similar situation exists in man where the Stanford–Binet IQ test is commonly used as a measure of intelligence, for questions have been asked regarding its suitability in all cultural situations (that is, environments) and populations. Furthermore, the degree to which *different* measures of intelligence give the same relative results between populations is open to debate. The mouse results are difficult enough to interpret, and show clearly that extrapolation to man is peculiarly difficult for traits associated with learning.

In conclusion, in spite of the results in Table 1 showing an association between morphology and the behavioral phenotype in mice, it seems from a more detailed consideration that this association does not necessarily hold for learning. In man, therefore, it seems likely that there would be little real association between somatotype and intelligence, but on the other hand, it would be expected that somatotypes may be associated with traits of lesser complexity from the behavioral point of view, such as those more directly associated with the skeletal, muscular, and vascular systems, which would mean traits of a sensory-perception type.

The mouse data referred to so far were all collected at a standard age. Results from younger mice generally showed that for all genotypes, weight and emotionality increased with age, but activity decreased. A similar situation was found for litter size, since mice from litter sizes less than six were less active and heavier than those from litter sizes greater than or equal to six, although there seemed to be no trend for emotionality (Rose and Parsons, unpublished). In other words, for these traits there seems to be some association between behavior and morphology, but in this case as a result of environment (age or litter size), rather than genotype.

For traits associated with learning, the situation again seemed more complex. Young mice showed a lower initial reaction to shock, but there was no litter size effect. Conversely, for the percentage of no-shock jumps, there was a litter size effect since the percentages were highest for litter sizes greater than or equal to six. For conditioned-avoidance learning generally over the 10 trials, litter size was found to have no consistent effect. There seemed to be an age effect in that younger mice tended to forget easier, especially after a long break (24 hours), but the effect was hardly significant. Once again, therefore, it seems difficult to say there is an association between morphology and traits with a high component of learning, although in this case morphology is altered by environmental means.

Extreme Environments and Genotypes

The importance of extreme genotypes in studying a trait in experimental animals has been stressed frequently. Thus, in outbred species, the use of inbred strains or of individuals that have been selected for extremes for a trait is common for quantitative traits including behavioral ones (Parsons, 1967a), and this approach is illustrated with inbred strains of mice in this chapter. The approach of studying extreme genotypes, in the sense of being largely homozygous, is of course not possible in man, since we have to work on the available population.

The geneticist, preoccupied with studying various genotypes, many of which are extreme, in experimental animals, seems to have paid less attention to the question of environmental variability, being mainly concerned with keeping the environment constant and often optimal for experimental organisms. There are exceptions, mainly in the area of plant breeding and in experimental work on some species of *Drosophila*. For example, in population cages of *D. pseudoobscura,* heterokaryotype advantage occurs at the more extreme temperature of 25°C as compared with 16.5°C (Dobzhansky, 1948). Other examples of heterokaryotype advantage in extreme environments in *Drosophila* include cold tolerance, mating speed and duration of copulation at high temperatures, and desiccation. Similarly, hybrids between inbred strains and other homozygotes tend to show an enhancement of heterosis in extreme environments in several species of *Drosophila*, mice, and plants such as *Arabidopsis thaliana, Nicotiana rustica,* and maize (see Parsons, 1971, and Parsons and McKenzie, 1972, for references). Such extreme-environment heterosis has been postulated to be associated with temperature sensitive and correlated enzymes, or the general poorer fitness of homozygotes compared with the corresponding heterozygotes, because of the breakdown of relational balance in the heterozygotes when forming homozygotes. It was postulated that extreme-environment heterosis could provide an explanation of the high level of polymorphism in natural populations, in addition to those already advanced in the literature, since it would not imply a high genetic load under relatively optimal environments. This again represents an example of genotype × environment interaction, since in moving the environment from optimal to extreme, the hybrids change relatively less than the homozygotes, leading to the observed homeostasis across environments for heterozygotes.

Because of the concentration of attention on genotypes, the study of behavior under extreme environments has been rather neglected. In *Drosophila,* the types of extreme environments that can be studied include extremes of temperature, desiccation, and competition. Rodents, however, being closer to man phylogenetically, seem to be worth more detailed study. Cooper and Zubek (1958) studied two lines of rats in which genetic differences in the rat's capacity to find their way through a maze had been accumulated by artificial selection, leading to "maze-bright" and "maze-dull" rats under a normal laboratory environment. Under a restricted environment, no differences between lines were found and both behaved at the same low level. Conversely, in a stimulating environment, there was a much larger improvement in the maze-dull than the maze bright rats. As pointed out in this study and by Bodmer and Cavalli-Sforza (1970), this could have implications in the determination of human IQ in restricted and stimulating environments. Manosevitz and Lindzey (1970) studied hoarding in various inbred strains in an enriched and standard environment, and found substantial strain × environment interactions. They also studied hoarding in a stress situation, which involved a 10-sec immersion in room temperature water 15–20 min before each daily trial. The general magnitude of the effect of treatment in the F_1 and F_2 generations was less than in the inbred strains themselves, which is not surprising in view of the evidence for the greater stability of heterozygotes as discussed above.

Other environmental extremes to which mice could be subjected are crowding, and high and low temperatures. It is known that under crowded conditions, adrenal weights of mice are high (Davis, 1966). A large adrenal gland leads to a high level of certain hormones, which have the effect of lowering reproductive rate. Such changes tend to regulate the population size. Under conditions of high population density, numerous aggressions occur and concurrently adrenal weight increases. There is some evidence for a similar situation in rabbits (Myers, 1966). Comparisons of different inbred strains and hybrids could well provide information of some importance on genotype × environment interactions in mice, and perhaps similar studies would be worth doing in mice collected from different populations in the wild, which would represent genetically more accurately the situation in human populations.

In the same way, because of the difficulty of studying extreme genotypes in man, the study of behavior under extreme environments could be worthwhile, especially if associated with various physiological and biochemical tests. Extreme environments could include the influence of drugs, alcohol, and temperature. Even so, we cannot go as far in studying humans, since differential mortality frequently has been observed in animals. Some of the issues discussed in this chapter may be advanced by an approach of this sort, for example, behavior under optimal and extreme environments in relation to socioeconomic class. The same could be studied in relation to somatotype and might provide additional information on the possible relationship between somatotype and behavior. Correlation matrices under optimal and extreme environments would be of interest; quite likely, levels of correlation might be lower under extreme environments. Overall comparisons between ethnic groups would be of interest due to known differential effects of certain drugs and presumably other environments on behavior in different ethnic groups. Generally, such studies could lead to the build-up of behavioral phenotypes under a multiplicity of environments, and this would probably be initially most successful for sensory-perception traits. It would also seem that additional insight could be obtained by carrying out parallel experiments in mouse and man simultaneously.

Mice and Men

Morphology and Behavior. Sheldon (1942; see page 79) discussed possible relationships between the somatotype and behavior. In spite of the high correlations he found, few further studies have been carried out (see Lindzey, 1967), and in fact Lindzey pointed out that there has been a reluctance of some psychologists to give serious consideration to the study of morphology and behavior: "The modal emphasis among psychologists in America has been upon learning, acquisition, shaping, or the modification of behavior, and not upon those aspects of the person and behavior that appear relatively fixed and unchanged."

Some associations between behavior and morphology can be cited in man. Thus the frail ectomorph cannot be expected to employ physical or aggressive responses with the same effect as the robust mesomorph; in other words height, weight, and strength put limits upon the adaptive responses an individual can make in a given environment. In women at least, linearity (ectomorphy) is negatively associated with the rate of physical and biological maturation. Individuals who are physically extreme in some sense, such as being excessively fat or thin, will be exposed to a somewhat different set of learning experiences than someone who is more modal physically or extreme in some other way. It should be pointed out that modality will vary between ethnic groups.

An extremely striking set of examples comes from data on the somatotypes of athletes (Carter, 1970). Thus almost all groups of championship athletes are rated high on mesomorphy, but the most mesomorphic are weight lifters, followed closely by Olympic Game track and field throwers, football players, and wrestlers. The least mesomorphic men are the distance runners. Women athletes range from the track and field jumpers and runners, who have the lowest mesomorphy, to the gymnasts, who have the highest. It is also of interest that champion performers at various levels of a particular sport exhibit similar patterns of body size and somatotype, but the patterns tend to become narrower as the level of performance increases, that is, extremes at the behavioral level correspond to extremes at the morphological level. Conversely, certain somatotypes found in nonathletes are not found at all in groups of champion athletes. It should be noted that extreme somatotypes in athletes may be made more extreme by training, but it would be unlikely that a nonathletic somatotype could be converted to an extreme athletic mesomorph merely by training.

Associations between morphology and temperament have been found in students with a very high correlation as was asserted in a striking form by Sheldon (1942). This has been confirmed more recently at a rather lower level of correlation (Child, 1950; Parnell, 1958; Walker, 1962; Lindzey, 1967). Among individuals showing criminal behavior, a number of surveys have shown an excess of mesomorphs (see Eysenck, 1964; Lindzey, 1967). Several investigators (see Heston, 1970) have found somatotype to be associated with schizophrenia since mesomorphs are underrepresented among schizophrenics, and ectomorphs are correspondingly overrepresented (see also Parnell, 1958). Paranoids, on the other hand are high in mesomorphy (Parnell, 1958).

Most morphological traits in man have a high heritability, as has been shown from twin studies and correlations between relatives (see for example Fisher, 1918; Clark, 1956; and Spuhler, 1962). The situation is rather more difficult to assess for behavioral traits because of the complication of possible environmental variation, but heritabilities of a number of traits are reasonably high (see Parsons, 1967a, for references). Unfortunately, in man, studies on sensory-perception traits, which can be regarded as "simpler" than learning or personality, are rarer than those on the latter. Since in some cases, we may be closer to the actual genetical and physiological basis of a sensory perception trait, Haldane's (1963) suggestion quoted by DeFries (1967), that qualitative (that is, single gene) and quantitative studies should be combined in work on human populations, clearly has merit. Haldane was referring to anthropometric data, but behavioral data are no different in principle. As he pointed out, this may lead to spurious correlations due to the presence of linkage disequilibria, however as an approach it seems worth exploiting further for those traits for which all members of a population can be assessed. Perhaps the results closest to Haldane's idea are certain sex-chromosome abnormalities that are known to have behavioral effects associated with morphological changes—in particular, changes in the gonads (see McClearn, 1970, for references).

In any case, whatever approach is used, the mouse data presented seem to argue for an association of morphology as assessed by weight and skeletal form with the simpler forms of behavior, but probably not with traits involving a large learning component. In man, as pointed out, it is these simpler traits for which associations would be expected to be found, but far fewer studies have been carried out with these than with learning.

The Mouse as a Prototype in the Study of Human Behavior. The experimental data discussed, in common with much work on the behavior genetics of the mouse (see Lindzey and Thiessen, 1970), are based on artifical laboratory strains of mice. In particular, the genetic architecture of the inbred strains probably bears little relationship to that of free-living populations of mice. Laboratory mice are normally tame and easily handled compared with wild mice. Inbred strains and hybrids among them have their place in behavior–genetic research, since they enable studies to be made on traits which will provide estimates of relative genotypic and environmental control, and hence may give us indications of traits worth studying in wild mice. In other words, they provide us with hints on phenomena and relationships that could be sought in the wild. But it must be remembered that an open natural population must cope with numerous conditions to which the closed and sheltered laboratory population is not exposed, and in fact it can be expected that genotypes would be developed in the laboratory that are inferior in viability in nature, since they are not exposed to the effects of the adverse and variable environments found in a natural environment.

Even so, in mice, it seems that we do have a species which could yield information directly relevant to the study of human behavior. There are three main groups of mice: (1) aboriginal mice, which so far as is known have never associated with man, (2) commensal mice, which have followed man around the world as scavengers, and (3) feral mice, which were once commensal with man, but are now not dependent on man for food and shelter. They all belong to the one species *Mus musculus,* and numerous varieties within the above three groups are known. There are therefore some analogies relevant to the study of man, for they represent a species divided into a number of populations, which from the morphological and coat color point of view are known to diverge. The various races of mice, inhabiting various different habitats and with presumed different behavior forms, could well provide us with a model through which inferences could be made about man, especially as many are associated with man. They have an added advantage since genetic manipulation can be performed with them. Little as yet has been done, but much could be done that may be of interest in studying behavior in man.

1. The behavioral profiles of races and subraces of mice could be studied, as well as the morphological and biochemical profiles.

2. Information on mating patterns in mice could be obtained; are they random, density-dependent for certain traits, or assortative? In man we know that mating is largely assortative (positive) for morphological, psychological, and sociological traits (Spuhler, 1962), and the same is likely for morphological traits in *Drosophila melanogaster* (Parsons, 1965).

3. Different races of mice can be studied under given environments, which is not possible in man. The environments could be both optimal and extreme.

4. Among other things, the study of mouse populations may provide hints as to likely processes that will occur due to overpopulation, especially with regard to behavioral changes.

5. The area where they may be the least beneficial is in the study of learning and reasoning, which are developed to their maximum in man, however the study of mice can provide some information.

It seems clear that the belief still held by some psychologists (Bruell, 1970) that it is possible to obtain "species-typical" estimates of behavioral parameters would be proven incorrect by such studies in mice. In man this is, of course, an oversimplification, since it is likely that just as different ethnic groups differ at the quantitative level for morphological features, they may differ for behavioral features. Unbiased evidence on behavior is more difficult to obtain, because of the effects of previous experience, hence we should probably be very careful in our interpretation of results on traits such as intelligence at this stage. On the other hand, differences between groups in man are known for simple sensory processes, and curiously enough, as pointed out by Spuhler and Lindzey (1967), the decades prior to 1900 probably saw more pertinent investigations of this type than occurred in the ensuing 50 years. Spuhler and Lindzey (1967) document a number of examples of racial differences in traits such as visual, auditory, olfactory, and tactile stimuli, and variations in taste and weight discriminations. Although there may be flaws in some of these data, they do suggest the possible existence of interesting and appreciable racial differences in behavior. Less complex processes studied include taste blindness to phenylthiocarbamide (PTC), and color blindness, which are under the control of major genes, and which vary in frequencies between racial groups in man. Just as differences between races in man have been quantified based on anthropometric traits, and polymorphic loci controlling blood groups, serum proteins, and enzymes, it is feasible that the same could be done for behavioral traits, especially for those traits measuring sensory perception. Between ethnic groups, there is a reasonable association between genetic distances for anthropometric traits and blood groups (Cavalli-Sforza and Edwards, 1964), and based on our argument we would expect this to occur for the simpler behavioral traits.

The Meaning of the Term "Race"

Since the term "race" has been used in this chapter, it seems desirable to give a working definition for it.

A race is a population in which the gene frequencies at some loci differ from one another. It is a quantitative and not a qualitative definition, as there are no biological isolating mechanisms between different human populations. Thus gene pools of different races have differing gene frequencies, and these are often maintained by positive assortative mating when races come together by migration, as is well-shown by minority immigrant groups in many countries.

It must be stressed that because the definition is quantitative rather than qualitative, the amount of difference needed to accept that we have two different races is completely arbitrary. Hence, there is the possibility of an infinity of classifications, as the literature shows, unless we say that populations differing by a certain arbitrary amount (probably measured as genetic distances) represent distinct races. Even the calculation of genetic distances is arbitrary, since they depend on the loci and/or the anthropometric measurements used to compute them, although it is known that there is often a reasonable correlation between distances calculated from gene frequencies and anthropometric measurements (Cavalli-Sforza and Edwards, 1964).

Distance computed on anthropometric traits are largely environment independent, especially if ages are taken into account, and gene frequency traits are completely environment independent. However can races or populations be distinguished based on behavioral traits which may be more environment dependent? For the simpler sense-perception traits of the type mediated by discrete and classifiable loci, it may be possible. As we approach traits involving learning and reasoning, this is less possible, since no two races can be said to exist in identical environments. The mouse experiments under discussion show this, since genotype × environment interactions were found to be the most important for traits incorporating a component of learning. In other words, the further we are from the direct action of genes, the greater the difficulty because of the likelihood of genotype × environment interactions, a problem presumably most acute for traits such as learning and reasoning. For example, if two races of man appear to differ in IQ or some such trait, I believe that this is meaningless from the interpretive point of view, since almost by definition, they will exist in different environments.

The other clear point is that there are good arguments for abandoning the term race, as it is clearly arbitrary, undefinable and without biological meaning. The term population, although suffering from many of the same difficulties, at least has a lesser emotional content.

Conclusions and Summary

1. The methods of biometrical genetics, as applied to plant breeding in particular, allow the estimation of the effect of genotype, environment, and genotype by environment interactions for quantitative behavioral traits in experimental animals. Because of the difficulties of defining environments and genotypes, the problem in man is much more difficult.

2. Three inbred strains of mice showed characteristic behavioral phenotypes: strains Ba and C57 were usually extreme, and C3H intermediate. The same was found for weight, and the incidence of minor skeletal variants. The similarity between weight and pattern of skeletal variation supports an association between the incidence of many skeletal variants and the size of structures associated with weight, and naively allows one to argue for an association between genotype, skeletal morphology, weight, and behavior. This may be reasonable, since skeletal variants are presumably associated with variants of the muscular, nervous, and vascular systems, and such variants would presumably have consequences at the behavioral level.

In man, this result supports postulated associations between somatotype and behavior described in the literature.

3. Most of the traits in mice showed dominance toward one or other extreme inbred strain in hybrids between the strains. These traits included open field activity, open field emotionality, exploratory activity, weight, and skeletal divergence. On the other hand, for learning in a conditioned-avoidance apparatus, heterosis was quite marked and was associated with lower variability in the hybrids compared with the inbred strains. The hybrids, therefore, show behavioral homeostasis for learning. Traits with a direct relation to fitness, of which learning is one, are expected to be subject to directional selection for higher fitness. They would be expected to show greater inbreeding depression, and consequently heterosis on crossing inbred strains, whereas the other traits studied would probably be subjected more to stabilizing selection showing less inbreeding depression and heterosis.

The behavioral homeostasis represents a form of genotype × environment interaction. If the problem of genotype × environment interaction is most marked for those traits related to fitness, such as learning, there are real difficulties in extrapolating to man, where neither genotype nor environment can be controlled. Simpler traits of the order of sensory-perception, therefore, should be more amenable to accurate study in man.

4. Correlation matrices between certain traits within strains and hybrids showed some consistency between strains and hybrids. In general, less extreme associations were found for hybrids than inbred strains, showing another form of genotype × environment interaction.

5. Genotype × environment interactions were found between trials in the conditioned-avoidance apparatus, for example, Ba and C3H tended to drop in learning ability after a rest, whereas C57 did not. The measure of learning is relevant, since if assessed as the percentage of no-shock jumps C3H is superior to C57, whereas based on the average times for all trials the reverse was found, so that different rankings can occur according to the method of assessing learning.

This points to even more complexities when attempting to extrapolate to man, for example, comparing results of different intelligence tests under differing types of previous experience.

6. Because of these complexities, the learning data in mice do not support any real association with weight and skeletal morphology, as found for simpler behavioral traits. Similarly, in man no real association would be expected between traits with a high learning component and somatotypes.

7. In experimental animals, genetic analysis is frequently based on extreme genotypes, and less frequently, extreme environments. Where extreme environments are studied, extreme genotype × environment interactions may occur such that inbred strains tend to be affected more than hybrids and extreme-environment heterosis tends to occur. It is considered that the study of behavior over many environments would be valuable in experimental animals. In man, where extreme genotypes cannot be bred, it is considered that the approach of using extreme environments should be explored more deeply and may add insight to a number of the issues raised.

8. The study of behavior in man is therefore partly one of unraveling genotype × environment interactions, and their estimation, a problem which becomes more acute as the learning component of a trait increases. Until this problem can be approached with greater precision, progress may be difficult, but a multidisciplinary attack in which experimental animals play a part may lead to insight. Basically, the function of experimental animals is to provide accurate and controlled data on relationships that could be investigated in man.

9. It is considered that possible associations between morphology and behavior in man should be explored further.

10. Some possible further extrapolations from mouse to man in the study

of behavior are considered, especially the possibility of extensive studies on wild mice.

11. Reasons indicating that the term race is not definable are given.

Acknowledgments

I am grateful to Mrs. Astrid Rose and Mr. Peter Close for helpful discussions. The research was supported by the Australian Research Grants Committee. Finally, I am very grateful to the members of the Committee on Basic Research in Education, who made my attendance at the workshop possible.

References

BERRY, R. J. (1963). Epigenetic polymorphism in wild populations of Mus musculus. *Genet. Res.* **4**, 193–220.

BODMER, W. F., and CAVALLI-SFORZA, L. L. (1970). Intelligence and race. *Sci. Amer.* **223**(4), 19–29.

BREESE, E. L., and MATHER, K. (1960). The organization of polygenic activity within a chromosome in Drosophila. II. Viability. *Heredity*, **14**, 375–399.

BROADHURST, P. L. (1967). An introduction to the diallel cross. *In:* "Behavior-Genetic Analysis," (J. Hirsch, ed.), pp. 287–304. McGraw Hill, New York.

BRUELL, J. H. (1964). Heterotic inheritance of wheelrunning in mice. *J. Comp. Physiol. Psychol.* **58**, 159–163. (a)

BRUELL, J. H. (1964). Inheritance of behavioral and physiological characters of mice and the problem of heterosis. *Amer. Zool.* **4**, 125–138. (b).

BRUELL, J. H. (1970). Behavioral population genetics and wild Mus musculus. *In:* "Contributions to Behavior-Genetic Analysis. The mouse as a prototype," (G. Lindzey and D. D. Thiessen, eds.), pp. 261–291. Appleton. New York.

CARTER, J. E. L. (1970). The somatotypes of athletes—a review. *Hum. Biol.* **42**, 535–569.

CASPARI, E. (1968). Genetic endowment and environment in the determination of human behavior: Biological viewpoint. *Amer. Educ. Res. J.* **5**, 43–55.

CAVALLI-SFORZA, L. L., and EDWARDS, A. W. F. (1964). Analysis of human evolution. *Proc. Int. Congr. Genet. 11th*, 923–933.

CHILD, I. L. (1950). The relation of somatotype to self-ratings on Sheldon's temperamental traits. *J. Personal.* **18**, 440–453.

CLARK, P. J. (1956). The heritability of certain anthropometric characters as ascertained from measurements of twins. *Amer. J. Hum. Genet.* **8**, 49–54.

COOPER, R. M., and ZUBEK, J. P. (1958). Effects of enriched and restricted early environments on the learning ability of bright and dull rats. *Can. J. Psychol.* **12**, 159–164.

DAVIS, D. E. (1966). "Integral Animal Behavior." MacMillan, New York.

DeFRIES, J. C. (1967). Quantitative Genetics and Behavior: Overview and perspective. *In:* "Behavior-Genetic Analysis," (J. Hirsch ed.), pp. 322–339. McGraw-Hill, New York.

DEOL, M. S., and TRUSLOVE, G. M. (1957). Genetical studies on the skeleton of the mouse. XX Maternal physiology and variation in the skeleton of C57B1 mice. *J. Genet.* **55**, 288–312.

DOBZHANSKY, T. (1948). Genetics of natural populations. XVIII. Experiments on chromosomes of Drosophila pseudoobscura from different geographic regions. *Genetics* **33**, 588–602.
EYSENCK, H. J. (1964). "Crime and Personality." Routledge and Kegan, London.
FISHER, R. A. (1918). The correlation between relatives on the supposition of Mendelian inheritance. *Trans. Roy. Soc. Edinburgh,* **52**, 399–433.
GRÜNEBERG, H. (1952). Genetical studies on the skeleton of the mouse. IV. Quasicontinuous variations. *J. Genet.* **51**, 95–114.
GRÜNEBERG, H. (1963). "The Pathology of Development: A Study of Inherited Skeletal Disorders in Animals." Blackwell Scientific Publications, Oxford.
GUTTMAN, R. (1967). Cross-population constancy in trait profiles and the study of the inheritance of human behavior variables. *In:* "Genetic Diversity and Human Behavior," (J. N. Spuhler, ed.), pp. 187–197. Aldine, Chicago.
HALDANE, J. B. S. (1963). *In:* "The Genetics of Migrant and Isolate Populations," (E. Goldschmidt, ed.), p. 43. Williams and Wilkins, Baltimore.
HESTON, L. L. (1970). The genetics of schizophrenic and schizoid disease. *Science,* **167**, 249–256.
HOWE, W. L., and PARSONS, P. A. (1967). Genotype and environment in the determination of minor skeletal variants and body weight in mice. *J. Embryol. Exp. Morphol.* **17**, 283–292.
JINKS, J. L., and FULKER, D. W. (1970). Comparison of the biometrical, genetical, MAVA, and classical approaches to the analysis of human behavior. *Psychol. Bull.* **73**, 311–349.
LINDZEY, G. (1967). Behavior and morphological variation. *In:* "Genetic Diversity and Human Behavior," (J. N. Spuhler, ed.), pp. 227–240. Aldine, Chicago.
LINDZEY, G., and THIESSEN, D. (1970). "Contributions to Behavior-Genetic Analysis: The mouse as a prototype." Appleton, New York.
McCLEARN, G. E. (1970). Behavioral genetics. *Ann. Rev. Genet.* **4**, 437–468.
MANOSEVITZ, M., and LINDZEY, G. (1970). Genetic variation and hoarding. *In:* "Contributions to Behavior-Genetic Analysis. The mouse as a prototype." (G. Lindzey and D. D. Thiessen, eds.), pp. 91–113. Appleton, New York.
MATHER, K. (1966). Variability and selection. *Proc. Roy. Soc. London Ser. B.* **164**, 328–340.
MYERS, K. (1966). Morphological changes in the adrenal glands of wild rabbits. *Nature* **213**, 147–150.
PARNELL, R. W. (1958). "Behaviour and Physique: An Introduction to Practical and Applied Somatometry." Arnold, London.
PARSONS, P. A. (1965). Assortative mating for a metrical characteristic in Drosophila. *Heredity* **20**, 161–167.
PARSONS, P. A. (1967). "The Genetic Analysis of Behaviour." Methuen, London. (a)
PARSONS, P. A. (1967). Behavioural homeostasis in mice. *Genetica* **38**, 134–142. (b)
PARSONS, P. A. (1971). Extreme-environment heterosis and genetic loads. *Heredity* **26**, 479–483.
PARSONS, P. A., and McKENZIE, J. A. (1972). The ecological genetics of Drosophila. *Evol. Biol.* **5**, 87–132.
ROSE, A., and PARSONS, P. A. (1970). Behavioural studies in different strains of mice and the problem of heterosis. *Genetica* **41**, 65–87.
SEARLE, A. G. (1954). Genetical studies on the skeleton of the mouse. IX. Causes of skeletal variation within pure lines. *J. Genet.* **52**, 68–102.

SHELDON, W. H. (with the collaboration of S. S. Stevens and W. B. Tucker). (1940). "The Varieties of Human Physique: An introduction to constitutional psychology." Harper, New York.

SHELDON, W. H. (with the collaboration of S. S. Stevens). (1942). "The Varieties of Temperament: A psychology of constitutional differences." Harper, New York.

SPUHLER, J. N. (1962). Empirical studies on quantitative human genetics. U.N./W.H.O. Seminar on "The use of vital and health statistics for genetic and radiation studies". Pp. 241–252.

SPUHLER, J. N., and LINDZEY, G. (1967). Racial differences in behavior. *In:* Behavior-Genetic Analysis. (J. Hirsch, ed.), pp. 366–414. McGraw-Hill, New York.

TRUSLOVE, G. M. (1961). Genetical studies on the skeleton of the mouse. XXX. A search for correlations between some minor variants. *Genet. Res.* **2**, 431–438.

WALKER, R. N. (1962). Body build and behavior in young children. I. Body build and nursery school teachers' ratings. *Monogr. Soc. Res. Child Devel.* **27**(3), 1–94.

DISCUSSION

LEONARD L. HESTON

University of Minnesota
Minneapolis, Minnesota

My training and major interest is in clinical psychiatry. I turned toward genetics because I was concerned with abnormal behavior and dissatisfied with the explanatory systems in vogue in psychiatry and psychology. I am thus a consumer of the sort of research described by Professor Parsons and not a competent critic of its technical aspects. My remarks will be aimed mainly at choices among research strategies.

The topic is "Genetic Determinants of Behavior (Human)." But the data presented by Professor Parsons come from animal research. Although this is understandable, it puts us on the horns of a chronic dilemma in human biology: To what extent is research on other animals applicable to man? Although the problem is old, I think that behavior genetics presents a somewhat different point of departure.

There can be no question that animals are worth studying in their own right. There can be no question that we have learned far more about general genetics from animals than we are ever likely to learn from man. Some animal work, for example, that of ethologists observing natural populations, has produced important insights. But as a consumer, I have found little help from recent studies of the behavior of laboratory animals. I look for lessons applicable to human behavior and find few of them. I look for new hypotheses, and find few of them, and none that seem central to human behavior. I find that basic problems that must be addressed in the study of a human population, assortative mating, for example, are defined out of most animal studies. To the extent that investigators intend their research to be applicable to man, different approaches may be more helpful. And we who deal with human behavior do need help. Because control of critical variables in the human population is impossible, we need help in developing a methodology. Which variables are likely to be important or most important? Is there perhaps a sequence of observations most likely to avoid hopelessly confounding our experiments? Perhaps by simulating the human condition, animal geneticists might find indirect methods of experimental control applicable to man. Now, of course, if one's aim were learning about the genetics of animal behavior, simulation of human conditions would mean doing animal experiments in irrational ways. But developing a methodology for doing human research is quite another goal, which I think is well worth pursuing.

Professor Parsons begins to come to grips with such issues. He suggests studying the mating patterns of wild populations. I agree, because this is a condition that must be met in human research. Inbred strains, on the other hand, seem largely irrelevant to the human condition. Yet they are almost universally used in animal behavior genetics. And I think we may go on to ask why rodents at all? Why not our primative primate ancestor the tree shrew? If the investigator is interested in human behavior, tree shrews, though no doubt more difficult to study, would probably be more pertinent. Professor Parsons also suggests the study of hybrids, which I heartily second. It seems to me that the study of hybrids is the best available method of defining and studying behavioral differences, if any, in human subpopulations. We have available now children of American soldiers of all colors by women of many countries of the world. We could probably learn much through the study of these populations, but I have no doubt we could learn much more if a methodological foundation were laid by animal research.

Another category of problem which Professor Parsons does not mention is that of trait selection. The definition of traits is a central problem in human behavior genetics, yet many of the traits studied in animals seem to me to have no human counterpart. For example, why has no one studied the open-field behavior of humans? Technically, this could be done. But what would we learn? Where would we go from there? The trait is of interest in the biology of mice, but man does not exhibit it in any recognizable form in his natural settings. If the investigator is interested in man, traits appearing in animals that are the same or analogous to human traits seem to me to be better choices. For example, what are the genetics of monogamous pair-bond formation? This trait might be studied in primate subspecies by using individual animals deviant for the trait in breeding experiments. Also, hybrids between monogamous and nonmonogamous groups should be informative. Other animals, including some much easier to handle than primates, could be studied in the same way. From such studies the human geneticist would get a model and a start toward a comparative science. I do not know whether it is possible for animal geneticists to do such research, let alone whether any want to. However, this sort of basic data is needed by human behavior genetics.

Let me carry the matter of deviant animals a little further. One of the main lessons of medicine is that we have learned most of what we know about normality from the study of abnormality. It seems to me that in man the extreme genotypes Professor Parsons seeks are usually pathological ones. Selecting animals for traits such as easy handling and health must have resulted in the loss of many informative genotypes and of course destroyed analogies between human and animal populations.

Professor Parsons suggests the use of extreme environments, which of course can be studied in animals much more readily than in man. The work that has been done relevant to behavior ordinarily involves some sort of social deprivation and we must ask if such experimental conditions have human counterparts, and if they do, what are they? I conclude I just do not know, largely because the experimental deprivation has been much more extreme than that occurring to man. Again, I wonder if the simulation of human environments might not yield more informative data. Perhaps one way to carry the matter further would be through simulated natural selection. Starting with wild populations and exposing them to different chronic environments corresponding, for example, to tropical, temperate and arctic ones, might tell us something about how man evolved.

Professor Parsons is not very hopeful about the study of the genetics of learning. It seems to me that learning is only a special case of gene–environment interaction. Sensory experiences from the environment are stored later to be retrieved, often integrated in novel forms. These are surely lawful processes and the laws must be based on gene action. The problem, of course, is that we know little about the physiology of learning. Yet leaving learning out of animal experiments seems to me to neglect an extremely important facet of human behavior. To the extent that mice do not provide means for studying learning, they are just not suitable models for human behavior genetics.

There is another sense in which I think human behavior genetics might be better served by animal work. Most of the analytic tools used in animal research are inappropriate to the objectives of human behavior genetics. The concern in man is fundamentally individual variation. Statistical description is not enough. Knowing the heritability of a trait tells me very little. It seems to me a very short time ago that demonstrating a genetic factor in schizophrenia was extremely important. In one sense, it was. The demonstration that genes are important undid a lot of nonsense and the nonsense was adversely affecting the lives of schizophrenics and their families. But what positive benefits accrued? Well, we cannot specify a single gene product associated with schizophrenia. Neither can we specify a single environmental codeterminant. We have a slightly better basis for genetic counseling and we have a better basis for further research. But the situation of schizophrenics and their families has not been improved very much. So I say we have enough of demonstrating heritability, even if we thereby win victories over prevailing modes of thought in psychology and psychiatry. These are Pyrrhic victories, if they take much of our time and energy. Indeed, to the extent that we allow them to take our time and energy, we are immature as a discipline.

Future victories seem to me to depend on getting down to proteins. To the extent that we aim to treat or intervene, we in human behavior genetics, do not have the methods of animal or plant breeders. Instead of culling out unwanted stocks and breeding for specific traits, our treatment must aim at the adjustment of environments. The scope for selection through genetic counseling is limited. In order to intervene through environmental adjustments, we will have to understand mechanisms, and proteins are our effector molecules. Again, the study in animals of protein variation associated with behavioral variation might get human behavior genetics off to a much faster start. Of course, protein variation is no royal road to knowledge. The simple demonstration of protein polymorphisms is of limited use. We must delineate associated physiological changes and physiology in this context means behavior. That is why I answer Dr. Thompson's earlier question about the place of psychologists in behavior genetics in the affirmative. We do need psychologists. We also, as I hope I have made clear, need more animal research, although research of a somewhat different kind.

COMMENT

NEWTON E. MORTON

University of Hawaii
Honolulu, Hawaii

Proposals to study behavioral traits in hybrid populations overlook the obvious fact that if there is an important environmental difference between the parental groups, this difference may well persist in the hybrid populations. Considering this fact, our studies of Hawaii have been concerned almost entirely with hybridity effects and to a very small extent with attempts to interpret differences among the parental groups, which for morbidity traits we regard as generally not rigorously explicable without experimental control. On this argument, the value of hybrid populations for behavioral traits may well lie in testing for nongenetic effects; for example, prenatal environment in reciprocal crosses could be assayed by electrophysiological response in neonates. There is also the process of selection and social pressure in the hybrid population acting in a variety of ways. It is certainly not possible to collect a strictly random sample of hybrids between Africans and Caucasians, because only a small and biased sample of the population participates in such crosses and (what is perhaps more important) is willing to participate in a behavioral study. I am greatly discouraged about the value of hybrid populations in the absence of reciprocal crosses which alone rescue such material from being a collection of uninterpretable data.

Chapter 6

Human Behavioral Adaptations: Speculations on Their Genesis

I. I. GOTTESMAN and L. L. HESTON
University of Minnesota
Minneapolis, Minnesota

It is presumptuous to talk about the evolution of any primate characteristics, let along the evolution of human behaviors among *echt* scientists. But like the rodent and the cobra, we may temporarily take pleasure from the excitement and fascination of the confrontation and take our chances with being consumed. As has often been noted, behavior leaves no fossils; it was paleontology that did so much to legitimize the scientific (as opposed to the philosophical) credence in Darwinian theory. Unfortunately, we behavioral evolutionists have no tool more powerful than analogous reasoning and little unassailable evidence. A few years ago, Bullock (1970) introduced his paper on the physiological basis of behavior to the XVI International Congress of Zoology with the following paragraph:

> The gulf between our present level of physiological understanding and the explanation of behavior as we see it in higher forms is wider than the gulf between atomic physics and astronomy and is indeed the widest gap between disciplines in science. But real understanding of behavior is the great

challenge of the future, not only for biology but for all sciences. It cannot wait for full development of the basic sciences on which it eventually rests, but must proceed, as it is proceeding, simultaneously on all levels [p. 451].

Although we can use such words as "amulet," we would like to present one further caveat to remind us of our vast ignorance of how natural selection operates to produce evolution (that is, in practice, not in principle). Li (1970), in discussing human genetic adaptation, said,

> Most of the selection models so far studied by geneticists are limited to one pair of genes. In fact, a general solution for selection with respect to two pair of genes has not yet been obtained and is still under active investigation, despite the liberal use of high-speed computers [p. 561].

Since virtually all behavioral traits of interest in man will be under both polygenic genetic and environmental control, our paper must of necessity consist largely of speculations about the evolution of human behavior. Hopefully the elucidation of a few principles of evolution found useful for understanding nonbehavioral traits will provide a frame of reference for a program oriented around basic research in education.

General Considerations: Evolutionary Outcomes and Kinds of Selection

Adaptedness is a product of evolutionary development which is maintained and can be improved by natural selection. Natural selection is the force underlying evolution; the essence of natural selection is the differential reproduction of genotypes best adapted to the demands of the environment. You may either accept the teleology involved here or spend a lifetime grappling with it (see Dobzhansky, 1970). Selection occurs in many different forms, with different consequences. First there is stabilizing or normalizing selection, which works to maintain the status quo of a population's gene pool; it is a conservative force, not to be ridiculed, rather like investing only in government bonds. Then there is balancing selection, which adds another technique for maintaining genetic variability; it permits genes to stay in the population even though they have bad effects in double doses because in single doses they appear to confer some kind of advantage in some environments. This kind of selection acts like a kind of premium paid for disaster insurance; it hurts to pay it, but it pays off if the disaster ever materializes. Of most importance to a discussion on the evolution of human behavior are directional selection and diversifying selection. The former causes the composition of the gene pool to change or shift in some particular direction so as to accommodate more efficiently changes which have occurred in the environment (it will also preserve fresh mutations which work better than the old ones even if the environment did not change). The latter, diversifying selection, seems to us to just be "multidirectional" selection, a kind that simultaneously favors two or more phenotypes in an environment with multiple niches. Multidirectional selection underlies such phenomena as sexual dimorphism and the formation of subspecies, incipient species, or new species.

It is immensely important when talking about the possible roles for selection pressures in molding the shape of a gene pool to remember that natural selection operates on the total organism (the phenotype), with indirect effects on the gene pool of the following generation. The Darwinian fitness of an individual (that is, the number of offspring he has) is the net result of the sum of his assets and his liabilities in a particular environment. The corollary of this proposition is that man is simultaneously being subjected to different selection pressures from many different selective forces. A widespread misapprehension about how natural selection works may stem from the wider familiarity of the public with artificial selection for one economically useful character at a time in domesticated animals, forgetting that the breeder can easily eliminate the animals which do not suit him.

A Brief Overview of Primate Phylogeny

Thinking in evolutionary terms requires some perspective of the time periods involved and the relative position of the phylogenetic branches.

With the development of methods for determining the sequence of amino acids in proteins, a new tool supplementing older ones was added to the study of evolution. Because a mutation at a structural gene locus may result in the substitution of one amino acid for another in the completed protein, tracing the variation in a protein through a group of organisms and counting the number of amino acid substitutions gives an estimate of the relative distance in time separating the species (an evolutionary protein clock). For example, human hemoglobin and chimpanzee hemoglobin are the same. Gorilla and human hemoglobin differ in 2 amino acids, men and monkeys differ in 12 amino acids, and men and horses in 43 amino acids (Wilson and Sarich, 1969). Figure 1 indicates the evolutionary paths that seem most likely for man and some of his closer primate relatives as well as rough estimates of divergence times.

Evolution of Brain Size and Tool Use

Thiessen said (1972) that "There is little investigative hope of constructing a phylogenetic tree to express the evolutionary trends of behavior. Evo-

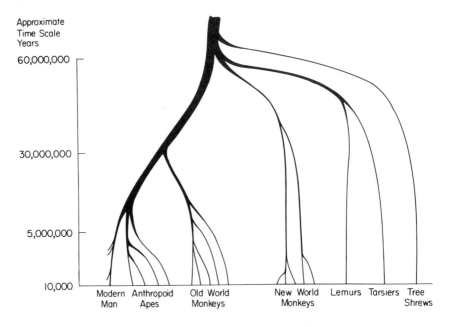

FIG. 1.

lution [of behavior] has not been progressive or linear and has not occurred at uniform rates. . . . There is more hope, it seems to me, in dealing directly with species-specializations and treating them as evolutionary reflections of ecological demands." We agree. One obvious line of evidence that must be pursued is the evolution of the brain.

Jerison (1963) has estimated the number of *adaptive neurons* in mammals of different sizes using information on brain and body weight; the method allows the eight major primate taxa to be distinguished from each other, but not closely related species or genera. Adaptive neurons are defined as those left over after basic "housekeeping," for example, moving muscles and maintaining visceral function. The results of the method are given in Table 1. It is obvious that the numbers of neurons have increased tremendously in hominid evolution. Keeping in mind the numerous speculations and approximations that have entered into the Jerison technique, it still manages to show a difference in the predicted direction between *erectus* and the anthropoid apes, despite similarity in brain size. It also demonstrates the predicted similarity between chimpanzee and gorilla despite a big difference in body weight.

The increase in "discretionary" neurons (roughly) parallels increasing ability to make and use tools, even though gigantic strides and sophistication

TABLE 1 Estimates of Adaptive Cortical Neurons in Mammals of Different Brain and Body Sizes[a]

Animal	Brain weight (gm)	Body weight (gm)	Number of adaptive neurons (billion)
Macaca (Rhesus)	100	10,000	1.2
Papio (Baboon)	200	20,000	2.1
Pan troglodytes (Chimpanzee)	400	45,000	3.4
Pan gorilla	600	250,000	3.6
Australopithecus	500	20,000	4.4
Homo erectus	900	50,000	6.4
Homo sapiens	1300	60,000	8.5
Elephant	6000	7,000,000	18.0
Porpoise	1750	150,000	10.0

[a] Adaptive neurons are those cortical neurons associated with the adaptive capacity of the brain. After Jerison, 1963. Reproduced by permission.

appeared after the level of 8.5 billion adaptive neurons was reached near the end of the Middle Pleistocene some 100,000 to 200,000 years B.C.[1] The archaeological record suggests that there was little improvement in the pebble tools used by Australopithecine at the beginning of the Lower Pleistocene some two to five million years B.C. until the flake tools of *Homo erectus* in the Middle Pleistocene, roughly a half million years B.C. Buettner-Janusch (1966) interpreted this to mean that the rather abrupt change in the "tool kit of man" (if it was abrupt) was not associated with a stepwise increase in adaptive neurons, perhaps because a sufficient threshold number had been reached permitting the adaptive capacity necessary for diversity and elaboration of tools. The degree of correlation between evolution of brain size and tool manufacturing is vague; the evolution of culture may have been as important or more so as a selection pressure favoring the increase in brain size.

> Once the neurological capacity to symbolize and to make culture evolved, the differentiation and rapid development of culture itself very likely put severe demands upon the brain. . . .This probably required elaboration of the cerebral cortex, a larger set of association neurons and interconnections between them [Buettner-Janusch, 1966, p. 352].

It is easy to imagine that even with tools at his disposal, early man required massive changes in social organization—the formation of a hunting group—which in turn demanded efficient communication, cooperation

---

[1] We ask forgiveness if these dates and others have again changed since our data sources were published.

among males and role specialization, planning ability, and longer term memory storage. To quote Buettner-Janusch again,

> The lineage of primates in which all these capacities were presumably developing would be under strong selection pressure to continue to develop and refine such traits, in an environment rapidly changing from forest to open bush and plains [p. 360].

Anthropologists disagree among themselves as to the relative importance of bipedalism, tool use, and social organization as selection pressures favoring increasing brain size (Washburn and Avis, 1958). We leave the intriguing data in Table 1 on the elephant and porpoise to the discretion of our anthropologist–geneticist–anatomist discussant, Professor Pollitzer.

Further discussion and references about the evolution of neocortex may be found in Diamond and Hall (1969) along with a very clear example of the results of convergent evolution on the visual system of squirrels and tree shrews (the "lowest" living primate), who are unrelated species occupying similar ecological niches. These animals have independently evolved visual systems that are virtually identical. Similar environments *are* able to produce similar organs. This provides a concrete example of the Markov chain principle as discussed by Li (1970); once similar endpoints have been attained by two populations, for many important purposes their past evolutionary history does not matter.

Ernst Mayr (1963) has applied the term *mosaic evolution* to the process whereby each organ and each system of organs has its own rate and pattern of evolution. Mosaic evolution characterizes the form of genetic response that follows a move into new adaptive zones; it supports the view that man became what he is today very, very gradually. Phylogeny in relationship to the evolution of behavior is discussed cogently by Hodos and Campbell (1969).

Within Species Behavioral Variability

All men belong to one species, but races of men or other Mendelian populations can be thought of in some respects as incipient species. *Homo sapiens* has failed to speciate for two main reasons.

> . . . man's great ecological diversity. Man has, so to speak specialized in despecialization. Man occupies more different ecological niches than any known animal. If the single species man occupies successfully all the niches that are open for Homo-like creatures, it is obvious that he cannot speciate. The second reason is that isolating mechanisms in hominids apparently develop only slowly. . . .The probability of man's breaking up into several

species has become smaller and smaller with the steady improvement of communication and means of transport. The internal cohesion of the genetic system of man is being strengthened constantly [Mayr, 1963, pp. 643–644].

From an evolutionary viewpoint, we are interested both in genetically conditioned homogeneity (species-specific characteristics) as well as genetically conditioned heterogeneity (non-species-specific characteristics). It can be hypothesized that the former evolved through parallel and convergent evolution while the latter evolved through divergent evolution.

Thiessen (1972), speaking as a comparative animal behavioral geneticist, has made a cogent case to the effect that traits related to fitness show a restriction of genetic variability (and low heritabilities). He suggested that polymorphisms observed in a species' gene pool may be functionally equivalent. From this he made the provocative suggestion that traits with high heritabilities may be "genetic junk." It has been observed by animal breeders (Lush, 1945) that artificial selection for trait uses up its additive genetic variance and leads to low heritabilities. These ideas may serve as points of discussion by this group. Although we agree with Thiessen in respect to other animals, in defense of our serious interest in within *Homo sapiens* variability, we must point out that we have no way of knowing in advance whether trait differences between populations reflect directional or diversifying selection, a transient polymorphic state of affairs, or nongenetic adaptability. We believe that a better understanding of human behavior may result from such concern with trait variation within our species.

Our species-specific curiosity and self-awareness make us want to know about the meaning and significance of our non-species-specific (and fascinating) diversity. Such a stance may also permit us to discern the directions in which man is continuing to evolve. One of the goals of our paper is to stimulate discussion about the circumstances that could have led both to similarities as well as to differences in the genetic bases for human behavioral traits within and between Mendelian populations.

Adaptability and Genotype–Environment Interaction

Given the well-worked-out example of the relationship between the gene for sickle cell hemoglobin and heterozygote advantage in a malarial environment, it is too easy to jump to the conclusion that other genetic polymorphisms are also maintained by some kind of selective advantage. Other examples, however, are exceedingly scarce. The genetic diversity of man has been amply demonstrated, but is hardly understood. We have good evidence based on enzymes and red cell antigens in humans that about 30% of all our genetic loci may be polymorphic; except for the one example of Hb^s, and possibly a few others, we do not know what maintains the remaining polymorphisms and do not understand the physiological function of the alleles involved. Do the kinds of phenotypic differences we see among races or Mendelian populations also imply selection pressures in their ancient histories with consequent changes in their genotypes? It turns out that it is very difficult to distinguish between changes due to behavioral and physiological adaptability and those due to changes in adaptedness via natural selection leading to gene pool changes (see Ayala, 1970, and Dobzhansky, 1968). A concise treatment of the difficulties may be found in Lorenz's (1965) essay *Evolution and Modification of Behavior*.

As an example of the problems, the increased height in Japanese children born to Japanese parents in the United States compared to those born in Japan is well documented (Greulich, 1957). It is a good example (assuming no selective migration) of a phenotypic change not associated with a genotypic one; it is an example of the reaction range concept (Gottesman 1963, 1968) with the improved pre- and postnatal environment in the Japanese–American promoting a changed phenotype. Height is an excellent trait for model building in that it is under both genetic and environmental control. The reaction range concept builds on the classical work of Clausen et al. (1948), who planted different races of plants together (genetic heterogeneity + environmental homogeneity) and transplants of the same plant in different environments (genetic homogeneity + environmental heterogeneity). Two important axioms of the reaction range concept are the fol-

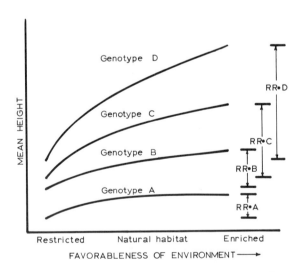

FIG. 2. The reaction range concept applied to height.

6. HUMAN BEHAVIORAL ADAPTATIONS

lowing: (1) Different genotypes may have the same phenotype (observed characteristic) and (2) Different phenotypes may have the same genotype (that is, for the trait under consideration). Figure 2 illustrates the concept with respect to human height (Greulich, 1957; Meredith, 1969; Morch, 1941) although it can be generalized to such traits as IQ score (see Gottesman, 1963). The units for both X and Y axes are only ordinal and not to scale.

Each curve in the figure can be construed as representing the phenotypic response of samples of individuals, homogeneous for four different levels of genetic potential for height, who have been reared in various trait-relevant environments (or niches) crudely characterized as restricted, natural habitat, and enriched. Curve Type A could represent a deviant genotype, for example, the one associated with the dominant gene for chondrodystrophic dwarfism, which has an incidence at birth of 1 in 10,000. The different environments to which such dwarfs have so far been exposed do not have much effect on their height; the mean height for 15-year-old cases (sexes combined) is only 120 cm. Curve Type B could represent samples of 13-year-old Japanese girls: in contemporary Japan they average 146.1 cm (= "natural" habitat); 13-year-old girls measured in postwar Japan (1950) only averaged 139.9 cm (= restricted environment nutritionally); 13-year-old Japanese girls born in the United States to Japanese parents averaged 150.5 cm (= enriched environment). The Reaction Range (RR B) for the genotype represented by 13-year-old Japanese girls under the range of environments sampled would be the largest value minus the lowest, or 10.6 cm. Curve Type C could represent the response of the genotypes of 15-year-old Japanese boys measured at the same times as the girls in B; we are dealing here with sexual dimorphism and a different genotype (*pace* women's lib) for height. Postwar boys averaged 151.1 cm; contemporary boys in Japan, 158.2 cm; and contemporary Japanese boys born in the United States, 164.5 for a reaction range of 13.4 cm, all attributable to environmental variations. Curve Type D could represent 15-year-old white American boys who average 168.7 cm (13-year-old white girls average 155.4 cm). Examples of the same phenotype with different genotypes are provided by some data on children of Japanese–American white matings (fathers always white); the 15-year-old boys averaged 164.7 cm while the 13-year-old girls averaged 151.5. It appears that the hybrids matched the American-born Japanese and were about halfway between contemporary Japanese and white children (under natural habitat conditions). Other genotypes could have been added to Fig. 2 for such diverse groups as the Mbuti pygmies and Nuer of Sudan, with adult mean heights of 144 cm and 184 cm, respectively. The thrust of the reaction range concept is that both heredity and environment are important in determining trait variation but in different ways, combinations, and degrees, some of which are amenable to dissection for some traits.

A particularly instructive example of adaptability in the face of selection pressure without, apparently, a genetic change is given by Harrison's (1967) work in Northern Ethiopia. Two populations of Amharic Ethiopians were shown to be essentially similar genetically by similar blood groups and because there were high migration rates between the two populations, who are only separated by a short geographical distance. However the homeland elevations involved are 5000 and 10,000 feet. Such a difference in altitude would be expected to exert differential selection pressures on the two groups. At 10,000 feet there was lowered partial pressure of oxygen and colder temperatures; at 5000 feet, malaria, dysentery, and measles were much more common. We will only report some of the morphological differences between the highlanders and lowlanders. The former were heavier and had greater chest circumference and antero-posterior and transverse chest widths. Harrison found that the parameters were not only modifiable during growth but also in adulthood; adult migrants to the highlands showed a morphology similar to the indigenous highlanders. The enlarged chests were due to hypertrophy of the intercostal muscles. Migrants to the lowlands lost some of their adaptability and were intermediate in morphology. The important lesson in these data according to the investigator was that they showed that these differing ecological niches did not require evolutionary change. Adaptability was all that was needed.

Men have worked on the surface of the moon, a very inhospitable ecological niche. We did not have to breed a new race or genotype for that niche; the adaptive potential of the *Homo* genotype, selected for plasticity and greatly extended and multiplied by technology, permits such phenotypic diversity. These kinds of examples can be seen many times over in Baker and Weiner's (1966) *The Biology of Adaptability* and lead us to counsel caution before automatically ascribing phenotypic differences in biologically based traits to genetic differences.

We now turn to the opposite sort of error. Some behavioral differences between Mendelian populations of *Homo sapiens* may be associated with genetic differences.

One of the many working hypotheses in a discussion of the evolution of human behavior is that cultural transmission is man's principal instrument of adaptation. At our present state of knowledge, this is only opinion. Although culture is transmitted nongenetically via learning, it has a genetic foundation that characterizes our species and which evolved genetically. With a few pathological exceptions, it can be argued, but not proved, that selection pressures were uniform across whatever races existed in the Middle Pleistocene for the general properties of educability, and the capacity to learn from positive and negative reinforcement (see Caspari, 1958; Dobzhansky, 1962). It is not empty diplomacy to talk simultaneously about genetic and cultural evolution of behavioral traits. To quote Dobzhansky (1969),

> Culture proved to be an adaptive contrivance more potent than any other which appeared in the whole evolutionary history of life. This does not make genetic development superfluous. However, genetic adaptation is shifted, so to speak, at two removes from the environments which the human species has to face. Genetic evolution enhances the efficiency and the open-endedness of the non-genetic, i.e., cultural evolution [p. 290].

As an illustration of this interaction, Dobzhansky sketched a scenario about the invention of the use of fire by *Homo erectus* in eastern Asia during the Middle Pleistocene.

> Here was an adaptive achievement of a highest order, symbolized in the myth of Prometheus. Did this race possess a special Promethean gene, which other races had to acquire before they too could use fire? This is unlikely. The inventors and their disciples had, however, a common genetic system which enabled them to learn and to transmit what they had learnt. It makes little difference to the argument if we suppose that the race *Homo erectus pekinensis* had a better, or only an equally good, genetic equipment for learning and for transmission of learned information as did other races of the same species [p. 290].

But do not let the story end there, for the case can also be made, and has, that natural selection can "shape" behavior just as it has shaped protective coloration in the famous example of industrial melanism in moths. Tinbergen (1970) has issued a number of warnings which we can paraphrase as follows:

1. Do not assume a behavior is without adaptive value unless you have ruled it out by observations, preferably in a natural setting.
2. Do not be too quick in blaming differences between groups on genetic drift.
3. Do not be too quick to attribute the presence of a character merely to pleiotropism.

The plain fact is we usually cannot choose among the alternatives presented.

The Evolution of Milk Drinking

Darwin in *The Descent of Man* (1871) suggested a strategy for making choices and anticipated modern developments in human genetics when he commented on the relationship between hair and skin color and immunity to tropical diseases. He had observed that white settlers in Africa died of malaria and yellow fever while natives did not, and that Sudanese recruited to fight in Mexico also escaped them. "That the immunity of the Negro is in any degree correlated with the colour of his skin is a mere conjecture," he said, and then proceeded to design a research project that was never completed. In the spring of 1862 he obtained permission from the army to elicit information from the surgeons stationed with troops in tropical areas about the hair color of Englishmen affected with dysentery, malaria, and yellow fever. He concluded his request with this prophetic comment,

> Theoretically the result would be of high interest, as indicating one means by which a race of men inhabiting from a remote period an unhealthy tropical climate, might have become dark-coloured by the better preservation of dark-haired or dark-complexioned individuals during a long succession of generations.

Another trait in which human populations differ is the concentration of the enzyme lactase. It is the only common trait known at both the biochemical and behavioral levels that contributes to "normal" variability in both. Although even here there is much that remains to be learned, lactase provides a reasonable model of divergent evolution. We owe much of our understanding to reviews of the subject by McCracken (1971) and Simoons (1970).

Lactase is an enzyme active in the villi of the small bowel and lactose is the main sugar in milk. Lactase splits the disaccharide lactose into the monosaccharides glucose and galactose. Monosaccharides can be absorbed into the portal circulation, but disaccharides cannot. In the absence of lactase, ingested lactose simply passes through the gut without providing nutrition. If too much is ingested, cramps and diarrhea result. The syndrome has been recognized in medicine only since 1963.

That the enzyme deficiency is a genetic and not an acquired trait produced by lack of dietary milk now seems likely. The evidence is provided by a study of Thai children living in an orphanage where milk was fed from birth. By age 2 all were lactose intolerant (Simoons, 1970). The 13 families reported in the literature (Ferguson and Maxwell, 1967; Fine *et al.*, 1968; Flatz and Saengudom, 1969; Welsh *et al.*, 1968; Welsh, 1970) are consistent with and suggest to us a two- or possible three-allele locus as the mode of transmission. One allele appears to be sufficient to maintain lactase production throughout life. Homozygotes for a second allele (*2-2*) cease producing lactase after infancy or early childhood. Thus tolerance for lactose appears to be a dominant condition. It is possible that a third allele may be associated with a rare recessive, usually fatal trait (*3-3*), called infantile

TABLE 2 Possible Genetics of Lactase System

Lactase genotypes	Phenotypic effect
1–1	Lactase present through life
1–2	Lactase present through life
1–3	Lactase present through life
2–2	Lactase deficient after infancy
2–3	Lactase deficient after infancy
3–3	Infantile milk intolerance (Rare, usually fatal)

milk intolerance; afflicted babies never produce lactase. On the basis of present evidence the most likely situation is as indicated in Table 2.

The evolutionary significance of lactase is suggested by the very striking differences in the geographical–racial distribution of phenotypes. In general, European populations digest and absorb lactose and thus can utilize milk as food. Asian, Amerindian, and African populations, on the other hand, are generally lactose intolerant. The proportion of tolerants is roughly 90–100% in northern Europe and 0–10% in most of the rest of the world. There are informative exceptions to the general distribution of lactose intolerance. African and Asian herders and cattle raisers are lactose tolerant. Such groups can be found living in areas adjacent to those occupied by a lactose intolerant population. Also, the Caucasian population of the southern rim of Europe has a high proportion of lactose intolerant persons. In general, groups utilizing milk for food tolerate lactose while groups who historically have not utilized milk for food do not tolerate lactose. A few examples of population prevalences are given in Table 3.

TABLE 3 Prevalence of Lactose Handling Phenotypes[a]

Population	N	Percentage tolerant	Percentage intolerant
Australia (Aborigines)	44	15	85
Australia (Europeans)	160	96	4
American Indian	24	0	100
American Caucasian	245	88	12
American Negro (Baltimore)	20	5	95
Chinese	71	7	93
Bantu	59	11	89
Thai	179	2	98
Finnish	134	82	18

[a] Data mostly from McCracken and Simoons.

Selection for lactose intolerance must have begun 10,000–12,000 years ago when human populations began domesticating milk-producing animals. Because the adult form of intolerance is not fatal and would only be disadvantagous when food supplies were very marginal (sour milk and some milk products such as yogurt or cheese are digested by intolerant persons) selection pressures must have been gentle. We may also note that selection favoring tolerance must have increased in populations where significant numbers had already become tolerant: The possession of a favorable trait increasing fitness leads to displacement of the other gene. Once some members of the population utilized milk as a food, the remaining members were at a relative disadvantage.

As in other examples of interaction between environment and genes, the more one understands about this specific phenomenon, the more difficult it becomes to separate genes from environment. In the case of lactase it appears that a cultural–technological advance, domestication of animals, was inexorably intermeshed with a change in gene frequency. At the same time, the cultural–technological advance must have accelerated the genetic change. The range of cultures and individual behaviors entailed by this genetic–environmental change is obviously extremely broad with ramifications into almost all aspects of life.

It appears that primitive man, like all mammals, must have been lactose intolerant after infancy. It is toleration for lactose that must have evolved. We may ask then what magnitude of selective advantage would have been required to change the frequency of a favorable dominant mutation to currently observed levels. Accepting the current prevalence of lactase deficiency in contemporary intolerant populations to be about 90% as opposed to about 10% in northwestern Europe, the corresponding frequencies of the gene for adult lactase production would be .05 in intolerant populations and about .60 in tolerant populations. With the help of a table provided by Lush (1945), we were able to work out an approximate selection intensity against homozygotes (*2-2*) required to produce a change in gene frequency from .05 to .60 in the 400 generations since domestication of sheep and goats. The selection intensity is approximately .01. The literal meaning of this number is that if lactose tolerant persons had an average of 1% more children per generation than lactose intolerant persons, the observed change in phenotype frequency could occur in the time available. Such subtle selection pressures would hardly have attracted attention. The value .01 is of reasonable magnitude for a selection coefficient. Table 4, reproduced from Lush, provides useful insights into the problems involved.

It is quite important for educators, dietitians, physicians, and public officials to apprehend that just because milk is good for babies it may not be good and in fact may be harmful for some children. The milk-break and

TABLE 4 Approximate Time Required for Selection to Increase the Frequency (q) of a Favored Gene by Various Amounts[a]

q to be changed from q_1 to q_2		Time, expressed in $1/s$ generations			Correction factor x
		Selection for a complete dominant ($h = 0$)	Selection when there is no dominance ($h = .5$)	Selection for a complete recessive ($h = 1.0$)	
q_1	q_2				
.01	.05	1.69	3.30	81.65	1.61
.05	.10	.81	1.49	10.75	.69
.10	.20	.95	1.62	5.81	.69
.20	.30	.72	1.08	2.21	.41
.30	.40	.68	.88	1.28	.29
.40	.50	.74	.81	.91	.22
.50	.60	.91	.81	.74	.18
.60	.70	1.28	.88	.68	.15
.70	.80	2.21	1.08	.72	.13
.80	.90	5.81	1.62	.95	.12
.90	.95	10.75	1.49	.81	.05
.95	.98	30.95	1.89	.98	.03
.98	.99	50.70	1.41	.71	.01
.99	.995	100.70	1.40	.70	.00
From answer in generations subtract:		x	$2x$	$x + 1/q_1 - 1/q_2$	

[a] From Lush, 1945, p. 126. Reproduced by permission.

school lunch programs may make many of our black, brown, red, and yellow children ill. Paige et al. (1971) reported that 20% of Baltimore Negro children drank less than half of the one-half pint of milk given them as part of the school lunch. The corresponding percentage of white children was 10%. The implications of milk intolerance for cross racial adoptions of children and for dietary treatment of stomach ulcer needs to be explored. Notorious examples can be cited of disadvantaged peoples in underdeveloped countries using powdered milk to whitewash their houses. Contrary to the advertising slogan, *not* everyone needs milk!

A few other traits deserve mention. The studies of Post (for example, 1964) have suggested that color blindness is more common in populations that have been longest removed from the hunter–gatherer stage of civilization. Similar relationships have been found in visual and hearing acuity and in the incidence of nasal septum deviation.

While we think there is much to be learned from the study of single locus and relatively simple traits, behavioral traits are mostly polygenic and selection acts on phenotypes. We have, we conclude, no wholly satisfactory models of selection for any polygenic traits in man, let alone behavioral ones.

The most conspicuous candidate is skin color, which appears to be due to 4–5 gene pairs acting in accordance with an additive polygenic model (Stern, 1970). We have theories seeking to account for the differences in skin color we observe in different populations but the critical physiological evidence to back the theories is lacking. Dark skin probably confers some protection against skin cancer and it may prevent overproduction of vitamin D. Dark-skinned Eskimos have a unique diet with ample vitamin D. They therefore did not need to evolve white skin. Although it seems evident that the differences among races are due to diversifying selection pressures, we cannot specify the pressures and hence the model is incomplete.

What Next?

How can our ignorance be remedied? Being aware of the evolutionary process leads us to ask questions about the evolution of human behavior. We have few answers and perhaps only now are prepared to look seriously. But we are painfully aware that other disciplines sharing the behavioral science niche likewise have no answers. How much further might we be in our understanding of human learning had all the man-years devoted to the laboratory white rat been spent with tree shrews? What might be learned if we admitted that races of men are Mendelian populations whose racial hybrids form natural experiments providing evidence (in certain circumstances) of genetic differences in behavior between the parental populations? What would happen if social scientists recognized aggression as a behavior with a long evolutionary history in our (and nearly all other) species; The ethologist would quickly point out that altruism has an equally long history and that perhaps man is subject to contradictory motivations. *Homo sapiens* in all our glory has evolved as a conglomerate of compromises; it is not a form of condescension to deal with members of our species via compromises. It is rather a cultural adaptation required by our genetic adaptedness.

Acknowledgments

We thank I. T. Diamond, F. Johnston, and D. D. Thiessen for helpful suggestions.

References

AYALA, F. J. (1970). Competition, coexistence, and evolution. *In:* "Essays in Evolution and Genetics in Honor of Theodosius Dobzhansky," (M. K. HECHT and W. C. STEERE, eds.), pp. 121–158. (Suppl. to Evolutionary Biol.) Appleton, New York.

BAKER, P., and WEINER, J. (1966). "Biology of Human Adaptability." Oxford Univ. Press, London and New York.

BUETTNER-JANUSCH, J. (1966). "Origins of Man." Wiley, New York.
BULLOCK, T. H. (1970). Physiological basis of behavior. *In:* "Ideas in Evolution and Behavior" (J. A. Moore, ed.). pp. 449–482. Natur. Hist. Press, Garden City, New York.
CASPARI, E. (1958). Genetic basis of behavior. *In:* "Behavior and Evolution" (A. Roe, and G. G. Simpson, eds.), pp. 103–127. Yale Univ. Press, New Haven.
CLAUSEN, J., KECK, D. D., and HIESEY, W. M. (1948). Experimental studies on the nature of species. III. Environmental responses of climatic races of Achillea. *Carnegie Inst. Washington Publ.* **581**, 1–129.
DARWIN, C. (1871). "The Descent of Man and Selection in Relation to Sen," 2nd ed. J. Murray, London, 1922.
DIAMOND, I. T., and HALL, W. C. (1969). Evolution of neocortex. *Science,* **164**, 251–262.
DOBZHANSKY, T. (1962). "Mankind Evolving." Yale Univ. Press, New Haven.
DOBZHANSKY, T. (1968). "On some fundamental concepts of Darwinian biology, evolutionary biology" (T. Dobzhansky, M. K. Hecht, and W. C. Steere, eds.), pp. 1–34. Vol. 2. Appleton, New York.
DOBZHANSKY, T. (1969). Evolution of mankind in the light of population genetics. *Proc. Int. Congr. Genet. 12th* **3**, 281–292.
DOBZHANSKY, T. (1970). "Genetics of the Evolutionary Process." Columbia Univ. Press, New York.
FERGUSON, A., and MAXWELL, J. D. (1967). Genetic aetiology of lactose intolerance. *J. Lancet, July 22,* 188–191.
FINE, A. E., WILLOUGHBY, E., McDONALD, G. S. A., WEIR, D. G., and GATENBY, P. B. B. (1968). A family with intolerance to lactose and cold milk. *Ir. J. Med. Sci.* **1**, 321–326.
FLATZ, G., and SAENGUDOM, C. (1969). Lactose tolerance in Asians: A family study. *Nature,* **224**, 915–916.
GOTTESMAN, I. I. (1963). Genetic aspects of intelligent behavior. *In:* The Handbook of Mental Deficiency: Psychological Theory and Research" (N. Ellis, ed.), pp. 253–296 McGraw-Hill, New York.
GOTTESMAN, I. I. (1968). Biogenetics of race and class. *In:* "Social Class, Race, and Psychological Development" (M. Deutsch, I. Katz, and A. R. Jensen, eds.), pp. 11–51. Holt, New York.
GREULICH, W. W. (1957). A comparison of the physical growth and development of American-born and native Japanese children. *Amer. J. Phys. Anthropol.* **15**, 489–515.
HARRISON, G. A. (1967). Human evolution and ecology. *In:* "Proceedings of the Third International Congress of Human Genetics" (J. F. Crow and J. V. Neel, eds.), pp. 351–359. Johns Hopkins, Baltimore.
HODOS, W., and CAMPBELL, C. B. G. (1969). Scala Natura: Why there is no theory in comparative psychology. *Psychol. Rev.* **76**, 337–350.
JERISON, H. J. (1963). Interpreting the evolution of the brain. *Hum. Biol.* **35**, 263–291.
LI, C. C. (1970). Human genetic adaptation. *In:* "Essays in Evolution and Genetics in Honor of Theodosius Dobzhansky (Suppl. to *Evolutionary Biology*)" (M. K. Hecht, and W. C. Steere, eds.), pp. 545–577. Appleton. New York.
LORENZ, K. (1965). "Evolution and Modification of Behavior." Univ. of Chicago Press, Chicago.

LUSH, J. L. (1945). "Animal Breeding Plans," 3rd Ed. Iowa State College Press, Ames.
McCRACKEN, R. D. (1971). Lactase deficiency: An example of dietary evolution. *Curr. Anthropol.* **12**, 479–500.
MAYR, E. (1963). "Animal Species and Evolution." Belknap Press, Cambridge, Massachusetts.
MEREDITH, H. V. (1969). Body size of contemporary youth in different parts of the world. "Monographs of the Society for Research in Child Development," Ser. No. 131, Vol. 34, No. 7. Univ. of Chicago Press, Chicago.
MORCH, E. T. (1941). Chondrodystrophic dwarfs in Denmark. "Opera Ex Domo Biologiae Hereditariae Humanae Universitatis Hafniensis," Vol. 3.
PAIGE, D. M, BAYLESS, T. M., and GRAHAM, G. G. (1971). Milk programs: Helpful or harmful to negro children. Address to Food and Nutrition Section, 99th Annual Meeting of the American Public Health Association.
POST, R. H. (1964). Hearing acuity variations among negroes and whites. *Eugen. Quart.* **11**, 65–81.
SIMOONS, F. J. (1970). Primary Adult Lactose Intolerance and the Milking Habit: A Problem in Biological and Cultural Interrelations, II A Cultural Hypothesis. *Amer. J. Dig. Dis.* **15**, 695–710.
SIMPSON, G. G. (1961). "Principles of Animal Taxonomy." Columbia Univ. Press, New York.
STERN, C. (1970). Model estimates of the number of gene pairs involved in pigmentation variability of the negro-American. *Hum. Hered.* **20**, 165–168.
THIESSEN, D. D. (1972). A move toward species-specific analysis in behavior genetics. *Behavior Genet.* **2**, 115–126
TINBERGEN, N. (1970). Behavior and natural selection. *In:* "Ideas in Evolution and Behavior," (J. A. Moore, ed.), pp. 519–542. Natur. Hist. Press, Garden City, New York.
WASHBURN, S. L., and AVIS, O. (1958). Evolution of human behavior. *In:* "Behavior and Evolution," (A. Roe and G. G. Simpson, eds.), pp. 421–436. Yale Univ. Press, New Haven.
WELSH, J. D., ZSCHIESCHE, O. M., WILLITS, V. L., and RUSSELL, L. (1968). Studies of lactose intolerance in families. *Arch. Int. Med.* **121**, 315–317.
WELSH, J. D. (1970). Isolated lactase deficiency in humans: Report on 100 patients. *Medicine*, **49**, 257–277.
WILSON, A. C., and SARICH, V. M. (1969). A molecular time scale for human evolution. *Proc. Nat. Acad. Sci.* **63**, 1088–1093.

DISCUSSION

WILLIAM S. POLLITZER
University of North Carolina
Chapel Hill, North Carolina

The central themes of this paper are (1) how did patterns of behavior arise in man and in different populations; (2) the problem of convergent versus divergent evolution; (3) the interaction of genetics and environment; and (4) our still vast ignorance.

Following Dobzhansky, the authors cite the kinds of selection: normalizing, balancing, directional, and diversifying. This last should produce cladogenetic, or splitting, evolution, in contrast to anagenetic evolution. In all probability, man arose from some combination of cladogenesis and anagenesis.

In the consideration of primate phylogeny, we have the new tool of molecular biology, amino acid substitutions. Wilson and Sarich suggest on the basis of albumin data that apes and man diverged five million years ago, six times more recently than the divergence of monkey and man; DNA and hemoglobin sequences tend to confirm the close kinship of apes and man. The assignment of *Ramapithecus* to 14 million years ago makes the absolute molecular dating questionable, but the relative dating reasonable.

Jerison's conclusions about adaptive neurons, those left over after basic "housekeeping," appears doubtful. It is based upon estimations of total brain volume from endocasts, on estimations of cortical neurons from brain size, and apparently on the estimation of adaptive neurons of present mammals from a comparison with Eocene mammals. Where E is brain size and P is body size,

$$E = kP^{2/3};$$

for contemporary mammals, k is .12, for Oligocene ones .06, and for Eocene ones .03. In the table taken from Jerison, there is indeed increase in the adaptive neurons in hominid evolution. But the estimated vast number of such neurons in the porpoise and the elephant remain hard to explain.

Quite likely some threshold number of neurons was reached for man's elaboration of tools and culture. Tool use among the Australopithecines implies relatively little brain was needed for this activity, perhaps less than for speech. One big gap in our knowledge is just how the use of the hands, freed by upright posture, stimulated brain development; if it did so, how could it be inherited; would mutations and selection alone be sufficient for the relatively rapid enlargement of the brain?

Diamond and Hall, working on the tree shrew, have given us new insights into the evolution of the neocortex unique to mammals. In contrast to the view that the association cortex arose in response to the selective advantage of increased opportunity for integration between modalities, their research suggests that it may have arisen as a primitive sensory area. That the gray squirrel shows elaboration of visual centers of the tree shrew rather than those of the rodent is an example of *convergence;* the expansion of the cortex in primates and carnivores is an example of *divergence*. The problem of the evolutionist is differentiating true phyletic relationships from the convergence or parallelism produced by adaptation to a similar environment.

Mankind is one species, and apparently shows no tendency toward further speciation. Man's spread over the earth and the increasing admixture of diverse populations prevents this trend, which is common in many other species. Gottesman and Heston say that characteristics which are species-specific evolved through divergent evolution. The age-old problem here is to differentiate the genetic heterogeneity of mankind from those traits which differ due to cultural factors.

The argument of Thiessen that traits with high heritabilities are "genetic junk" is not clear to me. Surely, if polymorphisms were exactly functionally equivalent, it should not matter in the evolution of the species which form one possesses. But, as in the well-known hemoglobin variants, they are certainly *not* equal in function—and they vary in frequency in different environments. In a few such cases we can gain insights into the kind of selection taking place. In our studies on the Gullah Negroes of Charleston, S. C., we found Hemoglobin S at approximately African levels, while blood type frequencies indicated white admixture—an example of selection due to malaria (Pollitzer, 1958). Workman has found indications of selection at other loci among Georgia Negroes.

The high degree of polymorphic loci in man is well documented by Harris; it is in line with Lewontin and Hubby's demonstration of the heterozygosity at many loci in *Drosophila,* and a continuation of the variability of that species shown by Chetverikov a half century ago. Further physiological investigations of man's polymorphisms should in time elucidate their evolutionary significance.

The example of increased height in children born to Japanese parents in the USA well illustrates the range of reaction of the genotype. The figure showing the four genotypes in three environments with their overlap has been used by Gottesman to illustrate IQ. The stunted chondrodystrophic dwarf with his narrow range of phenotype is the counterpart of Langdon Down's syndrome (better known as Mongolism).

In Harrison's study, two closely related Ethiopian populations dwelling at different altitudes exhibit quite different morphology. That the changes are physiological adaptations is shown by their occurrence in migrants from one environment to the other. Harrison sees this as a force reducing the intensity of diversifying selection and thereby maintaining the integrity of the human species. Is it possible that adaptations to cold and heat, such as those so ably demonstrated by Baker (1969) in the Andes, follow the same principle? If so, we are hard pushed to explained our inherited diversity.

While it is quite possible that selection pressures were uniform in mid-Pleistocene for the general capacity to learn, pressures could have varied then and subsequently for various specific abilities—abilities which might well have some genetic basis.

The interesting story of the correlation of milk drinking with lactose tolerance is presented as an example of divergent evolution. Those populations which lack the enzyme lactase—East Asians, American Indians, and some Africans—are the ones which have not had the milk-drinking habit. Several family studies indicate that the enzyme deficiency is a genetic trait; the selective advantage of lactose tolerance thus presumably first arose in people with a milk supply and relatively little else to eat. Bolin and Davis believe that some results alleged to support the inheritance theory have been misinterpreted, and they note in man and animals the decline in enzyme activity with the disappearance of the substrate challenge. Their own data on children in Singapore support the adaptation theory. Rosensweig recognizes a primary genetic control, but the possibility of adaptation induced by substrates other than lactose. Perhaps more family studies are needed. Even granting the genetic hypothesis, is there sufficient proof of a selective advantage in the lactose-tolerant persons? Would stomach cramps and diarrhea make people reproduce any less?

Color-blindness is another probable interaction of genetics and culture; it is more common in populations longest removed from the hunter–gatherer stage. Dr. Neil Kirkman has suggested an interesting hypothesis: Although hunters with normal vision would be at an advantage during most of the year, color-blind hunters who had learned to rely upon a sharpened sense of form would have an advantage at certain times. Thus, the best adapted population would contain a *few* color-blind individuals—an example of a frequency-dependent trait. Perhaps we need to search for many more examples of this kind.

How may our vast ignorance be remedied? An abundant supply of tree shrews and all other primates from lemurs through chimpanzees in the learning mazes might have helped a little. I view with caution the idea that the races of man and their hybrids can tell us much about genetic differences in behavior as control of the environment is imperative. The current research of my colleague at Chapel Hill, Dr. Robert Elston, may provide some new answers. He is currently embarked on a search in families for "major genes" underlying behavior traits, utilizing such "marker genes" as blood types and serum proteins.

My own research on Negro, Indian, and "triracial" isolates of the Southeastern United States, while concerned primarily with gene flow, has shed light upon the interaction of culture and genetics in the maintenance of their distinctive populations (Pollitzer, 1970). The "Haliwa Indians" arose from a surrounding Negro community and established their own Association and school. Phenotypically the "Indians" differed somewhat from the Negroes but blood typing showed the two subpopulations to be genetically quite similar. With the increase of friction between the groups, the Indians lowered their barriers to membership—and the experiment in microevolution came to an end. Our studies have also revealed some interesting data on assortative mating and differential fertility as possible evolutionary factors. Among the Haliwa, lighter women had slightly more children on the average; among the Seminole Indians of Florida, however, darker women had a slightly higher fertility. In the communities we have studied, skin color is fairly highly correlated between mates. Man's culture continues to direct the flow of genes and must be kept clearly in mind in any approach to the study of genetics and behavior.

At Chapel Hill, I serve both the Anatomy Department, where I am considered a liberal, and the Anthropology Department, where I am considered a conservative. Some in Anatomy apparently view man's behavioral differences as racial and thus genetic; some in Anthropology apparently see these differences as purely a matter of culture and view our biological inheritance as essentially uniform since *Homo erectus*. I have attempted to develop a balance rather than schizophrenia. I applaud the elucidation of every biochemical mechanism underlying behavior; if the lack of an inherited enzyme produces a pathological condition, it is reasonable to suppose that variations on its quantity may have an effect within the normal individual. But I acknowledge the enormous scope of man's environment, all the subtle influences of his culture in shaping his genotype into his phenotype. While the extent to which the components of IQ are genetic or environmental may make no practical difference to educators today, it could make a vast difference tomorrow. We are on the threshold of an age when genetic engineering may permit some alteration of the hereditary material or when we may know just what environment may best bring to fruition a particular genotype. Our task, then, is to discover both those genes underlying human behavior and their range of reaction.

References

BAKER, P. T. (1969). Human adaptation to high altitude. *Science,* **163,** 1149–1156.

HARRIS, H. (1967). Enzyme variation in man: Some general aspects. *In:* "Proceedings of the Third International Congress of Human Genetics" (J. F. Crow and J. V. Neel, eds.), pp. 207–214. Johns Hopkins Press, Baltimore.

LEWONTIN, R. C., and HUBBY, J. L. A molecular approach to the study of genic heterozygosity in natural populations, II. *Genetics,* **54,** 595–609.

POLLITZER, W. S. (1958). The negroes of Charleston (S.C.): A study of hemoglobin types, serology and morphology. *Amer. J. Phys. Anthropol.* **16,** 241–263.

POLLITZER, W. S. (1970). Some interactions of culture and genetics. *In:* "Current Directions in Anthropology" (Ann Fischer, ed.) *Bull. Amer. Anthropol. Ass.* **3,** 69–86.

ROSENSWEIG, N. S. (1971). Adult lactase deficiency: Genetic control or adaptive response. *Gastroenterol.* **60,** 464–467.

WORKMAN, P. L. (1968). Gene flow and the search for natural selection in man. *Hum. Biol.* **40,** 260–279.

Chapter 7   Biochemical Genetics and the
                  Evolution of Human Behavior

GILBERT S. OMENN[1] and ARNO G. MOTULSKY

University of Washington
Seattle, Washington

A large body of evidence from animal models, twin studies (particularly of identical twins raised apart), and family studies points to a prominent role of genetic factors in behavioral phenotypes in man (Hirsch, 1967; Fuller and Thompson, 1960; Shields, 1962; Erlenmeyer-Kimling and Jarvik, 1963; Rosenthal, 1970; Slater and Cowie, 1971.) The role of the genotype may be viewed first as one setting limits to nervous system function, the biological substrate for the range of normal behavior. In addition, abnormal genes predispose to or cause neurologic or psychiatric defects during fetal development and at various stages later in life. We have a strong faith in the generality of the mechanisms of gene action—that genetic information flows from the code of DNA via RNA messengers to protein products, with many regulatory steps affecting timing and magnitude of synthetic and degradative processes. All cells, including neurons and neuroglial cells, contain the same complement of DNA, but the regulatory processes of differentiation lead to different patterns of gene activation in different tissues. Evolution-

[1] Fellow of National Genetics Foundation

ary forces have acted on both the DNA complement and the processes of regulation. We have been assigned the formidable task of outlining the molecular basis of gene action, individual differences, and biochemical evolution and then trying to relate these biological processes to the structure and function of the human nervous system and to the evolution of human behavior.

We may ask what features of behavior are peculiarly human and cite language, upright posture, and increasing dependence on the technologies of our culture. Rensch (1970) maintains that complex human behaviors involving abstract conceptualization and foresight represent only extensions of the capabilities of other animals. Although much may be inferred in man from knowledge of the evolution of the brain and behavior in other species, it is likely that certain features of human behavior can be understood only by the study of man.

Table 1 contrasts the features of biological and cultural evolution (Motulsky, 1968). Biological evolution depends upon chance occurrences of mutations in the genome and the selection by environmental forces of those few mutations that serve to enhance the viability or fertility of the species. It must be emphasized that selection acts on the whole individual (Simpson, 1953; Franklin and Lewontin, 1970) not just on specific genes. However, the example of the protective effect of sickle hemoglobin against malaria infection (Allison, 1954; Motulsky, 1964) demonstrates that selection can rarely act upon mutations at single loci. Cultural evolution proceeds at a pace many orders of magnitude faster than the biological processes. Cultural

TABLE 1 Comparison of Biological and Cultural Evolution[a]

	Biological evolution	Cultural evolution
Mediated by	Genes	Ideas
Rate of change	Slow	Rapid and exponential
Agents of change	Random variation (mutations) and selection	Usually purposeful. Directional variation and selection
Nature of new variant	Often harmful	Often beneficial
Transmission	Parents to offspring	Wide dissemination by many means
Nature of transmission	Simple	May be highly complex
Distribution in nature	All forms of life	Unique to man
Interaction	Man's biology requires cultural evolution	Human culture required biologic evolution to achieve the human brain
Complexity achieved by	Rare formation of new genes	Frequent formation of new ideas

[a] After Motulsky, 1968. Reproduced by permission.

7. Biochemical Genetics and the Evolution of Human Behavior

forces include social customs, which change over generations, as well as technological advances, whose impact may be felt in only a few years (see Table 2).

We will describe two complementary approaches in this discussion: (1) reductionist analysis of brain function at many levels, with emphasis on features especially worthy of comparative study; and (2) an effort to integrate behavioral and cultural features of man in an evolutionary context. The hopelessness of understanding behavior from single analytical approaches can be compared to the task of seeking linguistic insights by a chemical analysis of a book! Nevertheless, reductionist explorations at many levels, seeking a convergence of conclusions from different types of data, are essential before reasonable integration is possible. We may hope that an evolutionary perspective will be helpful in avoiding blind alleys or false leads in each type of study.

Evolutionary Development of the Biological Substrate

Anatomical features have been inferred from extensive fossil records of man and other species. Little biochemical information can be generated from these sources. However, biochemical analyses of proteins of contemporary species seem to be consistent with the broad conclusions of the paleontologists.

Advances in protein biochemistry have permitted the determination of amino acid sequences of many homologous proteins and have justified the prediction that the amino acid sequence governs the conformational folding and biological activity of the protein (Anfinsen, 1959). By comparison of such sequences, it is possible to infer some of the evolutionary events at the genetic level of nucleotide sequences in DNA (Epstein and Motulsky, 1964; Dixon, 1966; Dayhoff and Eck, 1970). We must stress at the start that the time scale of the evolution of proteins is in the millions of years, analogous to the time scale of the paleontologist. Two quite different and complementary approaches may be taken to the evolution of macromolecules: The first and better described is the highly conservative evolution of specific proteins, based upon nucleotide substitutions in the structural genes and amino acid substitutions in the protein; the second, more important for subsequent discussion of major departures in evolutionary development of species, points to more drastic effects of gene and chromosomal duplication that might permit saltatory consequences and development of altogether new functions.

The evolutionary origin of the relationship between the DNA code of triplets utilizing only four nucleotide bases (adenine, thymine, guanine, cytosine) and the 20 amino acids of common proteins is still unclear. Simula-

TABLE 2 Evolution of Man

Mean brain volume (cc)	Time scale		Tool use	Life style	Arts and language
	Years ago	Generations ago			
400–550	1.7 million	85,000	Simplest stone & bone	Hunting & gathering	
900	600,000	30,000	More refined stone tools	Similar	
1300	50,000	2,500	Stone axes	Still hunters	Cave Painting Early languages
	30,000	1,500			
	10,000	500	Metal tools	Agriculture	Hieroglyphic, Iconic written languages
	8,000	400			
	6,000	300	More complex tools & vehicles for transportation	Cities & agriculture	Alphabetized languages
	3,500	175			
	300	15	Complex machinery	Industrialized centers	Printing
	30	1	Nuclear energy use	Atomic age	Radio, TV
	20		Computers	Post-industrial Age of "Aquarius"	

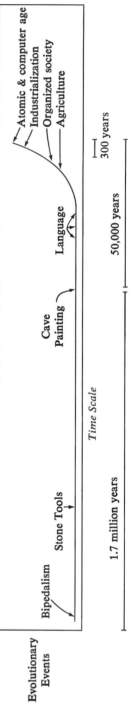

Evolutionary Events: Bipedalism, Stone Tools, Cave Painting, Language, Agriculture, Organized society, Industrialization, Atomic & computer age

Time Scale: 1.7 million years, 50,000 years, 300 years

tions of "primeval" atmospheric conditions have demonstrated that amino acids, purines, and pyrimidines can be generated from ammonia, hydrocyanic acid, methane, and carbon dioxide under ultraviolet and cosmic radiation (Calvin and Calvin, 1964). Hydrogen-bonded dimers readily form between adenine and uracil and between guanine and cytosine derivatives, without any requirement for enzymatic assistance. This inherent complementarity between purine and pyrimidine base pairs underlies the double-stranded helical structure of DNA. Mutation-induced single-base substitutions in the triplets sometimes cause no change in the amino acid specified (due to degeneracy in the code); usually they lead to replacement of one amino acid by another with fairly similar physical and chemical properties (Epstein, 1964), producing a variant protein. More drastic amino acid substitutions require two of the three bases in the DNA triplet to be replaced. Thus, the DNA code is highly conservative in its effects on protein structure, indicating either that the code arose after proteins were formed or that the structures of the two kinds of macromolecules converged. The DNA, of course, is arranged in chromosomes in cells.

Two very complicated processes, subject to all sorts of metabolic, hormonal, and physical regulation and to exquisite timing during development, are required to produce proteins from the genes. The first, called transcription, is the formation of complementary RNA messenger from the DNA sequence in specifically-activated genes in a given tissue. (In higher organisms, hunks of RNA larger than the actual messenger appear to be made first.) The messenger RNA then combines with an RNA-protein complex (the ribosome) to form the protein synthetic apparatus upon which amino acids transported specifically by transfer RNA molecules can be linked together into the polypeptide structure of proteins. This second process is termed translation of the genetic message into protein effectors. The transfer RNA molecules are specific for each amino acid, but have many crucial characteristics in common, including their tiny size of about 80 nucleotides (Holley et al., 1965). They are surely ancient evolutionary components of the life process.

Evolution of Allelic Gene Products

Changes in the DNA sequence occur spontaneously or upon induction by certain mutagenic agents. These changes are more or less random (depending upon the agent and the DNA and chromosomal structure). However, the nature of replication of the DNA and subsequent transcription and translation ensures that such a chance event will be perpetuated in the structure of the DNA and of the protein, with possible consequences in the function of the protein. The conservative nature of the relationship between the code and amino acids alluded to previously reduces the risk that these chance events will be damaging to the organism. In addition, natural selection acts to eliminate sufficiently deleterious changes in proteins essential for survival. On the other hand, some changes may not affect protein function too severely or, rarely, they may even improve the efficiency of the protein function in the usual environment or provide the adaptability to allow the organism to explore new environments. In this case, natural selection in favor of individuals carrying the new gene and protein combination may lead to accumulation of that new gene in the population. If the amino acid substitution is truly neutral in the functional and biosynthetic sense, its accumulation to a frequency above the low rate of such mutation must reflect the probabilistic processes of random genetic drift and effective population size.

When individual proteins, such as hemoglobins or cytochromes c, are compared among many species, sufficient homology of amino acid sequence is noted to "line up" the sequences and determine which sites remained unchanged during evolution, which sites allowed only some substitutions of similar amino acids, and which sites seemed to allow multiple or drastic substitutions while maintaining the overall function of these proteins (Hill et al., 1963; Margoliash, 1963). The number of amino acid differences (minimized by allowing gaps in the matching for maximum homology) as a function of the paleontological time scale can be used to estimate the rate of mutation—for several proteins listed in Table 3, roughly one effective (surviving) mutation per 100 amino acid residues per 1–10 million years. Rates may differ for different proteins or different species, and selection may markedly alter the effects for a specific protein; nevertheless, these allelic changes in genes for proteins that maintain their basic enzymatic or other activity can have little influence over the time scale of the evolution of man. The histones, basic proteins which combine with the DNA in chromosomes, are the most sluggish of all evolving proteins; comparable histones differ by only two amino acids (out of 101 residues) between the pea and the calf thymus (DeLange and Fambrough, 1968). It will be interesting to learn how homologous are such distinctive nervous system proteins as the S100

TABLE 3 Rates of Evolution of Homologous Proteins[a]

Homologous protein	Total amino acid residues in polypeptide chain	Millions of years per mutation per 100 amino acid residues
Histone IV	101	500
Cytochrome c	104	20
Hemoglobins	146	5.8
Fibrinopeptides A & B	40	1.1

[a] Based upon Dickerson, 1971.

and 14-3-2 proteins found mostly in glial and in neuronal cell populations, respectively, but each is immunochemically indistinguishable over many species (Cicero et al., 1970).

The slow evolution of homologous proteins with similar enzyme activity has incorporated at least one major development of complexity, however. Several examples, including the comparison of myoglobin and hemoglobin, point to the evolutionary development of allosterism of proteins (Monod, 1970). Allosterism refers to a thermodynamically stabilized capacity of a protein to alter its conformation and thereby its activity upon interaction with inducers, cofactors, ions, hormones, and inhibitors. The result is a regulation of protein function closely tied to physiological conditions of the tissue and to developmental needs. The implications of such capacities of proteins in the nervous system for cell–cell interaction, for postsynaptic responsiveness to neurotransmitters, for learning consolidation, and for other complicated behavioral processes are apparent, though still speculative.

Evolution by Gene Duplication

Major departures in evolutionary history must have required more drastic changes in the genome and in gene products than the amino acid substitutions we have been discussing. As Simpson has emphasized (1953), there is no basis for the notion of orthogenesis that evolution "progresses" steadily toward more complex or "higher" forms. Instead, features may become static, as the brain volume may have become in man, or regress, as olfactory structures clearly have. But how do new structures or new functions arise?

In a superb little book, Ohno (1970) has pulled together notions of the effects of gene duplications dating from Bridges and from Haldane (1932) and recent work of his own on evolution of vertebrate genomes, chromosome complements, and isoenzymes. Many striking chromosome changes in number are, in fact, highly conservative genetically, involving Robertsonian fusions and pericentric inversions. However, semisterility barriers introduced by inversions have probably been important in speciation (White, 1968), more so than point mutations accumulated in the genome. Tandem duplication of genes by unequal crossing over at mitosis or meiosis within chromosomes has produced several significant features:

1. Capacity for producing multiple copies of the gene product, particularly ribosomal RNA and possibly ribosomal proteins and transfer RNAs. The most interesting evolutionary question here is the maintenance of functional, nearly identical, yet redundant genes in the absence of apparent selective control. One explanation, which may be important for some central nervous system (CNS) processes, as well, is Callan's (1967) master–slave model, in

which only the master gene serves as the template for DNA replication after each cell division.

2. If the heterozygous state is advantageous, as for sickle hemoglobin, the incorporation of both alleles in a permanent form in the genome can be accomplished by having two loci. Otherwise, only a maximum of 50% of individuals can become heterozygous. Examples exist in the catostomid fish, whose esterase comprises a pair of variants, one active at 5° and the other active at 20°C, the range of temperatures through which the fish must survive (Koehn and Rasmussen, 1967). A problem of gene dosage can occur, in that twice the usual number of enzyme molecules may be formed, especially when a pathway of related enzyme functions is involved.

3. Another response to the gene dosage problem is the differential regulation of former alleles, now duplicated loci, in different tissues as tissue-specific isoenzymes. Lactate dehydrogenase (LDH) and fructose-diphosphate aldolase of the glycolytic pathway are examples of enzymes whose tissue-specific forms seemed to be well suited to the physiological needs of muscle and heart as extremes for LDH (Market, 1964) and muscle and liver as extremes for aldolase (Penhoet et al., 1966). The highly duplicated immunoglobulin system also reflects this solution to gene dosage compensation: Each plasma cell makes only one type of light and one type of heavy chain molecules.

4. Finally, we come to the major impact of gene duplication: the creation of a new gene product from a redundant duplicate of an old gene. As a redundant copy, the duplicate may absorb "forbidden" mutations that otherwise would have been eliminated by the conservative forces of natural selection, eliminated because of deficiency of the function of the old gene product. Once forbidden mutations begin to accumulate, there is the potential over long periods of evolutionary time for the appearance of useful new functions upon which natural selection will act favorably. Several instructive examples can be cited: (a) the pancreatic proteolytic enzymes trypsin and chymotrypsin (Neurath et al., 1967); (b) myoglobin and the hemoglobins (Ingram, 1963; Ohno and Morrison, 1966); (c) the light and heavy chains of immunoglobins (Baglioni, 1967); and (d) actin, the smaller of the muscle proteins that together make acto-myosin complexes, and the microtubule proteins of mitotic spindles, epithelial cilia, sperm tails, muscle sarcotubules, and axonal neurotubules (Renaud et al., 1968). The microtubular proteins bind colchicine and guanosine triphosphate, while actin retains the capacity of binding a nucleotide, ATP. Little is known yet of the amino acid sequence homologies and detailed functional comparisons of microtubular proteins, particularly in the nervous system. Recent evidence suggests that brain microtubular proteins have a half-life of only 4 days and contain non-

identical subunits of about 60,000 molecular weight (Dutton and Barondes, 1969; Bryan and Wilson, 1971).

Even though the amount of DNA and the number of chromosomes appear similar among the modern primates (Hsu and Benikschke, 1967), an outstanding example of very recent duplication is the haptoglobin locus (Barnicot et al., 1967; Buettner-Janusch, 1970). This hemoglobin-binding plasma protein is highly polymorphic in all human populations, yet most nonhuman primates seem to have no variation in this protein. Probably a partial duplication occurred subsequent to separation of pongid and hominid lines. It is possible that similar processes of duplication and of unrestricted evolution of redundant sequences occurred in the enlarging forebrain and that the resulting macromolecular products are involved in learning and memory storage and in language functions.

Reductionistic Description of the Human Nervous System

Anatomical Level. Several features have been identified as critical in the evolutionary development of the brain of man (see Tables 4, 5) (Washburn and Harding, 1970; Smith, 1970). The grossest change is a remarkably rapid increase in the volume of the brain, from 400–550 cm^3 in bipedal Australopithecus 2 million years ago, to double that size 600,000 years ago in ancestors skilled with stone tools, to about 1300 cm^3 in more recent and present-day man. The volume of brain varies considerably among individuals, of course, and some estimates are based upon single or only a few fossil skulls (Simons, 1971). Fossil and contemporary brain sizes of ungulates and carnivores indicate that the trend to larger mean brain size is accompanied by an increase in the variance, as though diversity were greatly favored (Jerison, 1970).

Underlying the rapid development in man of hand skills and social and linguistic skills is a striking relative enhancement in size and complexity of

TABLE 4 Anatomical Features of Human Brain Evolution

1. Absolute increase in brain size: 400–550 cc to 1300 cc[a]
2. Relative increase in forebrain — social & linguistic skills
3. Relative increase in cerebellum (3–4×) — hand skills
4. Regression of olfactory structures
5. Appearance of fetal ganglionic eminence
6. Slower maturation rate for neurogenetic processes

[a] Brain sizes represent samples from a probable range of sizes in any given population or time.

TABLE 5 Comparison of Some Mammalian Brains[a]

	Man	Chimpanzee	Macaque	Indian elephant	Rat	Mouse
Average weight of entire brain (gms)	1400[b]	435	80	4717	2.4	.2
Ratio of brain weight to body weight[c]	.02	.007	.05	.0015	.005	.015
Area of the cortex of one cerebral hemisphere (mm^2)	90,172	24,224	6940		16	
Number of cells per mm^3 of cortex[d]	10,500		21,500	6,900	105,000	142,000

[a] Information derived principally from Blinkov and Glezer, *The Human Brain in Figures and Tables,* Basic Books Inc., 1968. Modified from Smith (1970).

[b] Considerable variation occurs. Proportionate dwarfs may be as short as three feet in height, with brain weight of only 400 gm. Microcephalic individuals have normal height, small brain size, and mental deficiency. Nevertheless, rudimentary speech and social interaction develop.

[c] A commonly used comparative measure is $E/P^{2/3}$, where E is brain weight or volume and P is body weight or volume. Of course, different animals are built of differing proportions of fat, bones, connective tissue, etc. The ratio also varies with the age of the animals.

[d] This figure provides an indication of the volume of cortex available for the ramifying nerve cell processes.

both the forebrain and the cerebellum. Meanwhile, olfactory structures have regressed and other structures have presumably been left a more subservient role. The pioneering histologist Ramon y Cajal established that neurons are contiguous, not continuous, at synapses and that the neurons are the metabolic, structural, and physiological units of the nervous system. Evolutionary increase in cell number leads to a geometric increase in potential axodendritic connections. New fluorescent histochemical methods that outline fiber pathways of specific neurotransmitter agents (Hillarp *et al.,* 1966) offer powerful approaches to comparative studies of the connections between regions of the brain. In the morphogenesis of neural structures in man, two special features should be mentioned: (1) the fetal ganglionic eminence (Rakic and Sidman, 1969), a concentration of dividing cells, which go to form the basal ganglia and probably forebrain structures; analogous fetal cells beneath the lateral ventricles have not been recognized in other species and (2) a much longer time for maturation of the central nervous system in man. Unlike newborn apes and monkeys, who must be able to cling to their mothers, human newborns are delivered at a far less advanced stage of development, partly in evolutionary response to the narrowing of the bony birth-canal that accompanied bipedal locomotion. Presumably, such slow maturation is highly suited to molding of species-specific behaviors by cultural factors.

Neurophysiological Level. Many neuronal circuits appear to be genetically and developmentally "wired in" to function quasiautonomously in the breathing and sucking of the newborn, in the precise regulation of temperature, pH and osmotic pressure of the internal milieu, in sleep, and in other essential processes. These functions are primarily mediated in the brain stem, diencephalon, and limbic system, while greater plasticity is assumed to be a characteristic of cortical functions (McLean, 1970). In the cortex, probabilistic spatiotemporal configurations have been invoked to describe firing patterns and a capacity for "relearning" complex functions after ablation of specific areas. We may expect that the psychological correlates of cortical function will have a greater variety and greater variability of neurophysiological and biochemical properties than will the brain stem and limbic structures whose functions were well established much earlier in evolutionary time. Computer-averaged evoked cortical potentials (S. Fox, 1970) and pharmacologically manipulated electroencephalography (Fujimori and Himmich, 1969) may be potentially useful descriptive and comparative techniques.

Biochemical Level. Biochemical and neurophysiological studies have demonstrated that the old view of a stable set of quiescent neurons that could be stimulated to action must be revised. The brain constitutes 2–3% of body weight, yet consumes up to 50% of the resting energy and oxygen supply. The "resting" state of neurons is characterized electrophysiologically by intense rhythmic and spontaneous activity. Likewise, protein biosynthesis and transport of proteins, structural components, and other molecules through the axon of the neuron are continuous, active processes. Fertile areas for comparative neurochemistry include the following.

DNA Transcription. The result of differentiation of tissues is a selection of certain genes to be active in certain tissues, other genes to be active in other tissues, and some genes to be active in all tissues. DNA–DNA hybridization confirms that all cells contain the same DNA, while DNA–RNA hybridization confirms that only part of the genome is active in any tissue at any time (McCarthy and Hoyer, 1964). Some genes are redundant, coding for large amounts of ribosomal RNA needed for protein synthetic machinery. Appropriate methods can determine the proportion of "unique sequence DNA" genes present in single copies, that is, transcribed into RNA messengers. In most tissues, only 3–6% of this DNA is transcribed (Hahn and Laird, 1971; Grouse *et al.* 1972). In brain tissue, a remarkably higher proportion is transcribed—10–13% in the mouse and 20% in man (Grouse *et al.*, 1972). McCarthy and his colleagues are now testing different regions of the brain to see whether cortical regions utilize even more of the genetic complement than do brain stem or other regions. It will be interesting to determine whether stimulation by learning tasks or electrical means can increase the transcription activity even further. The theoretical limit is 50%, since only one or the other of the two DNA strands is transcribed along any portion of the double helix. Of course, the functions dependent upon the "extra" active DNA are not at all clear yet. Perhaps the diversity of transcription provides a complex variety of RNAs and proteins for recognition, integration, and memory-storage processes.

Compartmentalization of Protein Synthesis. Neurons appear to be more highly compartmentalized than other tissues in their capacity for protein synthesis, which is coupled to specialized neuronal functions (Shooter, 1970). Whether there are differences in the control of protein synthesis between regions of the brain or between man and other species is not clear. In large neurons, the cytoplasmic ribosomes are concentrated near the endoplasmic reticulum of the Nissl substance in the perinuclear region, the initial segment, and the axon hillock (Sotelo and Palay, 1968). A high proportion of these ribosomes are not attached to the membrane of the endoplasmic reticulum and may function directly in the synthesis of protein involved in axoplasmic transport to the nerve ending. Brain ribosomes require a high concentration of potassium ion (100 mM), suggesting a link with bioelectric phenomena and active transport of K^+. Brain mitochondria have their own apparatus for active protein synthesis, do not require an external source of ATP, are inhibited by the antibiotic chloramphenicol and not by ribonuclease. Mitochondrial protein synthesis is tightly linked to oxidative phosphorylation, increasing under conditions optimal for the latter and being inhibited when specific inhibitors of oxidative phosphorylation like rotenone and antimycin A are present. Finally, nerve ending fractions called synaptosomes carry on protein synthesis, with synergistic stimulation by appropriate concentrations of Na^+ and K^+. The ionic concentrations required for maximal incorporation of amino acids into protein also result in maximal sodium–potassium ATPase activity, K^+ uptake, and oxygen uptake. Ouabain, which inhibits the Na–K ATPase activity, markedly inhibits synaptosomal protein synthesis, further suggesting a close coupling of the synthetic activity with the ionic flux and energy metabolism in the nerve ending.

Specificity and Variety of Neurotransmitter Agents. Synaptic inputs can be either excitatory or inhibitory, and neurotransmitter function has been attributed to an increasing array of chemical agents. The evidence is good that such agents as acetylcholine, norepinephrine, dopamine, serotonin, and gamma amino-butyric acid are concentrated in synaptic vesicles, can be isolated and specifically taken up against a concentration gradient *in vitro* into isolated synaptosomes, and are probably released upon physiological stimulation across the synaptic cleft to excite (depolarize) or inhibit (hyperpolarize) postsynaptic membranes. The postsynaptic membrane has specific receptors for the action of these agents. Nevertheless, only a minority of all synapses can be assigned to even these incompletely proved transmitter agents; it is likely that a variety of amino acids (histamine, glutamate, aspartate, glycine) and possibly other compounds act as transmitters in certain fiber pathways (Snyder, 1970). The potential for evolutionary diversity and functional specificity is great.

Membranes and Macromolecules for Recognition Processes. The formation of cell–cell contacts, the specific migration of neuronal cell groups, and the processes of selective cell death may be mediated by macromolecules incorporated into the membrane structure of nerve cells. Complex glycoproteins are among the potential mediators (Barondes, 1970). Much of the work on this subject is still at the level of model systems, but rapid technical advances offer promise of substantive progress. We should emphasize that the diversity of specific proteins has been derived from a "simple" triplet code based upon only four different nucleotide base "letters." Thus, it is not unreasonable to expect sets of macromolecules to be able to perform complex memory storage and cell recognition functions, for the variety of intramembrane geometric arrays that could be formed with, for example, 10 different protein or carbohydrate-protein units in different combinations is enormous. It will be important in comparative biochemical studies to determine whether human cortical neurons have a greater variety of such units than do neurons of other species.

Discussion of complex molecular functions centers on the role of proteins, as will our discussion of the evidence for molecular evolution. Proteins have been termed "universal biological effector molecules" (Schmitt, 1967), since they act as enzymes for metabolic processes, as components of structural neurotubules and membranes, as recognition molecules for the neurotransmitters released across synapses, and possibly as electrogenic effectors for ion gating changes in the propagation of action waves down the axonal membranes. Monod (1970) has emphasized the capacity of proteins to act as molecular agents of structural and functional teleonomy—mediating oriented, constructive, and coherent activity through their ability to "sense" substrate or inhibitor concentrations, to carry in their structure the information for proper conformational folding as a response to such inputs, and to catalyze metabolic reactions or macromolecular interactions.

Although a biochemical basis for the characteristic human functions of cognition, language, and consciousness is beyond description at present, modern techniques of human genetics do allow us to demonstrate the individuality and uniqueness of different humans at a biochemical level and provide a basis to speculate about evolutionary mechanisms that must underlie the biological substrate of behavior.

Protein Polymorphisms

The ability to detect in the structure of proteins small differences, which reflect qualitative differences in the respective genes, permits an experimental approach to the question of human individuality. The basic technique in-

TABLE 6 Genetic Markers in Human Blood[a]

Genetic system	Probability that two randomly selected people have the same phenotype	Combined probability[d]
MNSs	.16	.16
Rh(CC^{w}cDEe)[b]	.20	.032
ABO(A_1A_2B)	.33	.011
Acid phosphatase	.34	.0037
Glutamic pyruvic transaminase (GPT)	.38	.0014
Kidd (Jk^{a}Jk^b)	.38	.0005
Duffy (Fy^{a}Fy^b)	.38	.0002
Haptoglobin	.39	7.9×10^{-5}
Gm(1,5)	.40	3.2×10^{-5}
Gc	.45	1.4×10^{-5}
Phosphoglucomutase (PGM)	.47	6.7×10^{-6}
Dombrock blood group	.55	3.7×10^{-6}
Lewis (Le^{a}Le^b)	.57	2.1×10^{-6}
P(P_1P_2)	.67	1.4×10^{-6}
Adenosine Deaminase (ADA)	.78	1.1×10^{-6}
Adenylate kinase	.82	9.0×10^{-7}
Pseudocholinesterase, E_2	.82	7.4×10^{-7}
Kell (Kk)	.84	6.2×10^{-7}
Lutheran (Lu^{a}Lu^b)	.86	5.3×10^{-7}
6PGD	.92	4.8×10^{-7c}

[a] Twenty blood genetic systems listed in order of their usefulness (that is, *MNSs* is the most useful) for distinguishing between two random samples of blood from western Europeans. Adapted, with permission, from Giblett, 1969, Table 17.1.

[b] Parentheses denote the antigens tested in a given system.

[c] This figure indicates that less than one in 2,000,000 people would be expected to have the same combinations of phenotypes in these 20 systems.

[d] Use of histocompatibility, immunoglobulin, and lipoprotein Lp antigens decreases the combined probability by another 3 orders of magnitude.

TABLE 7 Estimate of Number of Protein Polymorphisms in Man

Total nucleotide pairs in haploid human chromosome set	3 billion
Maximum number of genes (1 gene per 1000 nucleotide pairs)	3 million
Probable number of structural genes (2% of DNA)	60,000
Probable number of polymorphic genes (30% of structural genes)	20,000
Number of human polymorphisms known:	~50
Percent of polymorphic genes discovered (50/20,000)	.25%

volves electrophoresis of tissue extracts and specific staining for enzyme activity. Since electrophoretic mobility is based upon the net charge of the protein and may be altered by an amino acid substitution that changes net charge (about 30% of single-base substitutions), specific enzymes or other proteins can be compared in many individuals from a given species. Rare, variant proteins are found in single individuals, simply on the basis of continuing mutations. However, common variants (arbitrarily defined as greater than 1% gene frequency) require some selective advantage or random genetic drift in order to have accumulated, and are of great interest as "polymorphisms" both for questions of their origin and for application to the description of individuals. Using ABO and other red blood cell antigens, serum proteins, certain serum and red blood cell enzymes (Table 6), plus the histocompatibility, immunoglobulin, and lipoprotein antigens as testable polymorphic systems, the likelihood of finding two humans with identical results (except for monozygotic twins) is on the order of one chance in three billion (about the size of the world's population). And only a small fraction (much less than 1%) of the estimated number of protein polymorphisms has been discovered (Table 7).

Selective forces have been identified for the remarkable polymorphisms of sickle hemoglobin (up to 40% of individuals are heterozygous in parts of Africa) and glucose-6-phosphate dehydrogenase deficiency. Resistance of the heterozygote to early death from malarial infection seems to provide a definite example of natural selection in man (Allison, 1954; Motulsky, 1964). For all the other human polymorphisms, we do not know whether their presence represents the effect of past or current selective forces or is due to random genetic drift. Recent data of clines of gene frequencies (Selander, 1970) and models for gene disequilibrium and selection for whole regions of chromosomes (analogous to the genes locked into chromosomal inversions in *Drosophila*) (Franklin and Lewontin, 1970) make it plausible that many more polymorphisms are maintained by selection without introducing too severe a genetic load. However, many polymorphisms such as phosphatases, esterases, and peptidases, represent *in vitro* activities of enzymes which cannot be assigned specific metabolic reactions *in vivo*.

Current Studies of Enzyme Variation in Human Brain

We have recently embarked upon a novel application of biochemical genetic techniques and the notion of enzyme polymorphisms in the central nervous system of man. The rationale is as follows: Enzyme surveys in *Drosophila*, mice, and man indicate that about 30% of enzymes have common variants that can be detected by the electrophoretic screening method. Most of the electrophoretic variants that have been studied have a significant difference in quantitative enzyme activity, compared with the usual form of the enzyme (Table 8) (Motulsky, 1969). For example, *G6PD A+* has 85% and *G6PD A−* has 15% of the activity of the usual *G6PD B* form; and the three alleles of acid phosphatase, occurring in dimers, have relative activities of 100, 150, and 200. The acid phosphatase model is of special importance, for a quantitative survey in human populations suggests a normal distribution of enzyme activity; only with electrophoretic differentiation of the 6 dimeric phenotypes (*AA, AB, AC, BB, BC, CC*) can each subgroup be tested and be shown to have narrow ranges of enzyme activity (Fig. 1) (Harris, 1970). Unfortunately, the *in vivo* role of this interesting enzyme has not been elucidated.

We have selected crucial metabolic pathways in the brain and examined the relevant enzymes for the possibility of variant forms of the enzymes (Cohen *et al.*, 1970). A polymorphism of a rate-limiting enzyme, such as phosphofructokinase in the pathway of glycolysis (Fig. 2), would be highly significant even if associated with only a small difference in quantitative enzyme activity, since production of lactate at the end of the pathway and of ATP along the way would be affected. On the other hand, a small difference in activity of an enzyme normally present in concentrations well above rate-limiting activities could be expected to have no such consequences. Thus far, our attention has been directed primarily to the energy-generating metabolic processes of the nervous system. The brain is exquisitely sensitive to lack of oxygen or glucose, irreversible damage occurring within 5 minutes in man. The prime metabolic pathway for glucose utilization is glycolysis. We have screened all 11 enzymes of this pathway from hexokinase to LDH in some 150 human brain specimens. None of these enzymes has a common variant form. Only single, rare variants of phosphoglycerate kinase and of enolase were found (Table 9), presumably reflecting mutation. This negative finding may be highly significant (Omenn *et al.*, 1971), since similar lack of frequent variation was noted in our screening of mouse and monkey

TABLE 8 Structural Variation and Quantitative Activity of Enzymes

Enzyme	Relative activity of polymorphic structural loci						Author[a]
Acid phosphatase	P^A	: 100	P^B	: 150	P^C	: 200	Spencer et al., 1964
6-PGD	Pgd^A	: 100			Pgd^B	: 85	Parr, 1966
G6PD	Gd^B	: 100			Gd^A	: 80	Long, 1966
Adenylate Kinase	AK^2	: 100			AK^1	: 150	Modiano et al., 1970
Phosphoglucomutase							
Locus 1	PGM^1_1	: 100			PGM^1_2	: 100	Modiano et al., 1969
Locus 2	PGM^1_2	: 100			PGM_2^{pygmy}	: reduced	Santachiara et al., 1970
Galactose-1-Phosphate uridyl transferase	Gt^+	: 100			Gt^{Duarte}	: 50	Beutler et al., 1966
Pseudocholinesterase							
Locus 1	E^u_1	: 100			E^a_1	: 50	Simpson, 1968
Locus 2	E^-_2	: 0			E^+_2	: 30	Harris et al., 1963

[a] See Motulsky, 1969, for references cited.

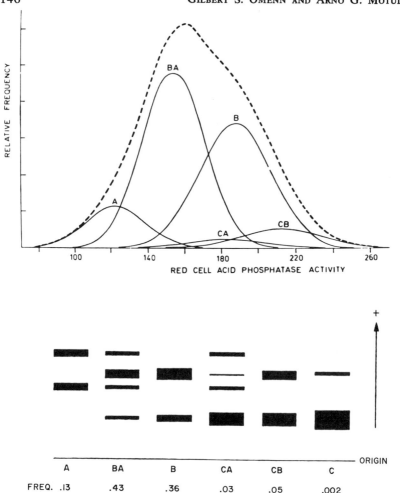

FIG. 1. Electrophoretic phenotypes and associated quantitative enzyme activity of human red blood cell acid phosphatase. (From Harris, 1970. Reproduced by permission.)

brain (Cohen *et al.*, 1972) and in screening of human erythrocyte enzymes of the glycolytic pathway by Chen and Giblett (1971). The only exception is the polymorphism of phosphohexose isomerase in the mouse. This enzyme is present in high activity relative to other enzymes in the pathway. It is possible that the very old evolutionary status of glycolysis and its central role as the primary pathway of glucose utilization in the brain have placed remarkable constraints on the tolerance for mutation-induced variation in the protein structure of these enzymes. Most of the glycolytic enzymes have

7. Biochemical Genetics and the Evolution of Human Behavior

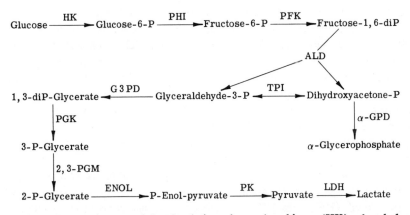

FIG. 2. Enzymatic steps of the glycolytic pathway: hexokinase (HK), phosphohexose isomerase (PHI), phosphofructokinase (PFK), aldolase (ALD), triosephosphate isomerase (TPI), glyceraldehyde-3-phosphate dehydrogenase (G3PD), phospho-glycerate kinase (PGK), 2,3-phosphoglycerate mutase (2,3-PGM), enolase (ENOL), pyruvate kinase (PK), and lactate dehydrogenase (LDH). In addition, α-glycerophosphate dehydrogenase (αGPD) is shown.

evolved tissue-specific isoenzyme forms—that is, different genes specify proteins with similar enzyme function in different tissues (for example, brain versus muscle). There are clinical consequences of such tissue variation within individuals. If an enzyme is deficient in erythrocytes, one would expect deficiency in other tissues only if the same gene specified that enzyme in the other tissues. Deficiencies of seven of the glycolytic enzymes have been identified as causes of hereditary hemolytic anemia in man. From their

TABLE 9 Electrophoretic Screening of Glycolytic Enzymes in Human Brain Tissue

Enzyme	Buffer system[a]	No. variant/Total alleles
Hexokinase	TP	0/300
Phosphohexoseisomerase	TC	0/300
Phosphofructokinase	TP + ATP (10^{-4}M)	0/240
Aldolase	TEB	0/600 (2 loci)
Triosephosphate isomerase	TC	0/300
Glyceraldehyde-3-phosphate dehydrogenase	TEB + NAD (10^{-4}M)	0/240
Phosphoglycerate kinase	TC	1/203
Phosphoglycerate mutase	TEB	0/300
Enolase	TP	1/300
Pyruvate kinase	TC	0/300
Lactate dehydrogenase	PHOS	0/600 (2 loci)

[a] Buffer systems: (TP) Tris-phosphate, pH 8.6; (TC) Tris-citrate, pH 7.5; (TEB) Tris-EDTA-borate, pH 8.6; (PHOS) Phosphate, pH 7.0.

electrophoretic and biochemical properties, only three (PHI, TPI, PGK) are likely to have the same form in brain as in red blood cells (Omenn et al., 1972). The original case reports of TPI and PGK deficiency did note prominent nervous system symptoms and signs. The PHI deficiency is not instructive, since the deficiency was mild even in the red blood cells. Deficiency of the other enzymes was not associated with any neurological abnormalities. Such tissue comparisons are important for another reason: if the same gene is responsible for the enzyme in all tissues, sampling of blood or skin or hair follicles may enable us to test for properties of the brain enzyme without needing to obtain brain tissue.

Another aspect of the conservatism of the glycolytic pathway is a comparison with the pentose–phosphate shunt. Here the first two enzymes have been studied, G6PD and 6PGD. These enzymes are controlled by the same gene in the nervous system as in other tissues and the same polymorphism known to exist in RBCs occurred in our brain specimens. We hope to extend study of this auxiliary enzyme pathway to additional enzymes to test the prediction that polymorphism is more likely in the less essential pathway.

A new polymorphism in man has been uncovered in our screening of the brain material (Cohen and Omenn, 1972). Malic enzyme, NADP-linked malate dehydrogenase, exists in a cytoplasmic form, which probably interconnects the Krebs cycle and gluconeogenesis, and in a mitochondrial form whose function is speculative, but may be involved in hydroxylation reactions. Studies in man and in monkeys demonstrate that the cytoplasmic and mitochondrial malic enzymes are controlled by different genes and vary and segregate independently. The mitochondrial malic enzyme in man has three phenotypes in starch gel electrophoresis, corresponding to gene frequencies of .7 and .3 for the two alleles.

The generation of high-energy phosphates is mediated first from stores of creatine phosphate in the nervous system. A striking variation in the activity of CPK with absent, intermediate, and intense activity in different specimens is suggestive of a difference in stability or kinetic parameters of a possible variant.

Other enzymes studied thus far include glycerol-3-phosphate dehydrogenase, rate-limiting for myelination; isocitric dehydrogenase, an NADPH-generating enzyme, which has disproportionately high activity in premature infants as compared to full-term infants or older individuals; and glutamic dehydrogenase and acetyl cholinesterase, enzymes involved in pathways affecting the neurotransmitters gamma amino-butyric acid and acetyl choline, respectively.

We intend to expand the study to other enzymes, particularly those involved in neurotransmitter metabolism and biosynthesis, with the expecta-

TABLE 10 Clinical Correlation of Isoenzyme Data for Glycolytic Enzymes

Glycolytic enzymes	Tissue-specific isozymes occur	Deficiency described in red blood cells	
		Hemolytic anemia associated	Neurologic signs associated
Hexokinase	+	+	0
Phosphohexoseisomerase	0	+	0
Phosphofructokinase	+	+	0
Aldolase	+	0	
Triosephosphate isomerase	0	+	yes
Glyceraldehyde-3-phosphate dehydrogenese	+	+	0
Phosphoglycerate kinase	0	+	yes
Phosphoglycerate mutase	+	0	
Enolase	?	0	
Pyruvate kinase	+	+	0
Lactate dehydrogenase	+	0	

tion that biochemical correlates of neural plasticity may more likely be found in such pathways than in the basic energy-generating processes, like glycolysis. In addition, study of monoamine oxidase, glutamic acid decarboxylase, and other such enzymes allows the marshalling of a second powerful experimental tool of the biochemical geneticist, pharmacogenetic analysis (Omenn and Motulsky, 1972). When certain drugs are given to a large number of people, the therapeutic response or incidence of side effects has a strikingly bimodal distribution, suggesting a major difference in the two groups of individuals. In several cases, the biochemical mechanisms underlying such differences have been demonstrated. For example, isoniazid, hydralazine, dapsone, and some sulfa drugs are acetylated rapidly by about half of the Caucasian population and slowly by the other half of the population. The rate of acetylation in the liver is determined by a single recessive gene (slow is recessive). Slow inactivators reach higher blood levels of active drug and higher risk of toxicity. There are known specific inhibitors for many of the critical brain enzymes, making it possible to screen many individuals for variants in susceptibility to inhibition by such drugs. Since these drugs have definite pharmacological and behavioral effects *in vivo*, such a genetic difference in response to these agents would allow direct manipulation of the appropriate neurotransmitter pathway (in mouse or monkey models and, with careful ethical controls, in patients receiving such drugs for therapeutic indications). The observations that a variety of psychopharmacological agents can modify affect, sleep, cognition, and sensory perception constitute a cornerstone of our assumption that such functions of the *mind* are mediated by the metabolic processes of the *brain* (Kety,

1967). It is likely that we have uncovered only the tip of an iceberg of specific enzyme–drug interactions that underlie the marked differences between individuals in their response to drugs and in their risk of side effects.

Study of polymorphic enzyme systems involving crucial metabolic processes in the brain seems a potentially fruitful approach to polygenic phenomena that underlie most behavioral traits. The electrophoretic screening method is capable of uncovering qualitative, structural differences in specific enzymes between individuals, without confusion by the alteration in quantitative activity in different parts of the brain or upon physiological stimuli. However, the interpretation of the physiological consequences of these qualitative enzyme differences will require careful measurement of the metabolic impact in individuals having the two different types of enzymes. In humans, such measurements must be carried out indirectly with radioactive tracers and with enzyme inhibitors; in model systems in mice or monkeys, more direct measurements may be feasible. The statistical notion that polygenic inheritance involves the equal and additive effects of a great many genes must be modified in light of metabolic interactions. Certain metabolic control points will be more important than others and much more important than enzyme reactions in minor pathways. Thus, it is possible that, even though a great many genes can interfere with normal brain development if completely deficient, the so-called normal range of development and function may be determined by a relatively few polymorphic genes sitting at rate-limiting steps in key metabolic pathways. Since we have already found no such polymorphism for glycolysis, it is likely that the rate-limiting points in the metabolic scheme of the brain that do have variation will be fewer than the potential number of sites. The fact that a normal or Gaussian distribution of some quantitative variable is obtained does not require a large number of genes for explanation. In fact, with just two alleles at each of two interacting loci or with three alleles at a single locus (acid phosphatase model), a quantitative distribution of some resulting trait can appear polygenic (Motulsky, 1969). The key feature of this model for congenital malformations or for classification of development as normal versus abnormal is the presence of a threshold in the quantitative sense, a threshold to which each allele can contribute and upon which various hormonal and environmental agents might act. The heuristic value of this point of discussion is to encourage the search for major gene mechanisms in polygenic traits and psychiatric disorders.

Approaches to Complex Behavioral Phenotypes

It is difficult to analyze phenotypes into meaningful "units" of behavior, in the sense that molecular evolution can be analyzed in units of proteins,

DNA, and chromosomal structures, or that the underlying biochemistry for the brain might be analyzed in terms of energy requirements, developmental switches, recognition phenomena, and possible electrochemical transformations. However, we have identified five operational approaches that may signify possible "handles" on certain aspects of modifiable behaviors at the level of integrated functions of the nervous system.

Sexually Dimorphic Behavior and the Effect of Fetal and Neonatal Hormones. One of the more difficult and especially timely questions about human behavior is the issue of male/female differences and the extent to which they reflect cultural impact of the assigned sex role or biological impact of the sex chromosomes and sex hormones. Young et al. (1964; Goy, 1970) have studied in the guinea pig, the rat, and the rhesus monkey the organizing or sex-differentiating action of fetal gonadal substances on behaviors beyond that which is primarily sexual. It was long ago established that mammalian fetuses lacking or deprived of fetal testes would undergo "indifferent" embryological development to normal female form and psychosexual orientation. Fetal testicular hormone is essential for differentiation of the Wolffian duct system into the male genital tract and for suppression of the Mullerian duct system. With female guinea pigs, prenatal injection of androgen produced a marked display of masculine behavior, as well as a lowered capacity to display feminine behavior (up to 92% later failed to come into heat) (Young et al., 1964). Conversely, castration of genotypic male rats led to significant postpubertal display of feminine behavior, measured as lordotic receptivity in response to mounting by intact males. Injection of testosterone propionate to the pregnant mother rhesus monkey caused genetic female offspring to later display behaviors distinctly shifted in frequency toward the male values. The measured behaviors are patterns of threatening, play initiation, rough-and-tumble play, chasing play, and immature double-foot-clasp mounting. In all cases, the differences between males and females are quantitative, not qualitative. The clinical implications of such studies for disorders of psychosexual identity in man are obvious. Money (1971) has found that hermaphroditic individuals raised in one sex, matched by individual of identical diagnosis raised in the other sex, typically differentiate psychosexually in concordance with the parentally-assigned sex. However, male transsexuals, some homosexuals, and occasional Klinefelter's syndrome patients with the XXY chromosomal anomaly tend to develop a gender identity resembling normal female, rather than normal male patterns (Money and Brennan, 1969). The distribution of these behaviors is shifted in males versus females, though a good deal of overlap exists.

The physiological bases for sex-specific patterns of behavior are uncertain, but certain hypothalamic regions are known to be excited or inhibited by the sex hormones, regions that might be involved in neural motivational systems. The potential for molecular biological exploration of such hormone-sensitive regions is at hand. For example, stereotactic implants of estradiol in the diencephalon of the female rat distinguish two hypothalamic centers sensitive to estrogen: destruction of the gonadotropin-regulating center in the anterior hypothalamus, or implantation of estrogen there, leads to gonadal atrophy; lesions in the basal tuberal-median eminence suppress mating behavior, but do not affect the gonads or interfere with the estrus cycle in the rat or cat (Lisk, 1962). Estrogen-sensitive oviduct and uterine preparations contain estrogen-binding cytoplasmic and nuclear proteins (Steggles et al., 1971). Similar binding sites probably exist in the sensitive regions of the hypothalamus and possibly in higher brain centers. In the Mongolian gerbil, implantation of minute amounts of testosterone in the preoptic region of the hypothalamus of castrated males can restore the species-specific and male-specific behavior of territorial marking, and such effects of testosterone can be blocked by simultaneous implantation of reasonable amounts of actinomycin D or puromycin (Thiessen, 1971). Testicular feminization, a syndrome in which genetic males with normal testes and normal production of testosterone fail to become masculinized because the target tissues fail to respond to the hormone, is now being studied in an animal model (Dofuku et al., 1971). These individuals appear and act as females. In the various studies in animals and in man, it may be possible to determine central nervous system mechanisms of sexually dimorphic behaviors.

Inborn Errors of Metabolism in Man. A striking array of enzyme deficiencies has been recognized as "experiments of Nature" in man. Some of these are associated with mental retardation or other behavioral abnormalities, others affect only red blood cells or other tissues, and some seem to have no detectable deleterious effects. Most are inherited as autosomal recessive traits, though a few are X-linked recessive traits manifested in males in hemizygous form. We have recently tabulated a variety of amino acidurias, carbohydrate and lipid and mucopolysaccharide storage diseases, and miscellaneous metabolic errors according to their impact on the nervous system (Omenn et al., 1972). A few of these syndromes are listed in Table 11. We distinguished a gross defect in mental development from more specific neurologic signs or psychiatric/psychological disorders, occurring without mental retardation or before mental deterioration. Some of these disorders are due to toxic effects of metabolites accumulated as a result of metabolic defects in other tissues, while other disorders are intrinsic to the nervous system. Among the latter, two of great interest are the Lesch–Nyhan syndrome and homocystinuria. The Lesch–Nyhan syndrome is comprised of hyperuricemia, choreoathetotic movement disorder, and a self-destructive, impulsive behavior. It is due to deficiency of an enzyme known as hypoxanthine-guanine phosphoribosyl transferase (HGPRT), involved in what was

TABLE 11 Inborn Errors of Metabolism and Predominant Phenotype

Syndrome	Enzyme	Mental retardation	Neurologic dysfunction	Phychiatric dysfunction	Intrinsic to CNS
Phenylketonuria (PKU)	phenylalanine hydroxylase	++++	0	0	No
Homocystinuria	Cystathionine synthetase	0/++	(vascular accidents)	??	Yes
Histidinemia	histidase	0/++	speech impairment in half of cases		?
Maple-syrup urine disease	Branch-chain ketoacid decarboxylase	++++	Ketotic coma	—	No
Variant	(Incomplete deficiency)	0	Episodic ataxia	—	No
Metachromatic leukodystrophy	Arylsulfatase A	0 (Secondary)	Motor & mental deterioration, age 2	—	Yes
Adult form	(8 cases)	0	0 (late)	"schizophrenia"	Yes
Lesch–Nyhan	HGPRTase	+++	++++	++++	Yes

previously discounted as a minor "salvage" pathway of purine metabolism. This enzyme turns out to have its highest activity in the body in the basal ganglia of the brain, allowing a correlation with the neurologic disorder of choreoathetosis. Just how to relate this metabolic disorder to an uncontrollable impulse to bite off the tips of the fingers or the lips is a challenge to investigators of impulsive or violent behavior disorders in man. The importance of the metabolic pathway is underscored by the finding that heterozygous females (cellular mosaics by the process of random X inactivation of this X-linked trait) have the expected 50% of normal activity for HGPRT in skin fibroblasts, but 100% of normal activity in blood cells (Nyhan et al., 1970). Presumably, all blood cell precursors lacking HGPRT activity were eliminated. We have no information on HGPRT-negative cells in the nervous system.

A second remarkable metabolic error intrinsic to the nervous system, as well as other tissues, is homocystinuria, due to deficiency of the enzyme cystathionine synthetase. As a result, cystathionine is not formed and homocystine and methionine accumulate. Cystathionine is a complex amino acid found normally in remarkably high concentrations in the brain, but its function is entirely unknown. A different inborn error, cystathioninuria, due to deficiency of the enzyme to break down cystathionine, seems to be unassociated with any major defects. Clinically, homocystinuria is characterized by vascular thromboses, skeletal anomalies, downward displacement of the ectopic lens of the eye, and—in only about one-half of all cases—mild to moderate mental retardation. There is considerable dispute whether affected patients or their sibs might have an increased incidence of schizophrenia; the evidence is not impressive, but the question is a sound speculation, based upon the hypothesis that methylated derivatives of normal neurotransmitter substances might be pathogenetically involved in schizophrenia or at least in experimental hallucinatory states. Why only one-half of cases have mental retardation is unclear. Perhaps the others have lower IQ than would have been their potential, but are still within the normal range. Perhaps the enzyme defect is different in different individuals.

A most important consideration in these rare recessive inherited metabolic disorders is the realization that even a disease with an incidence of only 1 in 40,000 births is associated with a heterozygous carrier frequency of 1%. For certain enzymes present in the brain and present in near rate-limiting activity, such a decrease to 50% of normal activity in the carrier might be a significant factor in predisposition to mild mental impairment or possibly to regionally-specific mental defects. There has been very little detailed psychometric study of such possibilities. The only definite finding comes, not from a metabolic error, but from the chromosome anomaly 45,X0 or Turner's syndrome, in which Money (1963) has demonstrated a striking defect in space-form relationships and in drawing ability. Anderson, however, has initiated a series of studies of manual dexterity and related specific functions in patients and carriers of the gene for PKU (Anderson *et al.*, 1968 and 1969). Since carriers for the long list of rare recessive diseases together make up a large percentage of normal individuals, any mild abnormalities that could be documented in such carriers might be useful in interpreting the range of normal behavior. Study of enzyme systems identified by syndromes of metabolic disorders seems complementary to our systematic approach, described earlier, to enzyme variation in the key metabolic pathways of the nervous system. A rational search for behavioral effects of another syndrome was reported by Scriver *et al.* (1970). Since the amino acid transport defect of cystinuria is present not just in kidney and intestine, but also in brain, it was reasonable to seek clinical consequences. The scattered cases of variable psychiatric disorder or mental retardation associated with cystinuria may not be a greater incidence than that due to chance, however.

For all such studies, more careful and more discriminating tests of psychological functions are needed. Some progress in this regard has been reflected in studies of brain-injured patients (Reitan, 1972).

Interracial Differences. The possibility of individual and racial differences in behavior before any obvious postnatal learning has been tested in a preliminary fashion (Freedman and Freedman, 1969). A matched series of infants (5–72 hours of life) of Chinese-American (Cantonese background) and of European-American background were evaluated on a Brazelton scale of 25 neurological and behavioral criteria. They were identical in scores of sensory and motor development, central nervous system maturity, and interest in their social environment. But there was a striking difference in scores of temperament, especially excitability/imperturbability ratings, such that the Chinese-American infants were less changeable, less perturbable, habituated more readily, and were consoled or calmed themselves more readily. The results are so consistent with a stereotype of adult behavior that further studies of this type will be of great interest.

There are definite differences in the gene frequencies for various polymorphic proteins in blood of oriental, negroid, and caucasoid populations (Harris, 1970; Giblett, 1969). There may be similar interracial differences in gene frequencies for enzymes in the brain and other tissues. Physiological differences might result from such polymorphic protein systems, but the impact of cultural forces on the biological substrate makes evaluation of behavioral phenotypes for interracial differences a most difficult task.

Polymorphisms of EEG Phenotypes and Possible Behavioral Correlates. Vogel has summarized a monumental, but unconfirmed, study of electroencephalographic (EEG) patterns in presumably normal individuals. The complex electrical potentials recorded from the scalp are determined almost

TABLE 12 Variants of the Normal Human EEG[a]

Rhythm	Genetic basis	Population frequency	Coment
Normal alpha (8–13 cps)	Polygenic		
Low voltage alpha	Auto Dom	7%	
Quick alpha (16–19 cps)	Auto Dom	.5%	
Occipital slow (4–5 cps)	??	.1%	?Psychopathy
Monotonous tall alpha	Auto Dom	4%	?Assortative Mating
Beta waves	Multifactorial	5–10%	Sex, age ?Assortative Mating
Frontal beta groups (25–30 cps)	Auto Dom	.4%	
Fronto-precentral beta (20–25 cps)	Auto Dom	1.4%	

[a] After Vogel, 1970.

entirely by genetic factors. Monozygotic twins share not only identical EEG patterns, but identical maturational transitions in the EEG patterns in adolescence and in later life. Analysis of pedigrees (Vogel, 1970) points to a polygenic mode of inheritance, with several specific variant EEG patterns inherited as Mendelian autosomal dominant traits.

Four percent of the population have the monotonous tall alpha pattern, determined by a single autosomal dominant gene, and another 5–10% have a beta wave pattern, with multifactorial determination. For both groups, limited data suggest that individuals of either of these variant EEG types tend to marry individuals of the same EEG type (Table 12). For another anomalous EEG pattern of less straightforward inheritance (posterior slow rhythm), there may be an as yet poorly characterized predisposition to psychiatric disorders. Apparently, discriminating psychometric analyses of individuals with different classes of EEG patterns have yet to be carried out. Also there is the potential to correlate the EEG patterns and any psychometric features with response to physiological (photic, auditory, sleep) stimuli and to pharmacological stimuli and to learn whether individuals of a given EEG pattern have distinctively different susceptibilities to various sedative or psychoactive drugs.

> *A desire to take medicine is, perhaps, the great feature which distinguishes man from other animals.*
> SIR WM. OSLER, *1891*

The Effects of Psychopharmacologic Agents. The potent effects of various drugs as sedatives, anesthetics, and central stimulants are well-established, though there is very little evidence on individual differences in susceptibility to desirable or toxic effects (Omenn and Motulsky, 1972). Several classes of drugs have proved effective in treatment of affective disorders; here certain clinical studies suggest that groups of patients may differ in their responsiveness or lack of responsiveness to monoamine oxidase inhibitors or to tricyclic antidepressants (Pare et al., 1962; Angst, 1961). Such patients manifested a similar pattern of response when treated during a subsequent episode of depression, as did relatives who were treated for depression. The bewildering array of "up" drugs used in all sorts of combinations by hippies and housewives alike impress the "street people" with the variety of response in different individuals. Always there is the suspicion that individuals who have "bad trips" may be predisposed to psychiatric difficulties. We have been reluctant to test L-dopa or other possible provocative agents in patients with a risk to develop Huntington's chorea, for the similar reason that we might induce psychotic symptoms in a predisposed patient and be unable to reverse the process. Finally, we should mention the current interest in hyperactive or hyperkinetic children and the recommendations (Task Force on Drug Abuse, 1971) that some 5–15% of young children be treated with amphetamines or methylphenidate (Ritalin). Here we are dealing with a potential culturally-decided behavioral modification program of generalized scale and frightening possible impact. The underlying behavior at issue is usually poorly characterized, the pharmacological basis for the treatment suspect, the biochemical actions of the drugs complex, and the metabolites of the drugs not readily detected by available chemical techniques. Nevertheless, the widespread use of amphetamines in the adult population and the acceptance of tranquilization of neurotic as well as psychotic individuals provides a cultural background suited to increasing modification of behavior with psychoactive drugs. This issue seems deserving of attention and control. The behavioral scientist has much to offer in studying the individual differences in mechanisms of response to these drugs and in providing a rationale basis for their use (Omenn, 1972).

The Central Role of Language in the Evolution of Man

Let us turn now from biochemical and physiological aspects of behavior to certain key features of the cultural evolution of man. All writers agree that symbolic, verbal communication in the media of language is the hallmark of *Homo sapiens*. Complex coordinating, representational, and cognitive functions of the human central nervous system are identifiable in other species. Animals, of course, may have elegant means of communication, too, but we assume that they lack the capacity to create subjective experiences, to carry out "subjective simulation" (Monod, 1970), to appreciate the notions of death and of self. Burial of the dead as an indication of such symbolic understanding appears in the fossil record very much later than the evidence of upright posture, apposed thumb, man-like jaw, and enlarged brain. However, *Homo neanderthalensis* had sufficient compassion to bury the dead and decorate the grave with clusters of brightly colored flowers (Solecki, 1971). Possibly such behavior reflects intellectual capacity greatly in excess of what was needed to cope with the environment of the time. The species of flowers found in burial sites deep in caves at Shanidar are now known to have medicinal properties (Solecki, 1971), but we cannot determine whether Neanderthal man was aware of such properties.

The time of appearance of "language," estimated as 10,000 to 50,000 years ago (Table 2), is altogether uncertain. Most modern linguists seem to ignore the issue. Nevertheless, a remarkable transition has occurred in the field of linguistics in the past decade or so. Previously, attention seemed to be riveted on the diversity of language and the possibility of tracing languages in cultures through what are now regarded as superficial aspects of vocabulary and grammar. Swadesh and others derived "evolutionary trees" of language interrelationships, analogous to the trees drawn by the paleontologist or the molecular taxonomist (Fig. 3). This technique of estimating "time-depth" of language relationships, called glottochronology or lexicostatistics (Swadesh, 1950; Lees, 1953; Gudschinsky, 1956), assumed that spoken language could be divided into core and general word lists, that the rate of retention of vocabulary items in the core is constant through time and in all languages (about 80% retained over 1000 years). By a simple calculation, from the percentage of true cognates between any two languages, the length of time that has elapsed since the two languages began to diverge from a single parent language can be estimated. Thus, the evolution of language appears similar to the evolution of proteins, alterations in spelling or form of words being analogous to the amino acid substitution in the proteins. Linguistic and blood group data, in fact, were used together by Watson et al. in an anthropological study in New Guinea, a large island where 2–3 million people are divided into some 400 language groups (Watson *et al.*, 1961). The major difference between the paleontological or molecular evolutionary trees and the trees of the glottochronologist lies in the time scale —millions of years for the biologists versus hundreds of years for the glottochronologist. It is this difference in time scale that reflects the impact of cultural evolution.

Chomsky and other modern structural linguists have described a basic unity in the midst of the diverse languages of man (Chomsky, 1965). There is no explanation of why any particular pattern of sounds signifies any given object or action (except onomatopoeia). Yet speech patterns of all languages operate on a few basic principles, and the semantic patterns may be similar if deep structures are deciphered. Deep structures consist of base

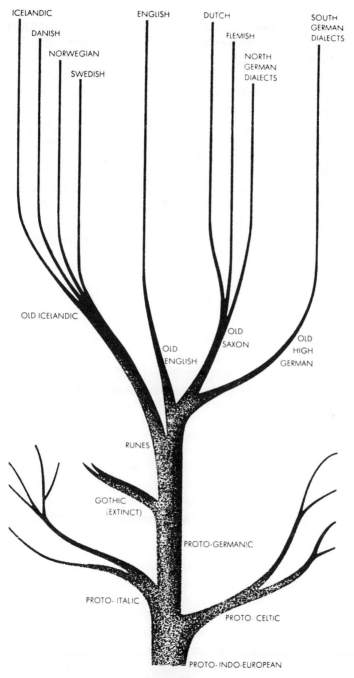

FIG. 3. Glottochronologic "tree" of related languages (From Hockett, "The Origin of Speech," © 1960 by Scientific American, Inc.).

phrase-makers upon which negative, passive, or question transformations, etc., act. The analysis employs a universal phonetic alphabet, starting with phones and advancing through phonemes, morphemes, and lexemes. Levels of grammaticality use base rules and transformation rules for syntactic, semantic, and phonological components (Chomsky, 1965). Such a unifying approach points to some biologically-determined potential of the species and sets a model for analysis of other features of man's culture.

As described by Lenneberg (1967, 1970), language has its roots in the physiological processes of cognition. Language-knowledge is viewed as an activity, rather than as a static storehouse of information, an activity of extracting peculiar relationships from the environment and interrelating these relationships. Examples drawn from neurological disorders show a parallel between acquired language disturbances and acquired disorders of such cognitive features as perceptual recognition. In the evolutionary context, it has been claimed that primates seem to have adequately developed motor systems for vocalization (Ploog, 1970) and visual-auditory perceptiveness for such clues to relationships (Sebeok, 1962).

There is controversy about the possibility that the laryngeal and pharyngeal anatomy has evolved parallel to the development of a capacity for language in the brain. Bryan (1963) claims that, although isolated larynxes appear identical, anatomical relationships and function of the epiglottis and soft palate and insertion of the base of the tongue making babbling and a variety of vocalizations easy only in man. Lenneberg (1970) agrees that structural changes in the vocal tract make the production of speech sounds uniquely possible in man, but insists that such modifications are not prerequisite for language capacity. Thus, children with deformed fauces can learn to understand English, even though their own speech is unintelligible. Also, children with congenital deafness, congenital blindness, or mesencephalic lesions that interfere with muscular coordination for speech can acquire language skills (Lenneberg, 1967).

Attempts to teach chimpanzees some form of human language suggest that the vocal tract difference is of some importance. Chimps Viki (Hayes and Hayes, 1955) and Gua (Kellogg, 1968), despite long efforts, acquired only a few words of barely intelligible English ("mama," "papa," "cup," "up," for example). The Gardners (1969), however, noted that this sociable animal, which forms close relationships to humans, tends to be silent and to vocalize only when excited. Instead, chimps use their hands extensively to communicate. On the hypothesis that gesturing by chimps might be a natural mode of expression, like bar-pressing for rats or key-pecking for pigeons or babbling for humans, they exposed Washoe to the American Sign Language gestures of the deaf and "taught" signs and rewarded learning. It is useful that some of the signs are iconic, while others are arbitrary. Washoe was an 8–14 month-old female at the start of the program. Within 22 months of the project, she could use 30 signs. From the time she had 8–10 signs in her vocabulary, she began to string two or more together and to transfer spontaneously a single sign to a wide class of appropriate referents.

Among the great many inherited and developmental syndromes affecting man, none seems to specifically alter language. However, there is interest in the speech impairment that accompanies half of the cases of histidinemia and in the possibility of metabolic abnormalities in some cases of reading impairment (dyslexia) (Childs, 1970).

It probably is not appropriate to view language itself as the evolutionary advance in the development of man; rather, language reflects some saltatory developments in complexity of cognitive processes. It is possible that elaborate macromolecular recognition mechanisms or novel transmitters and more complex synapses may underlie the advanced cognitive functions required for language. Although ablation or infarction of certain frontal and temporal cortical areas leads to aphasic defects in *speech,* we are uncertain whether specific anatomical structures in the brain can be assigned *language* function. Perhaps we have not adequately tested for such functions, however. For example, it has been a surprise that such a vague, diffuse, and varied function as affective state or mood could be localized to the limbic system, that stimulation or lesions in limbic structures can cause tameness or aggressiveness, hypersexuality, change in feeding or drinking or emotional expression, or recent memory impairment (Bullock, 1970). In the cerebral disconnection syndrome produced by complete section of the corpus callosum between the dominant and nondominant hemispheres (Gazzaniga and Sperry, 1967), disruption of inter-hemispheric integration produced remarkably little disturbance in ordinary daily behavior, temperament, or intellect. Writing and drawing with either hand are intact, indicating bilateral motor representation. Comprehension of both spoken and written language is intact. But information perceived or generated exclusively in the nondominant, right hemisphere could be communicated neither in speech nor in writing; it had to be expressed through nonverbal responses. Likewise, the separated minor hemisphere was incompetent in tasks of calculation. It is not clear whether these defects represent interference with the afferent side of the *speech* centers or with more basic language functions of the dominant hemisphere.

Two neuroanatomical substrates for integration of language functions have been postulated from recent studies of the defects in aphasia patients and of electrical stimulation during neurosurgery: (1) corticocortical connections between sensory association areas, as in the angular gyrus of the dominant temporo-parietal-occipital junctional region (Geschwind, 1968); and (2) the pulvinar and fibers related to certain nuclei in the left thalamus (Ojemann and Ward, 1971). Corticocortical connections and pulvinar size both are much more extensive in man than in certain nonhuman primates; however, no comparative descriptions of these structures are available for the chimpanzee. Cross-modal (that is, visual, auditory, tactile) transfers of input stimuli are crucial to naming objects and making reference to the environment.

The model of language as a species-specific universal behavior phenotype was extended by R. Fox (1970) to other species-specific "units" of behavior —kinship, courtship, and marriage arrangements; political behavior; associations of men which exclude women. Presumably there are definite limits to what the human species can do, to the kind of societies or cultures it can operate. No language seems conceivable that would violate the generative grammar rule of the universal language and be interpretable to man. Similarly, Fox argues that any behavioral patterns that were "gibberish" in terms of man's biological limits would cause a breakdown in social communication and be rejected. When infant baboons are raised in a zoo, they tend to mature and produce a social structure with all the elements found in the wild. Presumably, if a group of men and women were put into an experimental Garden of Eden without rules, they would produce a culture with the same basic properties as ours. The notion that we have a "wired-in" information processing capacity that responds specifically to certain kinds of inputs and responds with an element of timing in the life cycle (developmental stages) is consonant with the interactionist hypothesis of Piaget (1952) for general development of cognitive processes. It is conceivable that the evolutionary development of language reflected an analogous interaction of biological potential and cultural inputs. Two million years ago, ancestral men with brain sizes little larger than those of gorillas (then or now) were hunting, building shelters, making tools, treating skins, living in base camps, with well-established bipedalism and human dentition. Under presumed selective pressures for cultural adaptation and social communication, there may have been significant increase in the relative size, complexity of connections, and variety of transmitters and recognition molecules in the evolving neocortex. It is likely that our brains contain not only the capacity for culture, but also determine the forms of culture, through some universal grammar for both language and general behavior.

We may overemphasize the differences between culture and instinct. Stereotyped, instinctive mechanisms are highly efficient, but dangerously rigid. Ants can have societies, but not politics. Politics occurs only when members can change places in a hierarchy as a result of competition, as in gregarious, terrestrial primates. Yet, much of our behavior is "unconscious" or "automatic" in response to common environmental and developmental inputs—an iceberg of assumptions, values, and habits, plus the impact of the conscience or superego reflected in a sense of guilt, of having broken taboos or rules of the tribe. The capacity for imaginative thought and the need for self-control seem to have evolved biologically and culturally together. To what extent such features have become fixed in the biology of the species in the relatively short evolutionary time of man and to what extent they represent learned behavioral patterns remains controversial.

The written forms of language introduce additional considerations. It is remarkable that after 30,000 years or so of spoken language, iconic or hieroglyphic languages appeared in the short period of 2500 years in such widely separated peoples as the Sumerians, the Chinese, and the American Indians (4300 to 2000 B.C.). Then the Phoenicians and Hebrews and others adopted an alphabetized language. At least some Sumerian, Egyptian, and Chinese symbols do stand for words, which can be read in a spoken language. Certain Egyptian hieroglyphs were phonetic, not iconic, and completely phonetic alphabets developed from these hieroglyphs. On the other hand, the iconic efforts of American Indians cannot be read in a spoken language (Simpson, 1972). We are uncertain whether any important evolutionary conclusions can be drawn from the use of iconic or phonetic written forms of language in different cultures.

With the knowledge of man written into books, microfilm, libraries, and computers, the species has what might be called a "superbrain" (Rensch, 1970). Presumably a fertile group of men and women, a library, and materials would be sufficient for the reconstruction of human culture after a holocaust!

The Impact of Evolution of Man's Culture upon Man

When we realize that agriculture has been a part of man's life for only 10,000 years, that urbanization began some thousands of years more recently, and that industrialization is a phenomenon of the past few hundred years, we must admit that the pace of change in man's environment completely overwhelms the time scale of biological, evolutionary processes (Table 2). On the other hand, we find it difficult to evaluate whether or not such differences in life style require any remarkable change in the behavioral potential, the cognitive and affective functions of man. The possibility of selection is present, but its impact now must be small. Earlier development of man, by contrast, may have been dramatically enhanced by the drastic environmental changes of four successive periods of glaciation during the last million years of the Pleistocene epoch. *Homo erectus (Pithecanthropus)* and *Homo neanderthalensis* flourished during early interglacial periods and perished during glacial periods. Periodic decimation to small effective population size may have been crucial to the emergence of *Homo sapiens*. Now the number of our species has reached so high a level that the chance of any newly acquired hereditary traits being selected and fixed as a new species characteristic is small. In addition, the long generation time of man decreases the probability of significant change even further.

We may wonder how fragile our culture may be. Remarkable human civilizations in Egypt, Babylon, Rome, and Greece all but vanished. Political turmoil anywhere seems to diminish cultural values and functions. Likewise, religious dogma can be repressive; orthodox Christian ideas suppressed scientific inquiry for 1500 years. It is not clear whether the renewal of complex human culture should be attributed to lack of destruction of parallel civilizations at different stages of development or to basic capacities of remaining members of the species. For example, European Jews successfully returned to an agricultural life on a kibbutz in Palestine after some 2000 years of urbanized existence. All of these events, of course, occur in times that can mean little in the biological evolution of man.

The technology of our culture raises special possibilities. Man need not be dependent upon natural selection and upon the chance occurrence of mutations, so few of which might be advantageous. Artificial selection conditions and directed changes in the genome are present-day fascinations in the imaginative mind of man; they may become practical possibilities, intentionally or accidentally, in the future. We must understand much more of the "units" of behavior and their genetic and biochemical mediation to devise in a rational way any purposeful alteration of man's behavioral potential. Yet nonrandom mating with regard to intelligence and to a variety of social factors has probably occurred for a long time. Nonrandom mating is practiced on a huge scale by man. Other current practices, such as exposure to possibly mutagenic agents in the form of environmental pollutants, drugs, and radiation, can have little short-term positive genetic impact, for the reasons of population size and generation time given above. A negative impact from these agents becomes increasingly likely with urban population crowding. Such environmental agents are potentially more pervasive dysgenic forces than medical treatments for life-threatening illnesses that improve reproductive potential of these individuals. Ironically, the disease usually chosen to represent the dysgenic effects of modern medical care is diabetes mellitus, in which insulin therapy can carry patients with juvenile onset of the disease through the child-bearing period. The irony is derived from the hypothesis of Neel (1962) that diabetes may have represented a "thrifty" genotype in hunting and gathering societies, where food intake was more erratic and where delay in metabolism of carbohydrates and in mobilization of fat stores might have been protective against periods of poor food supply. It is interesting that American Indian tribes have exceedingly high prevalences of diabetes mellitus. Thus, diabetes might be a disease once favored by selection and rendered detrimental by "progress"!

Also, it is man and his way of life that made malaria an important disease and led to selection of sickle-hemoglobin, thalassemia, and G6PD deficiency in populations where malaria was especially prevalent. Livingstone (1958) traced these events to the "slash-and-burn" agriculture which opened the forest floor to stagnant pools. Such "technological advances" brought man into contact with the insect vectors of malaria; similarly, snails and rodents were attracted to settled populations and brought other epidemic diseases. The practice of single-crop agriculture also brought risks, since each cereal has its own limiting amino acids and propensity to protein undernutrition and endemic dysentery. Perhaps the most unusual vector for a specific disease is the culturally-based occurrence of kuru in the Fore language group of New Guinea, a degenerative disease of the nervous system caused by a slow virus and contracted only by the cannabalistic practice of eating the brains of worthy dead males.

In our society, the major cause of death in the child-bearing years is accidents. We might direct some attention to the predisposing factors in fatal accidents (clumsiness, epilepsy, aggressiveness, alcoholism, etc.) and test for disproportionate gene frequencies among those who are victims and instigators of the accidents.

Many models of cultural evolution exist in the products of our society, including some which may be viewed as technological extensions of central nervous system functions (Table 13). In fact, discussions of the evolution of the two-wheeled bicycle (Rensch, 1970) and of the MGB auto (Rowland, 1968) have been published! Of these, the computer bears the greatest interest, both for its simulation of human deduction and for the possibility that models could be devised which would undertake some kinds of synthetic, inductive "thinking" processes.

There is a potent desire in man to expand his awareness, his consciousness, his utilization of his brain's potential—by religious experience, by use of drugs, and by determined intellectual effort. We have little basis to assess how nearly completely that potential is realized or to compare how different individuals do so. Table 14 lists some approaches of genetic engineering and electrical and pharmacological manipulation that have been discussed in this context.

TABLE 13 Technological Extensions of CNS Functions

Vision	Microscope—Telescope—photosensitive transducers
Hearing	Stethoscope—Telephone receiver
Smell	Gas chromatograph
Information processing	Computers

TABLE 14 Deliberate Modifications of Brain and Behavior[a]

I. Affecting the development of nervous system structures
   A. Genes
      1. Selective fertilization by genotypes
      2. Cloning of desired genotypes in vitro or in foster uteri
      3. Introducing genes by viral transduction
   B. Gene expression
      1. Growth factors ⎫ at critical developmental periods
      2. Hormones ⎬
      3. Specific connections or transmitters

II. Nonprogrammatic modification of brains
   A. Surgical approaches
      1. Grafts = additions
      2. Ablations = subtractions
      3. Reconnections
   B. Electrical stimulation or interference, use of drugs, hormones, chemicals
      1. Generalized changes in efficiency
         a. Arousal systems
         b. Motivational systems
      2. Selective alteration of weighted factors in complex functions
      3. Input of artificial information
         a. Selective elicitation and suppression of behavior and subjective experience
         b. Selective reinforcement of behavior patterns
         c. Information for memory stores

III. Programmatic modification of brains
   A. Generalized enrichment or impoverishment ("cultural milieu")
   B. Modifying options and opportunities
   C. Reinforcing selected behavior patterns
   D. Shaping and selecting reinforcers
   E. More complex learning technologies

IV. Combinations of above with monitoring, telemetry, computer evaluation

[a] Based upon a table of G. C. Quarton.

Some biologists, evolutionists, and philosophers view the nature of man and of his consciousness as a complexity beyond human understanding (Eccles, 1967). While total understanding may not be possible, the potential to increase our knowledge of human behavior by both reductionistic analysis of brain function and integrative, comparative study of complex behavioral correlates offers excitement and challenge for the experimental exploration of the function and evolution of the nervous system.

Acknowledgments

This work was supported by Grants GM 15253 from the United States Public Health Service and IN-26 from the American Chemical Society and by a fellowship from the National Genetics Foundation to G.S.O.

References

ALLISON, A. C. (1954). The distribution of the sickle trait in East Africa and elsewhere and its apparent relationship to the incidence of subtertian malaria. *Trans. Roy. Soc. Trop. Med. Hyg.* **48**, 312–318.

ANDERSON, V. E., SIEGEL, F. S., TELLEGEN, A., and FISCH, R. O. (1968). Manual dexterity in phenylketonuric children. *Percept. Mot. Skills* **26**, 827–834.

ANDERSON, V. E., SIEGEL, F. S., FISCH, R. O., and WIRT, R. D. (1969). Responses of phenylketonuric children on a continuous performance test. *J. Abnormal. Psychol.* **74**, 358–362.

ANFINSEN, C. B. (1959). "The Molecular Basis of Evolution." Wiley, New York.

ANGST, J. (1961). A clinical analysis of the effects of Tofranil in depression. Longitudinal and follow-up studies. Treatment of blood-relations. *Psychopharmacol.* **2**, 381–407.

BAGLIONI, C. (1967). Homologies in the position of cysteine residues of the K and L type chains of human immunoglobulins. *Biochem. Biophys. Res. Commun.* **26**, 82–89.

BARNICOT, N. A., JOLLY, C. J., and WADE, P. T. (1967). Protein variations and primatology. *Amer. J. Phys. Anthropol.* **27**, 343–356.

BARONDES, S. H. (1970). Brain glycomacromolecules and interneuronal recognition. *In:* "The Neurosciences, 2nd Study Program," (F. O. Schmitt, ed.) pp. 747–760.

BRYAN, A. L. (1963). The essential morphological basis for human culture. *Curr. Anthropol.* **4**, 297–301.

BRYAN, J., and WILSON, L. (1971). Are cytoplasmic microtubules heteropolymers? *Proc. Nat. Acad. Sci. U.S.* **68**, 1762–1766.

BUETTNER-JANUSCH, J. (1970). Evolution of serum protein polymorphisms. *Ann. Rev. Genet.* **4**, 47–68.

BULLOCK, T. H. (1970). Operations analysis of nervous functions. *In:* "The Neurosciences, 2nd Study Program," (F. O. Schmitt, ed.), pp. 375–383.

CALLAN, H. G. (1967). The organization of genetic units in chromosomes. *J. Cell Sci.* **2**, 1–7.

CALVIN, M., and CALVIN, G. J. (1964). Atom to Adam. *Amer. Sci.* **52**, 163–186.

CHEN, H. S., and GIBLETT, E. R. (1971). Personal communication.

CHILDS, B. (1970). The genetics of reading disability. Personal communication.

CHOMSKY, N. (1965). "Aspects of the Theory of Syntax." MIT Press, Cambridge.

CICERO, T. J., COWAN, W. M., MOORE, B. W., and SUNTZEFF, V. (1970). The cellular localization of the two brain specific proteins, S-100 and 14-3-2. *Brain Res.* **18**, 25–34.

COHEN, P. T. W., OMENN, G. S., and MOTULSKY, A. G. (1970). Enzyme polymorphisms as genetic markers for analysis of brain function. *Amer. J. Hum. Genet.* **22**, 46a.

COHEN, P. T. W., and OMENN, G. S. (1972). Variation in cytoplasmic malic enzyme and polymorphism of mitochondrial malic enzyme in *Macaca nemestrina* and in man. *Biochem. Genet.* **7**, 289–301, 303–311.

COHEN, P. T. W., OMENN, G. S., and MOTULSKY, A. G. (1972). Restricted variation in enzymes of the glycolytic pathway in human brain. (Submitted)

DAYHOFF, M. O., and ECK, R. V. (1970). Atlas of Protein Sequences and Structures. National Biomedical Research Foundation, Silver Springs, Maryland.

DeLANGE, R. J., and FAMBROUGH, D. M. (1968). Identical COOH-terminal sequences of an arginine-rich histone from calf and pea. *Fed. Proc.* **27**, 392.

DICKERSON, R. E. (1971). The structure of cytochrome c and the rates of molecular evolution. *J. Mol. Evol.* **1**, 26–45.

DIXON, G. H. (1966). Mechanisms of protein evolution. *In* "Essays in Biochemistry," (P. N. Campbell & G. D. Greville, eds.), Vol. 2, pp. 148–204.

DOFUKU, R., TETTENBORN, U., and OHNO, S. (1971). Testosterone—"regulons" in the mouse kidney. *Nature New Biol.* **232**, 5–7.

DUTTON, G. R., and BARONDES, S. (1969). Microtubular protein: synthesis and metabolism in developing brain. *Science,* **166**, 1637–1638.

ECCLES, J. C. (1967). Evolution and the conscious self. *In* "The Human Mind," (J. D. Roslansky, ed.), pp. 1–28. Nobel Symposium. Amsterdam: North Holland Publ.

EPSTEIN, C. J. (1964). Relation of protein evolution to tertiary structure. *Nature,* **203**, 1350–52.

EPSTEIN, C. J., and MOTULSKY, A. G. (1964). Evolutionary orgins of human proteins. *Prog. Med. Genet.* **4**, 85–127.

ERLENMEYER-KIMLING, L., and JARVIK, L. F. (1963). Genetics and intelligence: A review. *Science,* **142**, 1477–1479.

FOX R. (1970). The cultural animal. *Encounter,* **42**, 31–42.

FOX, S. S. (1970). Evoked potential, coding, and behavior. *In:* "The Neurosciences, 2nd Study Program," (F. O. Schmitt, ed.) pp. 243–259.

FRANKLIN, I., and LEWONTIN, R. C. (1970). Is the gene the unit of selection? *Genetics,* **65**, 707–734.

FREEDMAN, D. G., and FREEDMAN, N. C. (1969). Behavioural differences between Chinese-American and European-American newborns. *Nature,* **224**, 1227.

FUJIMORI, M., and HIMWICH, H. E. (1969). Electroencephlographic analyses of amphetamine and its methoxy derivatives with reference to their sites of EEG alerting in the rabbit brain. *Int. J. Neuropharmacol.* **8**, 601–613.

FULLER, J. L., and THOMPSON, W. R. (1960). "Behavioral Genetics." Wiley, New York.

GARDNER, R. A., and GARDNER, B. T. (1969). Teaching sign language to a chimpanzee. *Science,* **165**, 664–672.

GAZZANIGA, M. S., and SPERRY, R. W. (1967). Language after section of the cerebral commissures. *Brain* **90**: 131–148.

GESCHWIND, N. (1968). Neurological foundations of language. *In:* "Progress in Learning Disabilities," (H. R. Myklebust, ed.), Vol. 1, pp. 182–198. Grune and Stratton, New York.

GIBLETT, E. R. (1969). Genetic Markers in Human Blood. Davis, Philadelphia.

GOY, R. W. (1970). Early hormonal influences on the development of sexual and sex-related behavior. *In:* "The Neurosciences, 2nd Study Program," (F. O. Schmitt, ed.), pp. 196–206.

GROUSE, L., CHILTON, M-D, and McCARTHY, B. J. (1972). Hybridization of ribonucleic acid with unique sequences of mouse deoxy ribonucleic acid. *Biochemistry* **11**, 798–805.

GROUSE, L., OMENN, G. S., and McCARTHY, B. J. (1972). Studies by DNA–RNA hybridization of the transcriptional diversity of human brain. *J. Neurochem.* (in press).

GUDSCHINSKY, S. C. (1956). The ABC's of lexicostatics (gluttochronology). *Word* **12**, 175–210.

HAHN, W. E., and LAIRD, C. D. (1971). Transcription of nonrepeated DNA in mouse brain. *Science,* **173**, 158–161.

HALDANE, J. B. S. (1932). The Causes of Evolution. Harper, New York.

HARRIS, H. (1970). "Principles of Human Biochemical Genetics,"pp. 135–140. Elsevier, Amsterdam.
HAYES, K. J., and HAYES, C. (1955). *In:* "The Non-Human Primates and Human Evolution," (J. A. Gavan, ed.), p. 110. Wayne Univ. Press, Detroit.
HILL, R. L., BUETTNER-JANUSCH, J., and BUETTNER-JANUSCH, B. J. (1963). Evolution of hemoglobin in primates. *Proc. Nat. Acad. Sci. U.S.* **50**, 885–893.
HILLARP, N.A., FUXE, K., and DAHLSTROM, A. (1966). Demonstration and mapping of central neurons containing dopamine, noradrenaline and 5-hydroxytryptamine and their reactions to psychopharmaca. *Pharmacol. Rev.* **18**, 727–741.
HIRSCH, J. (1967). "Behavior-Genetic Analysis." McGraw-Hill, New York.
HOCKETT, C. F. (1960). The origin of speech. *Sci. Amer.* **203**, 88–96.
HOLLEY, R. W., APGAR, J., EVERETT, G. A., MARQHISEE, M., MERRILL, S. H., PENSWICK, J. R., and ZAMIR, A. (1965). Structure of ribonucleic acid. *Science*, **147**, 1462–1465.
HSU, T. C., and BENIKSCHKE, K. (1967). "An Atlas of Mammalian Chromosomes." Springer-Verlag, Berlin and New York.
INGRAM, V. M. (1963). "The Hemoglobins in Genetics and Evolution." Columbia Univ. Press, New York.
JERISON, H. J. (1970). Brain evolution: New light on old principles. *Science*, **170**, 1224–1225.
KELLOGG, W. N. (1968). Communication and language in the home-raised chimpanzee. *Science*, **162**, 423–427.
KETY, S. S. (1967). Biochemical aspects of mental states. *In:* "The Human Mind," (J. D. Roslansky, ed.), pp. 141–152. North Holland Publ., Amsterdam.
KOEHN, R. K., and RASMUSSEN, D. I. (1967). Polymorphic and monomorphic serum esterase heterogeneity in Catostomid fish populations. *Biochem. Genet.* **1**, 131–144.
LEES, R. B. (1953). The basis of glottochronology. *Language* **29**, 113–127.
LENNEBERG, E. H. (1967). "Biological Foundations of Language." Wiley, New York.
LENNEBERG, E. H. (1970). Brain correlates of language. *In:* "The Neurosciences, 2nd Study Program," (F. O. Schmitt, ed.), pp. 361–371.
LISK, R. D. (1962). Diencephalic placement of estradiol and sexual receptivity in the female rat. *Amer. J. Physiol.* **203**, 493–496.
LIVINGSTONE, F. B. (1958). Anthropological implications of sickle-cell gene distribution in West Africa. *Amer. Anthropol.* **60**, 533–562.
McCARTHY, B. J., and HOYER, B. H. (1964). Identity of DNA and diversity of messenger RNA molecules in normal mouse tissues. *Proc. Nat. Acad. Sci. U.S.* **52**, 915–922.
McLEAN, P. D. (1970). The triune brain, emotions, and scientific bias. *In:* "The Neurosciences, 2nd Study Program," (F. O. Schmitt, ed.), pp. 336–348.
MARGOLIASH, E. (1963). Primary structure and evolution of cytochrome c. *Proc. Nat. Acad. Sci. U.S.* **50**, 672–679.
MARKERT, C. L. (1963). Cellular differentiation—An expression of differential gene function. *In:* "Congenital Malformations," pp. 163–174. International Medical Congress, New York.
MONEY, J. (1963). Cytogenetic and psychosexual incongruities with a note on space-form blindness. *Amer. J. Psychiat.* **119**, 820–827.
MONEY, J. (1971). Sexually dimorphic behavior, normal and abnormal. *In:* "Environmental Influences on Genetic Expression: Biological and Behavioral Aspects of

Sexual Differentiation," (N. Kretchmer & D. N. Walcher, eds.), Fogarty Intl. Center Proc., No. 2. U.S. Govt. Printing Off., Washington, D.C.

MONEY, J., and BRENNAN, J. G. (1969). Sexual dimorphism in the psychology of female transsexuals. *In:* "Transsexualism and Sex Reassignment," (R. Green and J. Money, eds.), pp. 137–152. Johns Hopkins Press, Baltimore.

MONOD, J. (1970). "La Hasard et la Nécessité." Editions du Seuil, Paris. Pp. 213.

MOTULSKY, A. G. (1964). Hereditary red cell traits and malaria. *Amer. J. Trop. Med. Hyg.* **13**, Part 2, 147–158.

MOTULSKY, A. G. (1968). Human genetics, society and medicine. *J. Hered.* **59**, 329–336.

MOTULSKY, A. G. (1969). Biochemical genetics of hemoglobins and enzymes as a model for birth defects research. *"Proc. 3rd Int. Congr. Congenital Malformations,"* pp. 199–208. Excerpta Medica, Hague.

NEEL, J. V. (1962). Diabetes mellitus: a "thrifty" genotype rendered detrimental by "progress"? *Amer. J. Hum. Genet.* **14**, 353–362.

NEURATH, H., WALSH, K. A., and WINTER, W. P. (1967). Evolution of structure and function of proteases. *Science,* **158**, 1638–1644.

NYHAN, W. L., BAKAY, B., CONNOR, J. D., MARKS, J. F., and KEELE, D. K. (1970). Hemizygous expression of glucose 6-phosphate dehydrogenase in erythrocytes of heterozygotes for the Lesch-Nyhan syndrome. *Proc. Nat. Acad. Sci. U.S.* **65**, 214–218.

OHNO, S. (1970). "Evolution by Gene Duplication." Springer-Verlag, Berlin and New York. Pp. 160.

OHNO, S., and MORRISON, M. (1966). Multiple gene loci for the monomeric hemoglobins of the hagfish (Eptatretus stoutii). *Science,* **154**, 1034–1035.

OJEMANN, G. A., and WARD, A. A., Jr. (1971). "Speech representation in the ventrolateral thalamus." *Brain* **94**, 669–680.

OMENN. G. S. (1972). Genetic approaches to the syndrome of minimal brain dysfunction *Ann. N.Y. Acad. Sci.* **205**, (in press).

OMENN, G. S., COHEN, P. T. W., and MOTULSKY, A. G. (1971). Lack of variation of glycolytic enzymes in brain. *Proc. Int. Congr. Hum. Genet. 4th. Abstracts,* Paris, Sept.

OMENN, G. S., COHEN, P. T. W., MOTULSKY, A. G. (1972). Perspectives for genetic analysis of normal and abnormal behavior in man. *Behav. Genet.* (In press).

OMENN, G. S., and MOTULSKY, A. G. (1972). Psycho-pharmacogenetics. *In:* "Human Behavior Genetics," (A. R. Kaplan, Thomas ed.) Springfield, Illinois.

PARE, C. M. B., REES, L., and SAINSBURY, M. J. (1962). Differentiation of two genetically specific types of depression by response to anti-depressants. *Lancet* **2**, 1340–1342.

PENHOET, E., RAJKUMER, T., and RUTTER, W. T. (1966). Multiple forms of FDP aldolase in mammalian tissue. *Proc. Nat. Acad. Sci. U.S.* **56**, 1275–1282.

PIAGET, J. (1952). "The origins of intelligence in children," (M. Cook, translator). Intl. Univ. Press, New York.

PLOOG, D. (1970). Social communication among animals. *In:* "The Neurosciences," (F. O. Schmitt, ed.) 2nd Study Program, pp. 347–361.

RAKIC, P., and SIDMAN, R. L. (1969). Telencephalic origin of pulvinar neurons in the fetal human brain. *Z. Anat. Entwicklungsgesch.* **129**, 53–82.

REITAN, R. M. (1972). Psychological testing of neurological patients. *In:* "Neurosurgery: A Comprehensive Reference Guide to the Diagnosis and Management of Neurosurgical Problems," (J. R. Youmans, ed.). Saunders, Philadelphia. (In Press).

RENAUD, F. L., ROWE, H. J., and GIBBONS, I. R. (1968). Some properties of the protein forming the outer fibers of cilia. *J. Cell Biol.* **36**, 79–90.

RENSCH, B. (1970). "Homo Sapiens: Vom Tier zum Halbgott." Vandenhoeck and Ruprecht, Gottingen. Pp. 231.
ROSENTHAL, D. (1970). "Genetic Theory and Abnormal Behavior." McGraw-Hill, New York. Pp. 318.
ROWLAND, R. (1968). Evolution of the MG B. *Nature* **217**, 240–242.
SCHMITT, F. O. (1967). Molecular parameters in brain function. *In:* "The Human Mind," (J. D. Roslansky, ed.), pp 113–138. North-Holland Publ. Amsterdam.
SCRIVER, C. R., WHELAN, D. T., CLOW, C. L., and DALLAIRE, L. (1970). Cystinuria: Increased prevalence in patients with mental disease. *New Engl. J. Med.* **283**, 783–786.
SEBEOK, T. A. (1962). Coding in the evolution of signalling behavior. *Behav. Sci.* **7**, 430–442.
SELANDER, R. K. (1970). Biochemical polymorphism in populations of the house mouse and old-field mouse. *Symp. Zool. Soc. London* **26**, 73–91.
SHIELDS, J. (1962). "Monozygotic Twins Brought up Apart and Brought Up Together." Oxford Univ. Press. London and New York. Pp. 264.
SHOOTER, E. M. (1970). Some aspects of gene expression in the nervous system. *In:* "The Neurosciences, 2nd Study Program, (F. O. Schmitt, ed.), pp. 812–826.
SIMONS, E. L. (1971). Relationships of amphipithecus and oligopithecus. *Nature*, **232**, 489–491.
SIMPSON, G. G. (1953). "The Major Features of Evolution." Columbia Univ. Press, New York.
SIMPSON, G. G. (1972). Personal communication.
SLATER, E., and COWIE, V. (1971). "The Genetics of Mental Disorders." Oxford Univ. Press, New York. Pp. 413.
SMITH, C. U. M. (1970). "The Brain. Towards an Understanding." Putnam, New York.
SNYDER, S. H. (1970). Putative neurotransmitters in the brain: selective neuronal uptake, subcellular localization, and interactions with centrally acting drugs. *Biol. Psychiat.* **2**, 367–389.
SOLECKI, R. S. (1971). "Shanidar: The First Flower People." Knopf, New York. Pp. 280.
SOTELO, C., and PALAY, S. L. (1968). The fine structure of the lateral vestibular nucleus in the rat. I. Neuron and neuroglial cells. *J. Cell Biol.* **36**, 115–179.
STEGGLES, A. W., SPELSBERG, T. C., GLASSER, S. R., and O'MALLEY, B. W. (1971). Soluble complexes between steroid hormones and target-tissue receptors bind specifically to target-tissue chromatin. *Proc. Nat. Acad. Sci. U.S.* **68**, 1479–1482.
SWADESH, M. (1950). Salish internal relationships. *Int. J. Amer. Linguistics* **16**, 157–167.
TASK FORCE ON DRUG ABUSE. (1971). Testimony to U.S. Senate Subcommittee on Juvenile Delinquency, Committee on the Judiciary, July 16.
THIESSEN, D. D. (1972). A move toward species—specific analyses in behavior genetics. *Behavior Genetics* **2**, 115–126.
VOGEL, F. (1970). The genetic basis of the normal human electroencephalogram (EEG). *Humangenetik* **10**, 91–114.
WASHBURN, S. L., and HARDING, R. S. (1970). The evolution of primate behavior. *In:* "The Neurosciences, 2nd Study Program," (F. O. Schmitt, ed.), pp. 39–47.
WATSON, J. B., ZIGAS, V., KOOPTZOFF, U., and WALSH, R. J. (1961). The blood groups of natives in Kaintanto, New Guinea. *Hum. Biol.* **33**, 25–41.
WHITE, M. J. D. (1968). Models of speciation. *Science*, **159**, 1065–1070.
YOUNG, W. C., GOY, R. W., and PHOENIX, C. H. (1964). Hormones and sexual behavior. *Science*, **143**, 212–218.

DISCUSSION

V. ELVING ANDERSON
University of Minnesota

In a review of molecular biology, Stent (1968) reflected on the development of that field. The "structural school," an early phase, studied the three-dimensional structure of proteins, guided by the basic assumption that all biological phenomena could be accounted for in terms of conventional physical laws. Another approach, the "information school," chose genetics as a focal point on the assumption that biology might make significant contributions to the physical sciences. As it turned out, the biological questions about information (transmission and translation) stimulated the breakthrough to fundamental insights about DNA. The results were fully compatible with basic physical principles, but could not have been obtained through physical questions alone.

Stent then turned his attention to the "one major frontier of biological inquiry for which reasonable molecular mechanisms still cannot be even imagined: the higher nervous system." If the parallel with molecular genetics is meaningful, *psychological questions* may be needed before any breakthrough showing that biological principles can account for higher nervous system function. He then concluded that "in the coming years students of the nervous system, rather than geneticists, will form the avant-garde of biological research."

At this point several strong objections can be raised. The excellent paper by Omenn and Motulsky provides clear evidence that molecular mechanisms *can* be imagined and tested, even though the results are still fragmentary. Furthermore, genetics can make important contributions to neurochemistry and developmental neurobiology. The genetic variability among individuals should be studied directly and not merely considered an unavoidable part of the error variance of measurement. In the search for simpler "model systems" for the study of behavior, special consideration should be given to those which permit combined biochemical and genetic analysis. In fact, it is entirely possible that genetics may provide the essential framework for studying this "last frontier" even as it did for molecular biology.

In this conference we have been discussing the extent to which research studies tend to be either reductionistic or holistic. The effective strategies in biochemical genetics are reductionistic, yet the results can be viewed in a broader evolutionary context which suggests hypotheses and helps to interpret the data. The paper under consideration presents reductionistic mechanisms framed before and aft with an evolutionary view.

This relationship is seen best in the major strategy which Omenn and Motulsky have utilized—looking for polymorphic variation in brain enzymes. Enzymes in the glycolytic (primary energy-producing) pathway turn out to show no common variations. By itself this would be an uninteresting observation, but the data make sense from an evolutionary view point. The energy-producing system apparently is so essential that any variations leading to a significant reduction in enzyme activity would be incompatible with life.

Screening for electrophoretic variants has been a major approach in biochemical genetics, and it is surprising that it has not been used extensively in neurochemistry earlier. The strategy has some practical advantages as well. By looking at qualitative differences, it is possible to study brain tissue some hours after death. The method will complement, but will not replace, the study of quantitative differences in enzyme activity. The identification of genetic factors in enzyme regulation will require data about levels of enzyme activity in different brain areas. A more serious limitation is that when polymorphic variations are detected, the relationship to behavior often will not be apparent.

The authors make the important point that some inborn errors of metabolism are intrinsic to the brain in the sense that the enzymes involved are normally produced only in brain tissue. Unfortunately such enzymes are the most difficult to study in the human. Some attention can be given to those enzymes important in brain function that can be tested in other tissues. Shih and Eiduson (1971) found that monoamine oxidase from adult rats showed more electrophoretic bands than the enzyme from fetal rats. Different patterns were found in brain, heart, and liver—an indication of isoenzymes.

For our purposes, it is more important to note that they were also able to analyze the enzyme in human serum and found several different patterns among the few subjects examined. Thus it is possible (although not yet established) that study of serum will give some indication of possible genetic variation affecting the brain.

We have chosen a different strategy, starting with those inborn errors of metabolism already known to affect human behavior (Anderson and Siegel, 1968). Such syndromes present a special case of psychopharmacology in which specific metabolites act as the equivalent of drugs. Eventually it should be possible to identify three levels of variation: (1) different mutant alleles affecting the primary enzyme system, (2) differences in secondary enzyme systems which handle the metabolites that accumulate when a pathway is blocked, and (3) genetic variation in brain response to the changed levels of metabolites. Thus the insights of psychopharmacology can be hybridized with those of pharmacogenetics to form what could be called "psychopharmacogenetics."

Using phenylketonuria as a model system for behavior genetics, we tried to select those behavioral measures that appeared appropriate. Affected children showed problems in manual dexterity as compared with control children matched for age, sex, and IQ (Anderson *et al.* 1968). Phenylketonuric children also made significantly more errors of omission on a Continuous Performance Test (CPT), a task which measures one aspect of attention or vigilance (Anderson *et al.* 1969). The CPT appears particularly interesting since it was a modification of one which Mirsky (1969) used to study children with petit mal epilepsy. Another form of the test was used with monkeys, and the data suggested that the effect of secobarbitol was largely on cortical areas, while chlorpromazine appeared to affect subcortical areas.

Unfortunately the more common human behavioral problems (such as schizophrenia and the affective psychoses) have not yet been resolved either biochemically or genetically. If a number of genetic loci with small effect are involved (a multigenic or polygenic hypothesis), then quantitative variation in specific metabolites may be the major biochemical finding. At some threshold, behavioral effects would become apparent. A parallel may be seen in the many systems involved in the control of serum glucose. Under this hypothesis a search for electrophoretically defined polymorphic variation would not be as productive as the analysis of levels of metabolites.

Another possibility is that of heterogeneity. There may be several (or many) rare inborn errors of metabolism that include psychotic disorder as a common manifestation. In this case a few individuals would show a marked deviation on specific biochemical tests, but their uniqueness would be hidden by any pooling of data.

Under either model (multigenic or heterogeneous) or a combination of the two, a research method of choice would be to study pairs of affected siblings. When siblings have reasonably similar pathology of behavior, the probability of similar genetic mechanisms is increased. Either a high correlation for a quantitative biochemical measure or a high concordance for a qualitative measure will provide valuable evidence for genetic factors.

It is fortunate that the biochemical genetics of the mouse has now advanced to the point that a nucleus of research workers has been identified and conferences on the topic have been organized (Paigen, 1970). In the mouse, 15–20 enzyme systems have been studied and the linkage relationships established. Furthermore, there is good evidence for genetic variation in the rate of synthesis, the rate of degradation, the activity of the enzymes, and the localization of enzymes within the cell (Paigen, 1971).

Such evidence can be very helpful in efforts to understand human biochemical traits with behavioral effects. Let me illustrate with the case of porphyria. Several of the more common types of porphyria have a dominant pattern of inheritance (Taddeini and Watson, 1968). Dominant traits have been extremely difficult to explain in biochemical terms, since the clinical manifestations occur in the heterozygotes. Heterozygotes for a gene producing a defective enzyme show intermediate levels of the enzyme, and this is generally enough to maintain normal functioning. In addition to this genetic puzzle in porphyria is the observation that among those with acute intermittent porphyria (AIP) about one-fourth show psychiatric signs (Wetterberg, 1967; Roth, 1968).

Porphyria involves disturbances in the pathway from δ-aminolevulinic acid (ALA) to heme. Recent evidence indicates that in AIP the hepatic conversion of porphobilinogen to porphyrins was less than 50% of that in controls (Strand *et al.* 1970). This appears to confirm the expected intermediate level in heterozygotes, but additional data from the mouse were essential for a more adequate interpretation. Inbred strains vary in the degree to which ALA synthetase is inducible. In six different strains, however, the levels of uroporphyrinogen synthetase (the enzyme presumably defective in AIP) were consistently low, showed no strain differences, and were not induced by substrate (Hutton and Gross, 1970; Gross and Hutton, 1971). Upon induction of ALA synthetase, the conversion of porphobilinogen to uroporphyrinogen became the ratelimiting step.

Although the nature of the biochemical defect in AIP must be verified, it may well turn out to illustrate the point made by Harris (1970, p. 252) that "dominant inheritance of a disease due to an enzyme deficiency is most likely to occur where the enzyme in question happens to be rate limiting in the metabolic pathway in which it takes part, because the level of activity of such enzymes in the normal organism will in general be closer to the minimum required to maintain normal function."

The behavioral manifestions in AIP remain to be explained, although some recent findings may provide leads. Porphobilinogen and porphobilin have been shown to inhibit the "miniature end-plate" potentials induced by potassium ion in rat phrenic nerve-hemidiaphragm preparations (Feldman et al. 1971). When ALA synthetase is induced in 13-day-old chick embryos, the concentration of serotonin in brains is also increased, presumably as an independent effect of the inducer (Simons, 1971). Simons suggested that a metabolite that induces clinical porphyria also could alter the serotonin level and thus account for the neuropsychiatric signs.

A different use of mouse data is seen in the case of the gene *pallid,* which causes reduced pigmentation, absence of otoliths, and congenital ataxia (Erway et al. 1970). It seems possible that a basic biochemical defect interferes with the mucopolysaccharide matrix on which the otoliths form in the inner ear. (If so, this would be a biochemical explanation for a sensory defect.) A phenocopy can be produced by a manganese deficiency in the diet of the pregnant female mouse. Later it was found that manganese supplement for the pregnant mouse will prevent the otolith defect in young mice homozygous for the *pallid* gene, although the pigment dilution is not affected (Erway et al. 1971). These observations reminded me of the suggestions by Ornitz (1970) that vestibular dysfunction might be involved in at least some cases of childhood autism.

The mouse is also providing an excellent model system for neurochemistry through the development of neuroblastoma cultures. When adapted to tissue culture conditions, the cells appear as mature neurons, and enzymes significant in neural functioning (acetylcholinesterase, cholineacetylase, and tyrosine hydroxylase) can be observed (Augusti-Tocco and Sato, 1969). Somatic cell hybrids of mouse neuroblastoma and L cells show electrically excitable membranes, another indication of gene function (Minna et al. 1971). More recently, different clones were tested for the presence of acetylcholine and catechols; clones were found with one or the other or neither of these, but not with both (Amano et al. 1972). To be sure, these neuroblastoma cultures cannot be considered as representative of normal neurons, but this is not a serious limitation. Many genetic variations are occurring, and lines differing in a variety of phenotypic features can be compared.

Genetics has yet another possible contribution to neurochemistry. For other types of biochemical dysfunction in man, it has proved instructive to select simpler systems for the study of genetic variation in the biochemical pathways involved. It would be helpful to pull together the present understanding of genetic variation (in simpler organisms) of the pathways involving compounds that serve as neurotransmitters in the mammalian brain or the glycoproteins proposed as recognition macromolecules. The behavioral implications could be disregarded for the moment. It may turn out that somewhat different enzymes and regulatory mechanisms are involved, but evidence about genetic variation should help in further studies of mammalian brain.

The discussion by Omenn and Motulsky about the role of language in human evolution is most interesting and provocative. I have been aware of the language problems in histidinemia and dyslexia, but had not considered the other linguistic problems they discuss. There is a potential difficulty in that the search for "deep structures" common to all languages appears to be based on a prior assumption of an underlying unity. In genetics we are conditioned to look for variability and then discover the similarities that exist. It would be hoped that these efforts to search for unity and diversity can mutually support each other.

There are two other evolutionary puzzles about neurochemistry that have interested me. The first involves the similarities between tyrosine and tryptophan metabolism. Both are involved in neurotransmitter biosynthesis, one pathway leading to norepinephrine and epinephrine, the other to serotonin. It is possible that a single enzyme may have the activity of both phenylalanine hydroxylase and of tryptophan hydroxylase (Barranger *et al.* 1972). Similarly, the activities of dopa decarboxylase and 5-hydroxytryptophan decarboxylase could not be distinguished immunologically (Christenson *et al.* 1972). The present results need verification, but it is clear that there are similarities (if not identities) which raise problems about enzyme regulation and about the evolutionary origin and modifications of the pathways.

The second question is based on the usual assumption that the phenotypes at one end of a distribution for behavioral traits have reduced fitness, thus producing a directional pattern of selection. There is the possibility, however, that for some behaviors both extremes may be at a selective disadvantage. For example, schizophrenia may involve overarousal, but underarousal also may lead to reduced efficiency (Kornetsky and Eliasson, 1969). If genetic factors are involved in a continuum for level of arousal, either extreme may be at a disadvantage, producing a balanced selection.

I am confident that the paper by Omenn and Motulsky will be viewed as a most important treatment of the problems and issues in this field. Part is factual and part is speculative, but this combination will stimulate the innovative research that is needed in the near future.

References

AMANO, T., RICHELSON, E., and NIRENBERG, M. (1972). Neurotransmitter synthesis by neuroblastoma clones. *Proc. Nat. Acad. Sci.* **69**, 258–263.

ANDERSON, V. E., and SIEGEL, F. (1968). Studies of behavior in genetically defined syndromes in man. *In:* "Progress in human genetics. Recent reports on genetic syndromes, twin studies, and statistical advances " (S. G. Vandenberg, ed.), pp. 7–17. Johns Hopkins Press, Baltimore.

ANDERSON, V. E., SIEGEL, F. S., TELLEGEN, A., and FISCH, R. O. (1968). Manual dexterity in phenylketonuric children. *Percept. Mot. Skills,* **26**, 827–834.

ANDERSON, V. E., SIEGEL, F. S., FISCH, R. O., and WIRT, R. D. (1969). Responses of phenylketonuric children on a continuous performance test. *J. Abnorm. Psychol.* **74**, 358–362.

AUGUSTI-TOCCO, G. and SATO, G. (1969). Establishment of functional clonal lines of neurons from mouse neuroblastoma. *Proc. Nat. Acad. Sci.* **64**, 311–315.

BARRANGER, J. A., GEIGER, P. J., HUZINO, A., and BESSMAN, S. P. (1972). Isozymes of phenylalanine hydroxylase. *Science,* **175**, 903–905.

CHRISTENSON, J. G., DAIRMAN, W., and UDENFRIEND, S. (1972). On the identity of DOPA decarboxylase and 5-hydroxytryptophan decarboxylase. *Proc. Nat. Acad. Sci.* **69**, 343–347.

ERWAY, L. C., HURLEY, L. S., and FRASER, A. S. (1971). Congenital ataxia and otolith defects due to manganese deficiency in mice. *Nutr.* **100**, 643–654.

ERWAY, L. C., FRASER, A. S., and HURLEY, L. S. (1971). Prevention of congenital otolith defect in pallid mutant mice by manganese supplementation. *Genetics,* **67**, 97–108.

FELDMAN, D. S., LEVERE, R. D., LIEBERMAN, J. S., CARDINAL, R. A., and WATSON, C. J. (1971). Presynaptic neuromuscular inhibition by porphobilinogen and porphobilin. *Proc. Nat. Acad. Sci.* **68**, 383–386.

GROSS, S. R., and HUTTON, J. J. (1971). Induction of hepatic δ-aminolevulinic acid synthetase activity in strains of inbred mice. *J. Biol. Chem.* **246**, 606–614.

HARRIS, H. (1970). "The Principles of Human Biochemical Genetics." North-Holland, Amsterdam.

HUTTON, J. J., and GROSS, S. R. (1970). Chemical induction of hepatic porphyria in inbred strains of mice. *Arch. Biochem. Biophys.* **141**, 284–292.

KORNETSKY, C., and ELIASSON, M. (1969). Reticular stimulation and chlorpromazine: An animal model for schizophrenic overarousal. *Science,* **165**, 1273–1274.

MINNA, J., NELSON, P., PEACOCK, J., GLAZER, D., and NIRENBERG, M. (1971). Genes for neuronal properties expressed in neuroblastoma x L cell hybrids. *Proc. Nat. Acad. Sci.* **68**, 234–239.

MIRSKY, A. F. (1969). Studies of paroxysmal EEG phenomena and background EEG in relation to impaired attention. *In:* "Attention in Neurophysiology," (C. R. Evans and T. B. Mulholland, eds.), pp. 310–322. Butterworths, London.

ORNITZ, E. M. (1970). Vestibular dysfunction in schizophrenia and childhood autism. *Compr. Psychiat.* **11**, 159–173.

PAIGEN, K. (1970). Closing remarks, Symposium on the genetic control of mammalian metabolism. *Biochem. Genet.* **4**, 237–242.

PAIGEN, K. (1971). The genetics of enzyme realization. *In:* "Enzyme synthesis and degradation in mammalian systems." (M. Rechcigl, ed.), pp. 1–46. Karger, Basel.

ROTH, N. (1968). The psychiatric syndromes of porphyria. *Int. J. Psychiat.* **4**, 32–44.

SHIH, JEAN-HUNG C. and EIDUSON, S. (1971). Multiple forms of monoamine oxidase in developing tissues: The implications for mental disorder. *In:* "Brain chemistry and mental disease" (B. T. Ho and W. M. McIsaac eds.), pp. 3–20. Plenum, New York.

SIMONS, J. A. (1971). Increase in brain serotonin in experimental porphyria. *Biochem. Pharmacol.* **20**, 2367–2370.

STENT, G. S. (1968). That was the molecular biology that was. *Science* **160**, 390–395.

STRAND, L. J., FELSHER, B. F., REDEKER, A. G., and HARVEY S. (1970). Heme biosynthesis in intermittent acute porphyria: Decreased hepatic conversion of porphobilinogen to porphyrins and increased delta aminolevulinic acid synthetase activity. *Proc. Nat. Acad. Sci.* **67**, 1315–1320.

TADDEINI, L., and WATSON, C. J. (1968). The clinical porphyrias. *Semin. Hematol.* **5**, 335–369.

WETTERBERG, L. (1967). "A neuropsychiatric and genetical investigation of acute intermittent porphyria." Svenska Bokförlaget (Norstedt), Stockholm.

Chapter 8   Gene–Environment Interactions and the Variability of Behavior

L. ERLENMEYER-KIMLING
New York State Psychiatric Institute
New York, New York

Introduction

The topic of this chapter is gene–environment interaction in the determination of behavior. As Haldane (1946, p. 147) once noted, "the interaction of nature and nurture is one of the central problems of genetics." Most of us, I think, would agree that it is a central problem of the study of genetics and behavior. We would probably agree also on the ubiquity of gene–environment interactions to be found in behavioral phenotypes (Lindzey *et al.*, 1971). It is remarkable, therefore, that so little systematic attention, either in research or theory, has been paid to the implications, extent, and meaningful analysis of interactions. For instance, a recent review (Lindzey *et al.*, 1971) that gives ample coverage of the literature relevant to behavior genetics in the past few years contains exactly one-half page (out of 40) devoted to the topic of interactions. This is not because the reviewers were remiss but because with a few notable exceptions, such as research by several investigators on audiogenic seizures (see Fuller and Collins, 1970;

Ginsburg, 1967), a series of studies by Norman Henderson (1968, 1970b), and theoretical discussions by Vale and Vale (1969) and by Harrington (1968, 1969)—workers in the field had given them little on which to report.

Perhaps we sometimes tend to be carried away by the complexities and the wide sweep of interaction possibilities. Perhaps for these reasons it seems better to refrain from fishing in such muddy waters. Yet, we do have quite an amount of information about gene–environment interactions from other areas of biology and medicine; we do have some models to serve us.

We have had information for a long time about differential genotypic responses to a variety of environmental conditions in plants, bacteria, and even Drosophila. We know that the embryological effects of teratogens and other intrauterine insults differ within and between mouse strains, and probably within and between human genotypes as well (Fraser, 1963). We are familiar with a long list of heritable susceptibilities to infectious agents (Cox and MacLeod, 1962) and a growing list of genetic conditions that are associated with adverse reactions to certain drugs (see Vesell, 1971) or special foodstuffs such as the fava bean (see Stamatoyannopoulos et al., 1966). Rh incompatibility of mother and fetus is clearly an interaction between the fetal genotype and the intrauterine environment provided by the mother's genotype. There is a lengthening list of metabolic errors that result in serious inabilities to cope with specific nutrients found in common foods, and many of these conditions, like phenylketonuria and galactosemia, have behavioral concomitants.

In behavior genetics itself, work on audiogenic seizures in mice offers a prototype for studies of the interactions of heredity and environmental factors: some strains being highly seizure-prone and others not; some being capable of seizure induction and others less so; some being sensitive over longer periods and others over shorter periods; etc. All of the complications of dealing with interactions are to be found in the audiogenic seizure research, but so are some of the uses to which analyses of gene–environment interactions may be put. For as Ginsburg (1958), Vale and Vale (1969), and others have emphasized, the study of the ways in which hereditary and environmental forces work together can provide one of the most powerful tools available for learning about mechanisms underlying behavioral processes.

What Is Interaction?

What do we mean when we talk about genes and aspects of the environment interacting? To many behavioral scientists, interaction means chiefly that environmental stimuli must impinge upon a biological substrate for a behavioral response to be emitted. These students of behavior believe that, barring major genetic deviations such as those involved in inborn metabolic

8. GENE–ENVIRONMENT INTERACTIONS AND THE VARIABILITY OF BEHAVIOR

or neurological dysfunctions, experiential factors mold the phenotype pretty much independently of genotypic factors. To geneticists, by contrast, the keynote of interaction is that different genotypes may respond differently to the same environmental conditions. Relationships between genes and environment can be of several kinds, however, and not all are consistently called interactions. My objective in this section will be to review briefly the several types of gene-involved relationships, to point to some of the overlapping between them, and to consider some of the difficulties that have arisen in attempts to classify interactions.

The Several Types of Gene Involvements. Genes can take part in three basic types of interactions besides those involving what we usually think of as environment. They are: interactions between alleles (dominance), between genes at different loci (epistasis), and between genes and cytoplasm (Mather and Morley-Jones, 1958). Of course, cytoplasm is itself a part of the nongenic environment, but usually these interactions are considered apart from the ones involving other environmental sources. Not a great deal is known about gene–cytoplasm interactions; almost nothing is known about such relationships and the development of behavior. Although we are concerned here only with interface points between genes and environmental factors, it must be remembered that interactions may be (in fact most likely are) going on at several levels at once. To take an obvious example, a phenylalanine loading test for heterozygote detection involves an interaction between an environmental manipulation (the administration of phenylalanine) and the product of an allelic interaction—the allelic interaction itself usually being undetectable except following exposure to the environmental treatment.

When we try to break down complex behavioral responses into components, we are likely to encounter epistatic interactions, or at least the sequential action of different genes that affect different parts of a behavioral chain. One illustration can be found in Rothenbuhler's (1967) work with honeybees. Nestcleaning, that is, disposal of diseased larvae from the nest, consists of two successive acts (uncovering of the cell containing the larva and removal of the larva). Each step is largely under the control of a different gene, but the environmental stimulus, presence of diseased (or otherwise-killed) larvae, is the necessary trigger for the behavioral sequence to occur. One may imagine that courtship, mating and fighting patterns in many species probably entail even more complex feedback relations among several genes and successions of cues from a rapidly altering environmental situation. Perhaps attention to multi-level interactions of this type would not usually prove to be highly rewarding, especially, if they led to infinite subdivisions of the behavior in question into smaller and smaller responses and movements, each of which might be part of several other behavioral patterns

(Scott and Fuller, 1963). In other contexts, however, examination of both intergenic and gene–environment interactions, and their interplay, might prove valuable. There are scattered indications, for instance, that heritabilities and dominance relationships are frequently altered over the course of learning processes. Do such changes, if they actually occur, merely reflect "progressive releases of the genetically-determined response from the effect of environmental stimuli irrelevant to it" (Broadhurst and Jinks, 1966, p. 471)? If heritability is decreasing, does the change reflect a progressively increasing importance of task-relevant variables compared to genetic variables? Or do the changes indicate that different genes or different groups of genes take over at various stages along the way? To my knowledge, attention to questions concerned with such multi-level interactions has so far been scant.

Two Relationships between Genes and Environment. Two types of relationships that occur between genes and environmental factors are frequently omitted from discussions of interactions. Both, though acting within the course of individual lifespans, have their main effects (usually) over the longer span of population time. These relationships are natural selection and gene–environment covariance.

The fact that gene–environment interactions form the basis for natural selection is, I think, quite obvious. Natural selection, of course, refers to the fact that different genes (or, more precisely, different alleles at given loci) are transmitted to successive generations in different frequencies. Differentials in transmission frequencies may be attributable to inequalities in either survival or reproductive rates (or both) among the carriers of different genes. Whichever may be the case, the source of the transmission differentials is to be found in the patterns of interaction, more or less favorable, that the genes in question form with various aspects of the environment. By creating new interactions, changes in the environment can also change previously existing differentials in reproductivity or viability. As observed by Caspari (1967), selection for coadaptive gene complexes, rather than for individual genes, is probably the general rule—a point which bears upon the questions of multi-level interactions raised in the preceding section.

There is ample documentation (see Part 1 in Hirsch, 1967) for the role of behavior as one of the important interaction products through which selection, stability, and change may be mediated. One point may be worth reiterating here. Gene–environment interactions by creating selection differentials may change previously existing environmental conditions and thereby eventually reach new selection levels as well. For example, in Ehrman's (1970) research, Drosophila males with rare genotypes are found to have a mating advantage over males that are common in the population, so that, through this selection differential, changes can be introduced in the genetic

8. GENE–ENVIRONMENT INTERACTIONS AND THE VARIABILITY OF BEHAVIOR

composition of the population (which in this instance can be regarded as an environmental parameter), with the selection differential, namely mating advantage, gradually diminishing as the population attains a balance between the initially rare and common genotypes. Analogies are to be found in the feedback chains linking human cultural developments, genetic factors, selection, and further pressures on the environment itself. For instance, the following hypothesis is suggested by Wiesenfeld (1967) in attempting to account for the relationship between sickle-cell trait, malaria, and agriculture:

> In the case of an intensely malarious environment created by a new agricultural situation, the variability of the normal individual is reduced and there is selection for the individual with the sickle-cell trait; this means that the nature of the gene pool of the populaton will change through time. This biological change helps to maintain the cultural change . . . [and] may allow further development of the cultural adaptation, which in turn increases the selective pressure to maintain the biolgoical changes [p. 1139].

While natural selection is often not mentioned at all in discussions of interactions, *covariance,* the second relationship between genes and environment mentioned above, is sometimes explicitly excluded. Covariance means that genotypes are differentially distributed across environmental conditions, the most obvious example being the ecological distribution in nature of species, subspecies, and population groups to those niches to which they are best adapted. Sickle-cell trait and certain other hemoglobinopathies that presumably confer protection against malaria occur mainly in regions where malaria was formerly endemic; adult lactose intolerance appears to be confined to populations in which dairy husbandry never developed—or, perhaps, it should be said that lactose *tolerance* appears mainly in populations that *did* develop the practice of using milk products. These are two reasonably well-established illustrations of the covariance between the environmental demands and human genetic variants.

Covariance is the result of selection based on gene–environment interactions that have taken place at some time. Several cautionary comments must immediately be appended to the foregoing statement. First, because we so very quickly move from natural to social selection and social implications when we touch upon questions relating to covariance, it becomes especially important to stress that both interactions and selection always in reality involve the phenotype rather than the genotype. This, we all know, applies at every point throughout the present discussion, but the noncongruence of phenotype and genotype, and the phenotypic basis of selection, are crucial concepts for our interpretations of the implications of empirically observed covariances. A second point has to do with the extent to which the basis for a given selection index correlates with a phenotypic character under study. If population subgroups had been differentially sorted on the basis of physical strength or eye color or nose shape, we would expect to find covariances between such subgroups and any characters that might correlate highly with the selection criterion. Covariance is restricted *only* to those correlated characters. Any other character that seems to be assorting differently into the subgroups delineated by the original selection criterion must be either random genetic drift (including founder effects) or sampling bias or experimental bias. The difficulty is that we are unable to set genetic correlations apart from chance or treatment effects in most of our empirical observations.

A third important question about covariance has to do with the maintenance of selection over generation spans by means of social mobility. As Haldane (1965, p. xcii) commented (probably with only partial accuracy): "If the sons of brahmans who could not learn the vedas and discuss philosophy had been expelled from their caste and made to sweep the streets, the brahmans might now dominate India completely." When rigid *non*mobile class or caste systems have been operative, the covariance of most behavioral characters with caste is probably negligible, for, as Haldane continued (probably with considerable accuracy): "In practice the efforts of members of every ruling group are largely devoted to preventing their children from falling in the social scale."

When social, rather than natural, selection is involved, it is exceedingly difficult to separate covariance and ongoing interaction effects. Two familiar teasers: (a) Are higher rates of schizophrenia found in lower socioeconomic classes because predisposed persons encounter greater environmental stresses in these classes (interaction) or because predisposed persons have down ward social mobility (covariance) (see Dunham, 1970)? (b) Do IQ and social class have a positive correlation because environmental factors relevant to intellectual development are differentially distributed over classes or because phenotypic selection in the form of social mobility has produced different clusterings of genetic factors in the different classes (see Gottesman, 1968)?

Neither of the foregoing examples necessarily presents mutually exclusive alternatives between covariance and ongoing interactions. It is highly probable that both types of relationships between genetic and environmental factors are continually operating as Thoday and Gibson (1970) found in their "model experiment" on environment, mobility, and "class" differences in Drosophila.

It is possible that, within a generally similar milieu, individuals may be free to choose specific niches, certain features of the environment as opposed to others, or variations in behavioral patterns. Many of these differences in self-placements may correlate, at least indirectly, with genetic differences. Some may have consequences for later behaviors or, most important, for behavioral development in the next generation. For instance, Polansky et al. (1969), studying poor Appalachian families, recently reported a positive correlation between mother's and child's IQ and also between mother's IQ and the adequacy of care given to the child. Investigators concerned with the effects of nutritional deficiencies or of perinatal complications upon intellectual development might well ask whether similar *within-group* correlations are to be found between maternal IQ and the nutritional adequacy of a child's diet or between maternal IQ and precautions taken during pregnancy to protect the health of the unborn child. The inference commonly encountered in behavioral, educational, and medical literature is that poor prenatal, postnatal, or later rearing conditions are the causes of low IQ (or various other unfavorable phenotypic outcomes). Without denying the significant, detrimental effects that such conditions impose upon development, we may also ask, however, whether the causal chain contains a parallel and equally important link, namely:

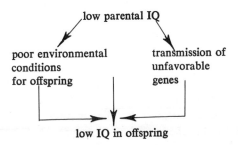

The point to be made here is that we run the danger, on one hand, of assigning too much weight to environmental variables if we neglect the possibility that certain genotypes are more likely to be found in certain environments as a result of selection based on phenotypic characters relevant to the ones that we may be studying. On the other hand, we must be equally alert to the opposite danger of overemphasizing hereditary influences by assuming that an observed covariance of genotype and environment necessarily bears upon the phenotype that we have under investigation.

Types of Gene–Environment Interactions. In general we do not have in mind natural selection or gene–environment covariance when we talk about interactions. What we usually mean is that genotypes (or strains, or populations) can be shown to react in different ways to the same environmental treatments. But how interactions are to be classified and what is to be done about them—on these points there is no solid consensus.

For many workers (see Haldane, 1946; Mather and Morley-Jones, 1958; Vale and Vale, 1969), the above description would be considered an adequate definition of interaction; any of the possible gene–environment relationships likely to be encountered in experimental data would be classified as interactions by these investigators. Others (see Broadhurst, 1967; Lubin, 1961), however, would insist on the further criterion that delineates nonadditive relationships along the lines of the analysis of variance model. Thus, for interaction to exist, according to these workers, the amount and/or direction of the differences between genotypes must change over the various environments under investigation, "at least one set of means must be nonparallel to the others [Lubin, 1961, p. 812]."

Three basic types of gene–environment relationships can be distinguished as follows: (1) *additive relationship* where phenotypic differences between genotypes remain constant in all observed environments; (2) *nonadditive relationship A* (Lindquist's (1953) ordinal interaction) where quantitative differences in the phenotypic values change with different environmental conditions but rank orders do not change; and (3) *nonadditive relationship B* (Lindquist's disordinal interaction) where the distinguishing characteristic is a reversal of phenotypic rank orders as the genotypes are moved from one environment to another.

Now, the tradition of equating interaction with nonadditivity grows directly out of the analysis of variance model that was originally designed to allow questions to be asked about main effects. Interaction terms were later incorporated into the model to handle the realities of the natural world where the main effects often do not add up in a simple fashion to account for all of the observed variance in the phenomenon under study. In spite of the provision for an interaction term, however, the analysis of variance model can create two sorts of difficulties for our understanding of the joint operations of genes and components of the environment.

First, significant interaction effects tend to be regarded by many experimenters as nuisance factors because the reason for using the model is still, in most cases, to look for main effects, not interactions. The interaction term is loosely hooked on to the model, and the idea generally is to try to shake it off. Therefore, upon encountering interactions in the data, investigators frequently attempt to remove them through scale transformations. Sometime this is effective for ordinal interactions, but never for disordinal cases (Lubin, 1961). If transformations fail, a more drastic solution may be offered by discarding parts of the data. Surprisingly, such procedures can be found in behavior genetics research just as they are in other areas of behavioral studies. Broadhurst (1967, p. 295) has pointed out, for instance, that two important assumptions are involved in biometrical methods of genetic analysis; these are that there be "no interaction between genotype and environment," and that the gene effects "be additive over the range of variations" studied! (Rather startling assumptions to be built into a method designed for use in a science of variations, but, fortunately, Broadhurst and Jinks (1966) have demonstrated that meaningful analyses of gene–environment interactions are possible with the biometrical methods after all.)

Some researchers (see Harrington, 1968; Lubin, 1961) argue that when significant interaction effects are turned up in the analysis of variance, the interaction term itself should be considered an important feature of the situation under study. Instead of attempting to *eradicate* the interaction term statistically, the aforementioned authors suggest that we try to *explain* it. Lubin (1961), who incidentally was speaking about nonadditive interactions generally rather than gene–environment relationships specifically, has stated the problem succinctly:

> To me it's far more important to determine the form of the equation relating the treatment effect to the block *(genotype, strain, group)* effect than to make accurate statistical inferences about the variance of the difference between two means. If a transformation eliminates the interaction, the inverse of the transformation specifies an equation which is a good fit to the raw data.

The first danger of the analysis of variance model and methods stemming from it, then, is that important interaction effects will be looked upon as trivial error variance or will be lost in statistical maneuvers.

The second difficulty is that, in analyses that include several different genotypes or several different environmental treatments for comparison, different types of relationships may emerge and may, in some instances, effectively cancel each other. For graphic illustration of a situation in which different kinds of relationships are to be found, I have plotted in Fig. 1 some of the data reported by Henderson (1970b) in a study of early experience effects on mouse behavior. There are 16 possible comparisons between strains. Among these are the following:

1. Disordinal interactions, involving rank order reversals, appear in three comparisons—between BALB and C3H, between C3H and A/J, and between C3H and RF.

2. Most of the interactions are ordinal, with quantitative changes only—BALB versus all strains except C3H, C57BL versus all except C3H, and DBA versus C57BL and C3H.

3. No comparison shows perfect additivity, but that between C57BL and C3H deviates only slightly from an additive relationship, with both strains showing nearly identical decreases in time to food goal in the enriched, compared to the standard environment.

4. One pair of strains, A/J and RF, gives identical means in both environments and fails to show any difference in behavior associated with the environmental treatments.

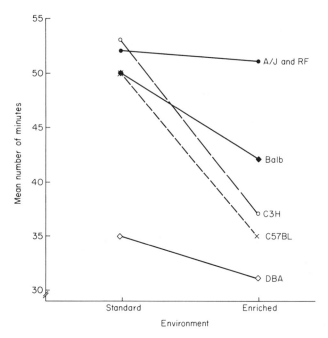

FIG. 1. Illustrative data from Henderson (1970b, Table 1) showing mean number of minutes required to reach food for six inbred strains of mice reared in standard and enriched environments.

In Henderson's study, a significant interaction effect was obtained in the analysis of variance. Had somewhat more of the strain comparisons shown additive relationships, however, the overall interaction term might have been nonsignificant even though some of the pairs showed markedly different responses to the environmental treatments. In summary, the analysis of variance model can mislead, and the limitation of the meaning of interaction to nonadditive relationships seems unwarranted.

In passing, it may be noted that Haldane (1946), in a now classical paper, included additive relationships among the possible types of significant interactions that may occur between nature and nurture. The additive relationships, in fact, account for a sizable proportion of such possibilities. Haldane bequeathed to us a formula, $(mn)!/m!n!$, to describe the number of theoretically possible types of interactions that might be found for m genotypes in n environments if all phenotypes differ from each other (that is, every genotypic–environmental combination produces a different phenotype than every other genotypic–environmental combination). These are, however, *theoretical* possibilities, whose full, impressive range may rarely be encountered in reality.

The purpose in classifying interactions is, or should be, to allow us to find reasonable ways to interpret mechanisms underlying the interactions. One classification-for-interpretation scheme has been offered by Vale and Vale (1969), who propose that additive and many ordinal interactions indicate that the same process underlies the phenotypic response to the environment in all of the genotypes under study. Disordinal interactions, however, are considered in this scheme to reflect differences in the processes underlying the response in different genotypes. The latter type of finding might occur when comparisons of crudely similar behavior (for example, maternal behavior) are made between species, when comparisons are made among genetically heterogeneous groups whose phenotypes may be similar in some circumstances but not in others (or between phenocopies and "hereditary" disorders), or when the phenotype being measured is not the same in all groups or all environmental conditions. It should be noted, though, that disordinal interactions need not per se imply differences in basic processes; for example, as Henderson (1968) suggests, if the relationship between emotional arousal and amount of prior stimulation should turn out to be U-shaped and if two genotypes have different optimal levels of stimulation, and if comparisons are made between only two or a very few levels of stimulation as is usual, then the effects of stimulation may appear opposite in the two genotypes, while comparisons at a larger number of treatment levels would show consistency in the relationship between treatments and the effect upon behavior. This point bears repeating: differential thresholds of sensitivity—whether we are concerned with sensory responses or with responses to drugs, alcohol, lack of sleep, etc.—can produce functions that look very dissimilar for different individuals over large ranges of intensity levels. With further extension of these ranges, however, different individuals may show similar (though widely displaced) treatment–response functions.

The foregoing comments do not detract from Vale and Vale's scheme as a first-approximation working base that may be useful in analyzing and understanding gene–environment interactions.

Just as the Twig Is Bent? (Illustrations from Early Experience Studies)

From Freud to Spitz and Bowlby right up to our most contemporary literature, it has been taken almost as an article of faith that the effects of experiences in early infancy are profound, enduring, and essentially universal for the members of the species. To a considerable extent, such assumptions are correct. Studies that have looked at genetic effects along with differences in early treatments, however, have some other things to show.

My purpose in discussing this research here is not so much to review the early experience concept as it is to call attention, through a brief scanning of data, to the kinds of consistencies and inconsistencies that are likely to appear when a considerable body of results on gene–environment interactions is at hand. As it happens, the work on early experience and subsequent behavioral development offers a number of comparisons on some of the same strains, the same phenotypic measures, and the same environmental treatments. Though not intended as an exhaustive coverage of the literature, the collection of sixteen studies referred to in Table 1 represents a good sampling of the available mouse research without selection for results.

Most of the studies include more than one measure of behavior; in some instances, the separate measures are supposed to be tapping the same phenotypic trait. We can look first for significant effects of early experience upon the later measures of behavior. We find that, out of a total of 40 measures, there are 37 in which at least one of the tested strains fails to display a significant difference between the experimental and control conditions. Looking at all strains × measures, we have a total of 162 opportunities in which to see significant effects of the early treatments. Actually, in 87 of these cases the early experience does *not* significantly influence performance on the subsequent behavioral test.

Reference to the original studies summarized here would plainly show us that all of the early treatments investigated do have very pronounced effects upon some behaviors in some strains. But the effects are far from universal. In fact, it seems that we have a better than even chance of not finding a significant relationship between an early treatment and a subsequent measure of behavior!

Interaction effects between genotypes and treatments can be examined for 33 behavioral measures on which two or more strains have been tested. In 14 of these comparisons, the treatment has an opposite influence in one or more strains compared to the other strains under study (last column of Table 1). Disordinal interactions (reversals of rank orders between two or more strains from the control to the experimental condition) occur in 14 out of the 33 comparisons. We can choose to lay stress upon these complicated relationships, or we can decide to emphasize that, in over half of the comparisons, when treatment effects occur, they tend to exhibit fairly regular patterns across genotypes.

There is a sufficient number of observations on some of the strains to permit closer examination by strain and type of early experience. Table 2 shows the number of behaviors measured and the number in which experimental and control animals differed significantly, for each of five strains and several types of experimental treatments. It can be seen that the C57BL strain responds to all types of treatments more frequently than do other strains—thus bearing out observations made by Ginsburg (1967), Henderson (1968), and others on the lability of the C57BL group. BALB shows generally low responsiveness, at least to the treatments considered here. The data are too scanty to allow careful comparisons to be made with regard to differential responsivities to specific treatments in different strains. Nevertheless, they suggest that, for some of the strains (C57BL, DBA, and C3H), general background variables, such as isolation, environmental enrichment, or cage illumination, may be less critical than more specific and possibly traumatic events, such as handling, shock, and noxious noise. The opposite may be true for BALB.

The behavioral measures that are most likely to reflect the influences of certain treatments also tend to differ among the various strains. For example, in the highly responsive C57BL strain, most of the behavioral measures (see Table 1) are substantially affected by the early experience treatments, but defecation scores are not greatly changed between controls and experimental subjects in most of the studies; C57BL's generally give low open-field defecation scores anyway. C3H, which shows treatment effects in only about 40% of the behavioral measures, seems especially unresponsive where measures involving learning (maze, avoidance, or water escape) are concerned, with only two of eight such measures showing an influence of the experimental manipulation. For DBA, on the other hand, maze-learning is the measure showing maximal response to the early treatments (in five out of six observations).

There is a large amount of literature on strain differences in behavior, quite a number of consistencies have been demonstrated in the relative phenotypic performances of several strains compared to each other, and some attempts have been made to construct behavioral profiles describing the relative strengths of various phenotypic characters within the different strains. The findings in the early experience studies tend to be consonant with the more general literature on strain differences. While two strains may sometimes reverse rank orders of performance as a result of treatments applied in infancy, such reversals are rarely found in behaviors on which one or the other of the strains usually scores particularly high or particularly low. Gene–environment interactions frequently appear to be chaotic, especially when seen within the confines of a single investigation, but Henderson (1968, p. 150) has noted that "most of these interactions are probably entirely consistent and interpretable when sufficient information is made available through the use of adequate designs and analysis techniques."

Before leaving this section, let me mention that the finding of nonsignificant treatment effects in a sizeable portion of measures of later behavior is by no means confined to mouse research or to studies of early experiences. Similar observations on manipulations during infancy can be made in studies of rats (see Levine and Broadhurst, 1963) and dogs (see Fuller, 1968), and there is one intriguing report (Kaufman and Rosenblum, 1967)

TABLE 1 Generality of Strain × Treatment Effects in Some Early Experience Studies in Mice

Ref[a]	Type of experience[b]	Behavioral measures[b]	Strains[c]	Treatment effects Not found[d]	Different directions[e]
(1)	variations in social vs. isolation rearing	1. fighting latency 2. sexual behavior	C5, B C5	B C5	yes no
(2)	handling vs. nonhandling	1. alley crossing 2. activity time 3. avoidance conditioning	P.m.b, P.m.g.	P.m.g. P.m.g. P.m.b.	yes no yes
(3)	infantile trauma (noise)	1. 30-day o.f., defecation, mean 2. 30-day o.f., defecation, day 10 3. stove pipe emergence, time 4. 100-day o.f., defecation, mean 5. 100-day o.f., motility, mean	C5, C3, D, J	C5, D, J D, J D C5 C3, D	no no C5 no D
(4)	handling vs. nonhandling	1. fighting latency	C5, D		yes
(5)	infantile trauma (noise)	1. 30-day o.f., defecation 2. 30-day o.f., motility 3. stove pipe emergence, time 4. 100-day o.f., defecation 5. 100-day o.f., motility	C5, C3, D, J	C3, J D, J D, (J ?) all J	no ? no ? C5
(6)	infantile trauma (noise)	1. maze errors 2. maze, time (g), mean 3. maze, time (g), day 13–15 4. maze, time (f), mean 5. maze, time (f), day 13–15 6. water escape, time, trend	C3, D, A/A	C3 A/A C3 all C3 all	no A/A no no no D

(7)	infantile trauma (noise)	1. maze, errors 2. water escape, time	(C3, D, A/A and three F_1 hybrids')	Three F_1 hybrids all	no no
(8)	shock vs. handling vs. nonhandling	1. defecation, o.f. 2. avoidance conditioning	C5		
(8a)	shock vs. handling vs. nonhandling	1. activity, o.f. 2. defecation, o.f. 3. runway emergence 4. avoidance conditioning	B	B B B B	
(9)	handling vs. nonhandling	1. fighting latency	C5, C3, C-A	C3, C-A	no
(10)	shock vs. handling vs. nonhandling	1. activity, o.f. 2. defecation, o.f.	(C5, B, C3, D, A/J, twelve F_1 hybrids)	C3, B, four C3F_1, three BF_1 C5, C3, three C3F_1, BDF_1	no B
(11)	deprivation of maternal care	1. defecation	B, J	B, J	yes
(12)	high vs. low illumination	1. defecation, o.f. 2. activity, o.f. (tested in high vs. low illumination; B's activity higher in low illumination testing)	C5, B	C5, B C5	yes yes
(13)	enriched vs. standard cages	1. hoarding	C3, J	C3	no
(14)	daily dosages CPZ vs. AMPH vs. saline	1. dominance	(random-bred Swiss, P.m.b.)	P.m.b.	no
(15)	enriched vs. standard cages	1. spontaneous alteration	(C5, B, C3, D, A/J, RF, and six F_1 hybrid averages')	all inbred, all F_1 except DF_1	C5,D, C3F_1, DF_3, RFF_1 vs. all others

Table 1 (Cont.)

	Type of experience[a]	Behavioral measures[b]	Strains[c]	Treatment effects	
				Not found[d]	Different directions[e]
(16)	enriched vs. standard cages	1. food-seeking (problem solving)	(C5,B,C3,D,A/J, RF & twelve F_1 hybrid averages)	D,A/J, RF, one A/JF$_1$ one RFF$_1$	F_1 of A/J female vs. all others

[a] References (in order): (1) King, 1957; (2) King and Eleftheriou, 1959; (3) Lindzey et al., 1960; (4) Ginsburg, 1963; (5) Lindzey et al., 1963; (6, 7) Winston, 1963, 1964; (8, 8a) Henderson, 1964, 1967a; (9) Ginsburg, 1967; (10) Henderson, 1967b; (11) Newell, 1967; (12) Dixon & DeFries, 1968; (13) Manosevitz et al., 1968; (14) Wolf and Rowland, 1969; (15, 16) Henderson, 1970a,b.

[b] Abbreviations (in cols. 2, 3): o.f. = open field test; maze, time (g) = time to reach goal; maze, time (f) = time from final choice to food cup; CPZ = chlorpromazine; AMPH = amphetamine.

[c] Strains: C5 = C57BL; B = BALB; P.m.b. = Peromyscus maniculatus bairdii; P.m.g. = Peromyscus maniculatus gracilis; C3 = C3H; D = DBA; J = JK; A/A = A/Alb; F_1's = hybrid generation crosses between the various inbred strains under study; C-A = C(Bagg) albino.

[d] Not found (col. 5): Experience effect not significant for the strain(s) shown.

[e] Different direction—yes = opposite effects of experience in the 2 strains studied; no = same directional effects of experience in the 2 or more strains studied; strain symbol(s) = opposite effects in the indicated strain(s) compared to other strains under study.

[f] No reciprocal cross data for F_1 hybrids.

TABLE 2 Types of Early Experience Treatments; Number of Significant Effects upon Behavioral Observations for Several Inbred Mouse Strains[a]

Type of experience[a]	C57B1 obs.	C57B1 sig.	BALB obs.	BALB sig.	C3H obs.	C3H sig.	DBA obs.	DBA sig.	JK obs.	JK sig.
Social-isolation	2	1	1	0	–	–	–	–	–	–
Enriched-standard	2	1	2	1	3	1	2	0	1	1
Subtotal	*4*	*2*	*3*	*1*	*3*	*1*	*2*	*0*	*1*	*1*
Handling	2	2	–	–	1	0	1	1	–	–
Handling-shock	4	3	6	1	2	2	2	2	0	0
Infantile trauma (noise)	10	7	–	–	18	8	18	9	10	4
Subtotal	*16*	*12*	*6*	*1*	*21*	*10*	*21*	*12*	*10*	*4*
Other (light)	2	0	2	1	–	–	–	–	–	–
All	22	14	11	3	24	10	23	12	11	5

[a] The strains were selected from studies in Table 1 based on frequency of observations.

on the effects of separation from mothers in pigtail monkeys (*Macaca nemestrina*). The offspring of the dominant female monkey failed to show the characteristic depression displayed by the other three pigtail infants—a possible genotype–environment interaction? Work on prenatal or preconception stimulation and on foster-rearing frequently also shows that one or more strains are not affected by the treatment (see DeFries *et al.*, 1967; Ressler, 1963; Thompson and Olian, 1961).

Parameters of Interaction

What are we measuring? When? In what circumstances? In whom? These are the questions that we are asking when we talk about gene–environment interactions. The question of genotype is, of course, basic to the discussion throughout this paper and need not be dealt with specifically here. The questions of behavioral phenotypes, time, and environmental conditions have, fortunately, received considerable attention from many other authors, so that I need only make a very few remarks about some points that seem, to me, most pertinent to the study of interactions.

Behavioral Phenotypes. Two questions arise about the choice of behavioral phenotypes for investigation. First, are we really measuring what we think we are, or are we measuring "noise" from interfering responses? If we want to compare learning processes in two groups or in two different environments, are we getting at the same phenotypic levels in both groups, both environments? If we are comparing learning in two groups *and* two environments, is our measure uncontaminated with competing behaviors in all four cells (or, at least, is the type and amount of contamination constant over cells)? There are numerous illustrations in which apparent strain differences in learning, activity levels, social behaviors, memory, emotionality, etc., have turned out to stem from differences, for instance, in fearfulness or in the motivational aspects of the task in the circumstances peculiar to the testing situation (see Fuller, 1967; Henderson, 1968; Ross et al., 1966). Obviously, this is an especially serious problem in research on human behaviors, where the same testing conditions may tap different functions in different subjects or groups of subjects. Thus, Clark et al. (1967) have shown that the "differences in perceptual function between psychiatric patients and control (subjects) found with traditional psychophysical procedures can be attributed to differences in response bias rather than to differences in sensory sensitivity [p. 41]." The implications of such findings are enormous and disturbing.

The second problem has to do with the relevance of our behavioral measures to the organisms under study. Consider, for example, the study in which the customary rat-type measure of emotionality, that is, open-field defecation, was applied to cats; *Felis domesticus,* having a very different response style, supplies no data in this situation, as could have been predicted by anyone who knew the animal. The fact is, however, that many investigators have only the vaguest notions about the natural behaviors of their experimental subjects. Whitney (1970), among others, has recently called attention to the arbitrary nature of the operational definitions assigned to many of our traditional laboratory measures and the dangers of drawing analogies between species based on superficial resemblances in behavioral variables.

Discussions of these and other problems relating to the choice and interpretation of behavioral phenotypes may be found in Ginsburg (1967) and Thompson (1967).

Time. A good deal of attention has been given to critical periods when events must occur if a particular response (behavioral or physiological) is to develop, and to sensitive periods when the organism is maximally vulnerable to specific types of treatments. We know of a great many behaviors for which different genotypes show different sensitive periods—outstanding examples being those found in audiogenic seizure research. Fuller and Collins (1970) note that there are even genotypic differences in the diurnal rhythm of susceptibility to seizures. One point not often mentioned is that sensitive periods need not be confined to a single interval of time in the life span. Many disease susceptibilities, for example, appear to show periods of heightened vulnerability occurring at several different times over the lifespan.

Some seeming dissimilarity among different sets of gene–environment interactions disappears when time factors and developmental rates are taken into account (see Henderson, 1968). But sometimes the opposite is true, and we find discrepancies emerging *only* when observations are taken at different points in time. Fuller and Clark (1968) have demonstrated that the time elapsed between an environmental treatment and the measurement of behavior can be an important variable. Similarities in responses observed shortly after the application of a particular treatment may diverge with time, as individual differences in recovery or retention rates gradually take over.

Environment. As mentioned earlier in this discussion, it is especially critical in attempting to understand gene–environment interactions that we specify the range of stimulus intensities examined. We need to consider the possibility that extensions of the observed range may reveal regularities in the response functions of different genotypes that are obscured under more restricted ranges of measurement.

There is one final point to be mentioned here about the choice of environmental treatments for meaningful analyses of behavior and genetic variables. This point is closely tied to one made above about the selection of behaviors that are relevant to the organism under study. It is simply that we must also question the meaning of environmental conditions imposed experimentally or seen in field observations in terms of the evolutionary history of the species. A good deal of work in the behavioral sciences is as flawed by the neglect of the kinds of environments that the subject species may be expected to encounter naturally as it is marred by inadequacies in the choice of behavioral phenotypes for study.

Man is a special problem. What shall we say is man's "natural" environment? From what baseline can we speak of deprivation, enrichment or inadequacy of stimulation during infancy and early childhood? We may not have answers to these questions, especially when we reckon with the fact that man is a genetically diverse animal adapted to many different environments. Nevertheless, I would quarrel with those who claim that no environments are universally good or bad. Surely a vermin-infested slum is a *bad* environment for any child, though there may be some environments that are relatively worse and some genotypes that manage relatively better than others in the same bad surroundings.

Concluding Remarks

Not long ago, most of the theoretical positions subsumed by the behavioral sciences found at least one common meeting ground: genetics could be safely ignored because heredity had little, if anything, to do with behavior. Environment was counted the all-important force in behavioral development

—though the bond of unity among theorists quickly dissolved when it came to specifying what the significant aspects of environment might be.

Nowadays, the nature–nurture controversy is often declared to be a thing of the past. "Everyone," says David Rosenthal (1968), "agrees that all human behavior is a function of both heredity and environment. . . [p. 78]." Perhaps *everyone* does not agree, for the same volume in which Rosenthal's enthusiastic note is sounded also contains a more skeptical point of view: "I would emphasize . . . the relative lack of scientific information concerning the genetic basis for human behavior [Haller, 1968, p. 225]." But if refusals to credit geneticists with having compellingly demonstrated their claims do persist, at least it may be said that outright refusals to credit genetic factors with *any* influence on behavior appear in the psychological and psychiatric literature with increasing rarity. Indeed, it is more and more common for contemporary discussions of both animal and human behavior to include some reference to interactions between genes and environment. Moreover, in recent years several leading proponents of behavioral theories heretofore conspicuously lacking in attention to any biological differences among individuals have seen fit to take notice of genetic factors, declaring further that they themselves had long held interactionist views about behavior! (These were evidently very privately held views that were strictly guarded against in the serious business of research and theory making.)

All of these should be encouraging signs. Yet paper tributes to the contributions of genes can scarcely be said to point to a revolution in the established environmentalist traditions that have so long dominated the behavioral fields. Nor do they indicate accommodation. It is only necessary to observe that, when they occur, acknowledgements of heritable effects are usually tucked into the general introductory remarks or the closing caveats of an article to realize that the implications of genotypic diversity have penetrated neither thinking nor action levels in behavioral studies. Admitting or not that heredity does have something to do with behavior after all, most students of behavior continue in the comfortable assumption that genetic principles and methods can still be largely ignored.

Dobzhansky (1962) and others have cited a variety of explanations for the emergence in former years of an anti-hereditarian bias. These ranged from historical reasons rooted in some of the earliest philosophical heritage of the social sciences, to the perversions of social Darwinism and its noxious offshoots, to misapprehensions about curability and inevitability, to emotional responses having to do with one's own self-determination. All of these background ideas were alike, of course, in that they represented statements of basic ignorance about gene–environment interactions. All posed alternatives: either genetic fixity, with phenotypic expression being insusceptible to change in response to environmental factors, or limitless environmental plasticity, with heredity being inconsequential in behavioral development.

That state of confusion seems to be chronic, for when we examine many of the modern treatments of nature and nurture we are likely to find them retaining the notion of opposed forces, teams that rarely go into play simultaneously. Thus, we find that environment is said to operate "irrespective of genetic constitution," "In spite of genetic limitations," or "without regard to heredity." Quite often, references to the interaction of heredity and environment turn out to mean nothing more than "there must be a genotype, that is, organism, upon which environment can act." Lacking is an appreciation of the enormous amount of genetic variability existing in human and animal populations and the individuality of the reaction ranges (Gottesman, 1968) of each of these variants. In short, the very essence of the gene–environment interaction concept has been missed. The nature–nurture controversy has not really died or even faded away; with a sprinkling of a few pleasant words about heredity for modern flavor, the nurture side of the argument thrives in quiet complacency.

To a large extent, developments in behavior genetics have not been conducive to dispelling the confusion. As noted earlier in this paper, there have been comparatively few attempts to take up the challenges of exploration and explanation in connection with gene–environment interactions and behavior. Part of this neglect can be accounted for by the fact that behavior geneticists, as a group, have often kept busy just in the effort to gain from entrenched environmentalists some enduring recognition of the need to reckon with heredity in behavioral studies. Unfortunately also, discussions of the implications of genetics for behavior sometimes convey the impression that endlessly proliferating interactions between hereditary and environmental variations can only result in a morass of disorderly individual differences. That is a discouraging prospect and one which is certainly overdrawn! Students of behavior may well fail to see any possibilities of discovering meanings in the chaotic state implied. With work on plants and lower organisms and with examples from developmental embryology as a frame of reference, workers in behavior genetics, however, should be able to think of gene–environment relationships in terms of underlying mechanisms in which some order is to be found.

I have tried to demonstrate in the data from the early experience literature examined briefly here that gene–environment interactions are numerous and that treatment effects are frequently reversed in direction for different genotypes. At the same time, I have tried to emphasize that not every strain × treatment combination produces a discernible difference in behavior compared to (a) the untreated members of the same strain or (b) similarly treated members of a different strain. Thus, environmental treatments very often do not produce any effect in some genotypes—at least not any change in the behaviors studies—strains sometimes do not differ among themselves, and, when they do differ, they more often show quantitative deviations from each other than sign reversals in performance. Moreover, it is possible to see some (admittedly crude) patterns of responsivity among some of the strains included in the several studies reviewed here. Thiessen (1965) and Abeelen (1966) have called attention to the consistencies in relative performances of several mouse strains when studied in a number of investigations on various behavioral measures that presumably tap a common phenotypic domain. Ginsburg (1967) and others have commented upon the further finding that some strains (for example, the C57BL types) are consistently more labile, more responsive to (at least certain kinds of) treatments than other strains. The data from the early experience studies tend to show both performance level and responsivity consistencies across strains, as well as consistencies in the behavioral phenotype which do and do not maximally reflect treatment effects within each strain. Diversity in plenty is certainly there, but, as Henderson (1968) and Vale and Vale (1969) have stressed, basic regularities can be found among the various sets of interactions with sufficiently fine-grained analysis, and sometimes even with a very coarse net.

Genotypic uniqueness is a fact (Hirsch, 1962). So, too, probably, is the uniqueness of total environmental complexes encountered by each individual. Nevertheless, it may be reasonable to suppose that many genotypic–environmental encounters do not produce interactions so unique that they differ appreciably from interactions formed in the encounters of many other genotypes and environments. Data reported by Broadhurst and Jinks (1966), for instance, strongly suggest that stability of behavioral development is under genetic control and that the genes which confer greater stability (that is, resistance to environmentally induced variations) tend to show dominance. Their discussion of the evolutionary significance of such behavioral stability during development proposes that gross individual differences in adult reactions to stimuli might be highly disadvantageous in many natural populations. Early stability, it is further suggested, might thus afford a comparatively homogeneous baseline of adult reactivity, from which a considerable amount of behavioral plasticity could then emerge. The hypothesis is of interest in that it offers a possible explanation for the fact that interactions between genotypes and environments do not seem to represent quite so much buzzing confusion as their separate diversities might indicate.

The idea of limits to the amount of phenotypic variation attained through gene–environment interactions has been stated most lucidly by Vale and Vale (1969). They say:

> If the canalization concept may be applied to behaviors, there would appear to be a property of development that acts to reduce the phenotypic expression of behavioral uniqueness. This balance is necessary if a population

is to exploit a limited species range while maintaining the genetic diversity imperative for evolution. If every genetic and every environmental difference produced important phenotypic differences, it is difficult to see how any population could reproduce and survive, so morphologically, biochemically and behaviorally different would be the individuals composing it.

In fact, it has been observed (e.g. Dobzhansky, 1955; Lerner, 1954, p. 6) that morphological variance in natural populations is smaller than would be expected considering genetic segregation and differences in environment."

The point has been made by Vale and Vale (1969) that the interactions of nature and nurture are often to be understood in terms of basic mechanisms underlying the shared behavioral response. Ginsburg (1967) has also stressed the usefulness of analyzing gene–environment interactions as a lever for the "meaningful investigation of problems of behavior at every level, from the molecular, through the organismic, to the population" (p. 153). In other areas of genetics, interactions are used, for example, to explore the timing and mechanisms involved in the development of specific characters (Caspari, 1964), to investigate maternal–fetal responses (see Fudenberg and Fudenberg, 1964; Howell and Stevenson, 1971; Levine *et al.,* 1941) influencing developmental patterns, and to explore the action of specific environmental agents upon metabolic pathways (Fraser, 1963). And Harris (1970), in a quite different context, has commented that one of the important applications of research in human genetics will lie in the possibilities of modifying or tailoring the environment according to the individual needs of persons with different genetic constitutions. In short, rather than regarding heredity–environment interactions as nuisance variables, many people are looking for ways to take advantage of them as a type of research stratagem.

Schizophrenia is a case in point where closer attention to gene–environment interactions should be a minimal requirement for all future research designs. If most of us who pursue the etiological *ignes fatui* of schizophrenia really believe that some kind of interactional phenomenon is involved (see Rosenthal and Kety, 1968, entire proceedings, *The Transmission of Schizophrenia*), then why are we so often found to be following our separate tracks, nature or nurture, as of old? In choosing to concentrate on one side or the other, we have options of sampling and design that would permit us to include at least gross analyses of gene–environment interactions. Some progress toward an interactional approach to schizophrenia has already begun to appear with Heston's (1966) study of adoption in children of schizophrenic mothers and the prompt follow-up by other investigators (Kety, Rosenthal, Wender) in making use of adoptee samples (see review of these studies in Rosenthal, 1971). In another elegant attack upon the heredity–environment problem, Rosenthal (1971) and colleagues are comparing children of schizophrenic parents reared in kibbutzim and in their own homes with control children in both types of rearing situations. In all of these studies, the idea is to separate genetic from rearing variables, the biological transmission of genes from the possible intrafamilial transmission of psychopathology. Clearly, future work will have to take into account subtler aspects of the environment (because, as Heston's work has demonstrated, the proportion of predisposed children manifesting schizophrenia is the same whether they are reared by their schizophrenic parents or by others), but a pattern for the dissection of interactions has now been set by these investigations. Other types of programs which concentrate on the prospective study of individuals presumed to be at risk for the later manifestation of schizophrenia (see Anthony, 1968; Erlenmeyer-Kimling, 1968; Mednick and Schulsinger, 1968) may be in a position to examine interactions between genotypes and environmental stresses over various periods of development. In such studies it may be possible, moreover, to use observed interactions to test specific hypotheses about, for example, neurophysiological or biochemical pathways in which aberrations occur.

No studies of gene–environment interactions are going to be easy to do, and the methodological problems, especially in connection with human behavior, are obviously immense. The study of behavior, however, has not been at all well-served thus far by *apartheid* tactics between environmentalists and geneticists. Rather than ignoring gene–environment interactions or being overawed by them, we will have to cope with them and learn to put them to our service in the understanding of the "hows" of behavior.

References

ABEELEN, J. H. F. van (1966). Effects of genotype on mouse behaviour. *Anim. Behav.* **14**, 218–225.

ANTHONY, E. J. (1968). The developmental procursors of adult schizophrenia. *In:* "The Transmission of Schizophrenia," (D. Rosenthal and S. S. Kety, eds.). Pergamon, Oxford.

BROADHURST, P. L. (1967). An introduction to the diallel cross. *In:* "Behavior-Genetic Analysis," (J. Hirsch, ed.). McGraw Hill, New York.

BROADHURST, P. L., and JINKS, J. L. (1966). Stability and change in the inheritance of behaviour in rats: A further analysis of statistics from a diallel cross. *Proc. Roy. Soc., Ser. B,* **165**, 450–472.

CASPARI, E. (1964). The problem of development. *Brookhaven Lecture Series No. 35,* (April 15), BNL 976 (T-411).

CASPARI, E. (1967). Introduction to Part 1 and remarks on evolutionary aspects of behavior. *In:* "Behavior-Genetic Analysis," (J. Hirsch, ed.). McGraw Hill, New York.

CLARK, W. C., BROWN, J. C., and RUTSCHMANN, J. (1967). Flicker sensitivity and response bias in psychiatric patients and normal subjects. *J. Abnorm. Psychol.* **72**, 35–42.

COX, R. P., and MacLEOD, C. (1962). Relation between genetic abnormalities in man and susceptibility to infectious disease. *In:* "Methodology in Human Genetics," (W. J. Burdette, ed.). Holden-Day, San Francisco.

DeFRIES, J., WEIR, M., and HEGMANN, J. (1967). Differential effects of prenatal maternal stress on offspring behavior in mice as a function of genotype and stress. *J. Comp. Physiol. Psychol.,* **63,** 332–334.

DIXON, L., and DeFRIES, J. (1968). Effects of illumination on open-field behavior in mice. *J. Comp. Physiol. Psychol.,* **66,** 803–805.

DOBZHANSKY, T. H. (1962). *"Mankind Evolving."* Yale Univ. Press, New Haven.

DUNHAM, H. W. (1970). Social class and mental disorder. *Brit. J. Soc. Psychiat.* **4,** 76–83.

EHRMAN, L. (1970). Simulation of the mating advantage in mating of rare Drosophila males. *Science,* **167,** 905–906.

ERLENMEYER-KIMLING, L. (1968). Studies on the offspring of two schizophrenic parents. *In:* "The Transmission of Schizophrenia," (D. Rosenthal and S. S. Kety, eds.). Pergamon, Oxford.

FRASER, F. C. (1963). Methodology of experimental mammalian teratology. *In:* "Mammalian Genetics," (W. J. Burdette, ed.). Holden-Day, San Francisco.

FUDENBERG, H. H., and FUDENBERG, B. R. (1964). Antibody to hereditary human gamma-globulin (Gm) factor resulting from maternal-fetal incompatibility. *Science,* **145,** 170–171.

FULLER, J. L. (1967). Experiential deprivation and later behavior. *Science,* **158,** 1645–1652.

FULLER, J. L. (1968). Genotype and social behavior. *In:* "Biology and Behavior: Genetics," (D. C. Glass, ed.). Rockefeller Univ. Press and Russell Sage Foundation, New York.

FULLER, J., and CLARK, L. (1968). Genotype and behavioral vulnerability to isolation in dogs. *J. Comp. Physiol. Psychol.* **66,** 151–156.

FULLER, J. L., and COLLINS, R. L. (1970). Genetics of audiogenic seizures in mice: A parable for psychiatrists. *Sem. Psychiat.* **2,** 75–88.

GINSBURG, B. E. (1958). Genetics as a tool in the study of behavior. *Perspect. Biol. Med.* **1,** 397–424.

GINSBURG, B. E. (1963). Causal mechanisms in audiogenic seizures. *In:* "Psychophysiologie Neuropharmacologie et Biochimie de la Crise Audiogene." *Colloq. Int. Centre Nat. Rech. Sci.* No. 112, 227–240.

GINSBURG, B. E. (1967). Genetic parameters in behavioral research. *In:* "Behavior-Genetic Analysis," (J. Hirsch, ed.). McGraw-Hill, New York.

GOTTESMAN, I. I. (1968). Biogenetics of race and class. *In:* "Social Class, Race and Psychological Development," (M. Deutsch, I. Katz and A. R. Jensen, eds.). Holt, New York.

HALDANE, J. B. S. (1946). The interaction of nature and nurture. *Ann. Eugen.,* **13,** 197–205.

HALDANE, J. B. S. (1965). The implications of genetics in human society. *In:* "Genetics Today," (S. J. Geerts, ed.). Proc. Int. Congr. Genet., (The Hague, Netherlands, Sept. 1963). Pergamon, Oxford.

HALLER, M. H. (1968). Social sciences and genetics: A historical perspective. *In:* "Biology and Behavior: Genetics," (D. C. Glass, ed.). Rockefeller Univ. Press and Russell Sage Foundation, New York.

HARRINGTON, G. M. (1968). Genetic-environmental interaction in intelligence: I. Biometric genetic analysis of maze performance of Rattus norvegicus. *Develop. Psychobiol.* **1,** 211–218.

HARRINGTON, G. M. (1969). Genetic–environmental interaction in "intelligence" II. Models of behavior, components of variance, and research strategy. *Develop. Psychobiol.* **1,** 245–253.

HARRIS, H. (1970). "The Principles of Human Biochemical Genetics." American Elsevier, New York.

HENDERSON, N. D. (1964). Behavioral effects of manipulation during different stages in the development of mice. *J. Comp. Physiol. Psychol.* **57,** 284–289.

HENDERSON, N. D. (1967). Early shock effects in BALB/C mouse. *J. Comp. Physiol. Psychol.* **64,** 168–170. (a)

HENDERSON, N. D. (1967). Prior treatment effects on open field behavior of mice —a genetic analysis. *Anim. Behav.* **15,** 364–376. (b)

HENDERSON, N. D. (1968). The confounding effects of genetic variables in early experience research: Can we ignore them? *Develop. Psychobiol.* **1**(2), 146–152.

HENDERSON, N. D. (1970). A genetic analysis of spontaneous alterations in mice. *Behav. Genet.* **1,** 125–132. (a)

HENDERSON, N. D. (1970). Genetic influences on the behavior of mice can be obscured by laboratory rearing. *J. Comp. Physiol. Psychol.* **72,** 505–511. (b)

HESTON, L. L. (1966). Psychiatric disorders in foster home reared children of schizophrenic mothers. *Brit. J. Psychiat.* **112,** 819–825.

HIRSCH, J. (1962). Individual differences behavior and their genetic basis. *In:* "Roots of Behavior," (Bliss, ed.). Harper, New York.

HIRSCH, J. (ed.). (1967). "Behavior-Genetic Analysis." McGraw-Hill, New York.

HOWELL, R. R., and STEVENSON, R. E. (1971). The offspring of phenylketonuric children. *Social Biology,* Suppl. **18,** 19–29.

KAUFMAN, I. C., and ROSENBLUM, L. A. (1967). Depression in infant monkeys separated from their mothers. *Science* **155,** 1030–1031.

KING, J. A. (1957). Relationships between early social experience and adult aggressive behavior in inbred mice. *J. Gen. Psychol.* **90,** 151–166.

KING, J. A., and ELEFTHERIOU, B. E. (1959). Effects of early handling upon adult behavior in two subspecies of deermice, Peromyscus Maniculatus. *J. Comp. Physiol. Psychol.* **52,** 82–88.

LEVINE, P., BURNHAM, L., KATZIN, M., and VOGEL, P. (1941). The role of isoimmunization in pathogenesis of erythroblastosis fetalis. *Amer. J. Obstet. Gynecol.* **42,** 925–937.

LEVINE, S., and BROADHURST, P. L. (1963). Genetic and ontogentic determinants of adult behavior in the rat. *J. Comp. Physiol. Psychol.* **56,** 423–428.

LINDQUIST, E. F. (1953). "Design and Analysis of Experiments in Psychology and Education." Houghton, Boston.

LINDZEY, G., LYKKEN, D. T., and WINSTON, H. D. (1960). Infantile trauma, genetic factors, and adult temperament. *J. Abnorm. Soc. Psychol.* **61,** 7–14.

LINDZEY, G., WINSTON, H. D., and MANOSEVITZ, M. (1963). Early experience genotype, and temperament in mus musculus. *J. Comp. Physiol. Psychol.,* **56,** 622–629.

LINDZEY, G. LOEHLIN, J., MANOSEVITZ, M., and THIESSEN, D. (1971). Behavioral genetics. *Ann. Rev. Psychol.* **22,** 39–94.

LUBIN, A. (1961). The interpretation of significant interaction. *Educ. Psychol. Meas.* **21,** 807–817.

MANOSEVITZ, M., CAMPENOT, R. B., and SWENCIONIS, C. F. (1968). Effects of enriched environment upon hoarding. *J. Comp. Physiol. Psychol.,* **66**, 319–324.

MATHER, K., and MORLEY-JONES, R. (1958). Interaction of genotype and environment in continuous variation: I. Description. *Biometrics,* **14**, 343–359.

MEDNICK, S. A., and SCHULSINGER, F. (1968). Some premorbid characteristics related to breakdown in children with schizophrenic mothers. *In:* "The Transmission of Schizophrenia," (D. Rosenthal and S. S. Kety, eds.). Pergamon, Oxford.

NEWELL, T. G. (1967). Effect of maternal deprivation on later behavior in two inbred strains of mice. *Psychon. Sci.,* **9**, 119–120.

POLANSKY, N. A., BORGMAN, R. D., DeSAIX, C., and SMITH, B. J. (1969). Mental organization and maternal adequacy in rural Appalachia. *Amer. J. Orthopsychiat.* **39**, 246–247.

RESSLER, H. (1963). Genotype-correlated parental influences in two strains of mice. *J. Comp. Physiol. Psychol.,* **56**, 882–886.

ROSENTHAL, D. (1968). The genetics of intelligence and personality, *In:* "Biology and Behavior-Genetics," (D. C. Glass, ed.). Rockefeller Univ. Press and Russell Sage Foundation, New York.

ROSENTHAL, D. (1971). A program of research on heredity in schizophrenia. *Behav. Sci.* **16**, 191–201.

ROSENTHAL, D., and KETY, S. S. (eds.) (1968). "The Transmission of Schizophrenia." Pergamon, Oxford.

ROSS, S., NAGY, Z., KESSLER, C., and SCOTT, J. (1966). Effects of illumination on wall-leaving behavior and activity in three inbred mouse strains. *J. Comp. Physiol. Psychol.* **62**, 328–340.

ROTHENBUHLER, W. C. (1967). Genetic and evolutionary considerations of social behavior of honeybees and some related insects. *In:* "Behavior Genetic Analysis," (J. Hirsch, ed.). McGraw-Hill, New York.

SCOTT, J. P., and FULLER, J. L. (1963). Behavioral differences. *In:* "Methodology in Mammalian Genetics," (W. J. Burdette, ed.). Holden-Day, San Francisco.

STAMATOYANNOPOULOS, G. FRASER, R., MOTULSKY, A. G., FESSAS, P. H., AKRIVAKIS, A., and PAPAYANNOPOULOU, T. H. (1966). On the familial predisposition to favism. *Amer. J. Hum. Genet.* **18**, 253–263.

THIESSEN, D. D. (1965). Persistent genotypic differences in mouse activity under unusually varied circumstances. *Psychon. Sci.* **3**, 1–2.

THODAY, J. M., and GIBSON, J. B. (1970). Environmental and genetical contributions to class difference: A model experiment. *Science,* **167**, 990–992.

THOMPSON, W. R. (1967). Some problems in the genetic study of personality and intelligence. *In:* "Behavior-Genetic Analysis," (J. Hirsch, ed.). McGraw-Hill, New York.

THOMPSON, W. R., and OLIAN, S. (1961). Some effects on offspring behavior of maternal adrenalin injection during pregnancy in three inbred mouse strains. *Psychol. Rep.* **8**, 87–90.

VALE, J. R., and VALE, C. A. (1969). Individual differences and general laws in psychology: A reconciliation. *Amer. Psychol.* **24**, 1093–1108.

VESELL, E. S. (ed.). (1971). Drug metabolism in man. *Ann. N. Y. Acad. Sci.* **179**, (Whole volume).

WHITNEY, G. (1970). Timidity and fearfulness of laboratory mice: An illustration of problems in animal temperament. *Behav. Genet.* **1**, 77–85.

WIESENFELD, S. L. (1967). Sickle-cell trait in human biological and cultural evolution. *Science,* **157**, 1134–1140.

WINSTON, H. D. (1963). Influence of genotype and infantile trauma on adult learning in the mouse. *J. Comp. Physiol. Psychol.* **56**, 630–635.
WINSTON, H. D. (1964). Heterosis and learning in the mouse. *J. Comp. Physiol. Psychol.* **57**, 279–283.
WOLF, H. H., and ROWLAND, C. R. (1969). Effects of chronic postnatal drug administration on adult dominance behavior in two genera of mice. *Develop. Psychobiol.* **2**, 195–201.

DISCUSSION

W. R. THOMPSON

Queen's University
Kingston, Ontario, Canada

I think that none of us would doubt that the problem of gene–environment interaction is very central to behavior genetics as well as to the separate disciplines of genetics and psychology. Yet there seem to be no clear notions of how the interaction can be analyzed so as to guide empirical research in fruitful directions.

In her valuable contribution, Dr. Erlenmeyer-Kimling has dissected conceptually the idea of gene–environment interaction, and made explicit the basic theoretical components involved; and, further, she has illustrated these factors with contemporary studies in the area of behavior genetics.

I shall now underline some aspects of her paper and supplement the illustrative studies she has reviewed with some additional data of my own.

In general, it is certainly worth emphasizing again to the behavioral sciences community that environmental stimuli do not impinge on an organism that is, in some sense, neutral, but on one that already possesses certain limitations and dispositions. Not only do all individuals (with the exception of identical twins) differ genetically from each other, but single individuals themselves change over time through growth, maturation, and experience.

This is not to say, of course, that we may never have any general laws of behavior. The old aphorism *scientia non est individuorum* still applies. But we must expect that genotype and age will enter as major modifying variables into any equations we educe to relate environment and behavior.

Dr. Erlenmeyer-Kimling quite rightly stresses the critical distinction between genotype–environment covariance and genotype–environment interaction. It is certainly not easy, however, to distinguish these in natural populations. Thus in the class structure in most societies, we find a strong relation between IQ and class membership. However, there is still some disagreement as to what this relationship means. The long-continued debate between Halsey (1958), on the one hand, and Burt (1961) and Conway (1958) on the other deals precisely with this issue: Do members of the low socioeconomic classes possess low IQ's by virtue of poor environment (an interaction) or by virtue of a process of natural selection that engenders downward mobility for those innately possessing low innate ability (covariance)? In reviving this issue in the popular press, Herrnstein (Atlantic Monthly, September 1971) opts for the latter position. There is, of course, nothing very new or startling in this suggestion. Many others during the 1950s and earlier have put forward the same point of view as Dr. Fuller and I did in our text in 1960.

Whatever one's beliefs regarding the generalities of the matter, however, it is quite clear that not everyone is convinced that social-class—much less racial—differences may be innate and relatively fixed. Consequently, it certainly seems highly advisable that model experiments be undertaken dealing with complex basic problems that emerge from any consideration of the biological determinants of the structure of society. For example, Dobzhansky and Spassky (1967) at Rockefeller University and Thoday and Gibson (1970) at Cambridge have used fruit flies as subjects. It will be of great interest to see what kind of congruence emerges between the results of experimental work with simple organisms and the results of research on natural human populations using techniques such as the Multiple Abstract Variance Analysis (MAVA) developed by Cattell (1960) and the biometrical methods of Jinks and Fulker(1970).

Dr. Erlenmeyer-Kimling presents a lucid discussion of the basic types of interaction possible: additive ordinal, nonadditive and disordinal, nonadditive. Of course, it is not always so easy to explain the cases that are found to occur empirically. Merely to state that a particular kind shows up as a statistically significant effect is not very satisfying. What is of much more interest is to understand exactly what processes underlie each, the range of environments and genotypes for which they hold and, if possible, the evolutionary or genetic significance of each. It is certainly quite true that early experience, for example, will not appreciably affect all genotypes. Of those genotypes which are affected, some may change in one direction, others in the opposite direction. One must turn to the parameters that underlie such interactions in order to make testable hypotheses. I would now like to focus on each of the three parameters discussed in this paper.

The Phenotype Involved. It is clear that one of the greatest problems facing the behavior geneticist—indeed, the psychologist in general—is knowing just what entity with which he is dealing. Behavior is complex, in the sense that it is usually made up of many components, each of whose contribution to the whole may vary over time, over test situation and over individuals. We are all familiar with Searle's analysis of Tryon's (1940) results. Searle (1949) showed that, although we might select bright and dull strains, as defined by a score on a particular maze, there were many reasons why an animal came to be classified as bright or dull, many of these probably having little to do with intelligence as usually conceptualized.

The problem of equating the phenotype is exactly the same when comparing age groups. For the last 5 years, I have been concerned with the ontogeny of learning, and note that it is seldom clear whether the objectively defined task or treatment which we may impose on an adult animal will mean the same thing for it as it does for a young animal. Consequently, any score differences obtained (or not obtained) need not reflect a general superiority of one age over the other. Rather it may represent the fact that animals at different stages during development may simply employ different strategies to cope with various environmental contingencies and that each of these, from the standpoint of that age, may be highly adaptive.

For example, we found that the degree of retention of a conditioned emotional response is almost absent in 15-day-old rats. However, it was large in three other older-aged animals, particularly in 23-day-old rats. Apparently young Ss do not classically condition readily. The question remains, however, have they in fact learned anything at all from their exposure to the shock-tone treatment used? An answer to this question was provided by a comparison of groups receiving unpaired treatments (that is, "stimulation") with groups which received no treatment. The young had in fact learned something—namely to be less emotional in general. This effect appeared to increase with subsequent testings. Just the opposite held true for the adults. They showed an immediate sensitization effect which diminished with time.

The point then is clear: an animal is apt to exhibit many behavioral patterns (or phenotypes) in a given situation. Hence it is always dangerous to draw conclusions of any great degree of generality. In psychology, we may like to think in terms of simple hierarchies of ability or mental health or intelligence. But such concepts may be inappropriate and much less suitable than a concept like adaptivity whereby the individual copes with whatever circumstances he faces by the best means available to him, whether these be physiological or behavioral.

Time. Unlike many morphological characters, behavior may change drastically from one moment to the next. The term I have used before to describe this property of behavior is fluidity. Changes in the organism occur due to development, to light–dark cycles and to learning; and for any particular trait, the heritability and the manner of genetic transmission are liable to vary greatly.

Again, examples of this are very numerous. However, I would like to supplement those put forward by Dr. Erlenmeyer-Kimling with one of my own. Mr. J. Kluger and Dr. G. J. S. Wilde (unpublished) at our University have studied the heritability of visual-motor coordination and pitch discrimination using MZ and DZ twins. The methods were based on those of Jinks and Fulker. For performance of MZ's and DZ's on a pursuit-rotor task, the plot of heritability (Holzinger's H and Nichols HR) over time traced out a U-shaped function, being rather high at the start and ending of training and low in the middle. On the other hand, the same heritability estimates for pitch discrimination appeared to show cylical decelerating changes. Except for day one of training, however, heritability values for pitch were nonsignificant. Although the samples used were small, these data illustrate a problem in human studies which is of interest. When heritability values do show changes, it is always difficult to know the cause. Kluger has favored the view that the nature of the task changes with time, in the sense that different kinds of abilities and skills are demanded at different points in time; hence the genetic factors involved may vary accordingly.

The Environment. As Dr. Erlenmeyer-Kimling suggests, the behavior-geneticist should always specify the range of stimulus intensities to be examined. It is, of course, desirable that a wide range be sampled. I am reminded, in this connection, of a study by one investigator who tested a great variety of organisms belonging to various species, genera, and classes in a multiple T-maze. Such an enterprise could hardly be expected to produce results of any great interest or value. A maze is a most "unnatural" setting for many, perhaps most, organisms. One might also apply this criticism to many other behavioral tests commonly used by psychologists. Use of such methods does not provide suitable research designs for the behavior geneticist whose interest is surely variation in respect to individuality, and in respect to environments.

Finally, one must pay attention not only to the matter of present environments, but also to past environments—particularly those occurring early in life. The effects of restriction on later behavior are well documented. The extent to which such effects are due to interaction between early and later environments, however, is not so well understood. Fuller and Clark

(1966) have emphasized this kind of explanation, and, though I cannot accord it as much weight as they do, I do think that such interactions are important. The point is made by a study in our laboratory done by Margaret McKim (1971), which dealt with the effects of prenatal maternal restriction or enrichment (that is, environmental stimulation) on offspring open-field ambulation at 23 days. Complete cross-fostering was carried out, and following parturition all offspring were housed in standard (control) cages. Highly significant effects were produced by enrichment given either pre- or postnatally. Control and restricted treatments did not differ greatly. There were also some interesting interactions between different combinations of treatments. Some of the largest effects were found when opposite treatments followed each other, that is, an enriched prenatal maternal environment followed by rearing by a previously restricted mother, or, conversely, a restricted prenatal maternal environment followed by rearing by a previously enriched mother.

It is obvious then that the sequences of environments to which an animal is subjected interact in striking ways and the *disparity* between them rather than the *order* of the sequence may be the most significant factor in determining some behaviours.

In conclusion it seems clear that a consideration of the gene–environment interaction issue suggests that we have perhaps an embarrassment of riches by way of experimental questions to work on. What is less clear is the choice of where the best payoffs are. This is always the most difficult problem in science. We cannot tell in advance what endeavors will turn out to be the most fruitful. As Michael Polanyi (1958) has pointed out, we are faced with the dilemma of the snark hunter when advised by the Bellman that the best way to spot a snark was by observing its peculiar habit of eating dinner the following day. However, I would guess that work in the next decade will yield some major advances. During the 1960s, I think our main concern was with problems of methodology. We have seen the development of MAVA by Cattell (1960) and his colleagues; the factor analytic approach of Vandenberg (1965); the adoptee methods of Rosenthal (1970) and others in psychopathology; the application of Fisherian and Matherian methods by Burt and Howard (1956) to the study of intelligence, by Broadhurst and Jinks (1961) to the study of various traits in animals; and, more recently, the sophisticated extensions of these by Jinks and Fulker (1970) into a generalized statistical methodology that seems capable of giving exact solutions to a variety of genetic psychological and sociological problems.

Thus we now have the proper tools. And these tools are of a character, not only to help us in answering questions we have already thought of, but in asking new ones that have not as yet occurred to us.

References

BROADHURST, P. L., and JINKS, J. L. (1961). Biometrical genetics and behavior: reanalysis of published data. *Psychol. Bull.* **58**, 337–362.

BURT, C. (1961). Intelligence and social mobility. *Brit. J. Stat. Psychol. XIV*, 3–24.

BURT, C., and HOWARD, M. (1956). The multifactorial theory of inheritance and its application to intelligence. *Brit. J. Stat. Psychol.* **9**, 95, 131.

CATTELL, R. B. (1960). The mutliple abstract variance analysis equations and solutions for nature-nurture research on continuous variables. *Psychol. Rev.* **67**, 353–372.

CONWAY, J. (1958). The inheritance of intelligence and its social implications. *Brit. J. Stat. Psychol. XI*, 171–190.

DOBZHANSKY, T., and SPASSKY, B. (1967). Effects of selection and migration on geotactic and phototactic behavior of Drosophila. *Proc. Roy. Soc., Ser. B*, **168**, 27–47.

FULLER, J. L., and CLARK, L. D. (1966). Genetic and treatment factors modifying the postisolation syndrome in dogs. *J. Comp. Physiol. Psychol.* **61**, 251–257.

HALSEY, A. H. (1958). Genetics, Social structure and intelligence. *Brit. J. Sociol. IX*, 15–28.

JINKS, J. L., and FULKER, D. W. (1970). Comparison of the biometrical genetical, MAVA, and classical approaches to the analysis of human behavior. *Psychol. Bull.* **73**, 311–349.

KLUGER, J., and WILDE, G. J. S. Wilde is in Queen's department. Kluger was his doctoral student. Ph. D. thesis not published.

McKIM, M. (1971). The behavioral effects of prenatal enrichment and restriction. Honours B.A. Thesis, Queen's University. (unpublished)

POLANYI, M. (1958). "Personal knowledge." Routledge and Kegan, London.

ROSENTHAL, D. (1970). "Genetic theory and abnormal behavior." McGraw-Hill, New York.

SEARLE, L. V. (1949). The organization of hereditary maze-brightness and maze-dullness. *Genet. Psychol. Monogr.* **39**, 279–325.

SNEDDEN, D. S., SPEVAK, A. A., and THOMPSON, W. R. (1971). Conditioned and unconditioned suppression as a function of age in rats. *Can. J. Psychol.* **25**, 313–322.

THODAY, J. M., and GIBSON, J. B. (1970). Environmental and genetical contributions to class difference. A model experiment. *Science*, **167**, 990–992.

THOMPSON, W. R. (1966). Multivariate analysis in behavior genetics. *In:* "Handbook of Multivariate Analysis. Experimental Psychology," (R. B. Cattell, ed.), Rand McNally, Chicago.

TRYON, R. A. (1940). Genetic differences in rats. "39th yearb. Nat. Soc. Stud. Educ." Part I, 111–119. Public School Publ. Co., Bloomington, Illinois.

VANDENBERG, S. G. (1965). "Methods and goals in human behavior genetics." Academic Press, New York.

COMMENT

ERNST CASPARI

University of Rochester
Rochester, New York

It has been mentioned that environmental factors may influence behavior if they act before the birth of the individual. It should be added that environmentally induced modifications acting on parents may be transmitted through more than one generation. This phenomenon, known for some forty years, has been designated "dauermodifications." The phenomenon has not been thoroughly studied and its place in the overall picture of genetics is unknown (see Caspari, 1948).

A "dauermodification" affecting brain structure and behavior has been recently described by Zamenhof *et al.* (1971). These workers subjected female rats to a low protein diet and found that their offspring were reduced in size and in number of cells in the brain, as measured by DNA content. This effect on the brain could not be reversed by improved nutrition after birth; the sensitive period for the specific effect of low protein diet on the number of cells in the brain extends from one week before impregnation through the first half of pregnancy. The reduced number of brain cells results in behavioral defects, extensive descriptions of which have not as yet been published. The reduction of cell number (DNA content) in the brain is further transmitted to the progeny of the affected animals, even though they had received standard food after birth. The effect is transmitted through the female but not through the male parent, suggesting transmission through the cytoplasm of the ovum.

It is obvious that effects of this type are of practical importance, since they show that harmful effects of environmental factors cannot always be remedied immediately upon improvement of the environment. It further shows that gene–environment interactions may be even more complex than suggested in our models. Transmission of an acquired behavioral character from mother to offspring would ordinarily be attributed to social and cultural factors. This seems unlikely in an effect involving diet and cell number in the brain.

References

CASPARI, E. (1948). Cytoplasmic inheritance. *Adv. Genet.* **2**, 1–64.

ZAMENHOF, S., van MARTENS, E., and GAUEL, L. (1971). DNA (cell number) in neonatal brain: Second generation (F_2) alteration by maternal (F_0) dietary protein restriction. *Science,* **172**, 850–851.

COMMENT

NEWTON E. MORTON
University of Hawaii
Honolulu, Hawaii

Fisher (1918) considered three essentially different models of assortative mating according to whether the phenotypic correlation is primary, the genotypic correlation is primary (inbreeding), or there is a primary association for another trait that happens to be correlated to the one in question. These three assumptions lead to essentially different predictions for correlation between relatives. No one has yet suggested an experimental design that would make it possible to distinguish these models in human material, and so this subject of assortative mating remains what it was in 1918, a theoretical exercise with no practical consequences.

Applied to IQ, for example, we could ask whether parental correlation is due to a primary association for educational level, social class, ethnic group, intelligence per se, or other factors, and would be forced to admit that the causation is extremely complex and, in my view, unanalyzable, so that none of Fisher's formulas would be applicable. Of course, if it were an important problem to estimate heritability, some assumptions would have to be made about the genetic consequences of assortative mating. If, however, this is not an important problem, assortative mating may well be neglected. There seems no point in measuring something if we cannot make any inferences subsequently.

Reference

FISHER, R. A. (1918). The correlation between relatives on the supposition of Mendelian inheritance. *Trans. Roy. Soc. Edinburgh* **52**, 399–433.

Chapter 9

The Meaning of the Cryptanthroparion[1]

E. TOBACH

The American Museum of Natural History
New York, New York

Introduction

In a general sense, to many people, the cryptanthroparion is the inherited destiny of each person—a destiny Grecian in its relentless, predetermined power to write the scenario of the individual's life. The earliest human thought, in all of the five cradles of civilization, featured such a concept. Later, in the early days of science as it is thought of today, there were many proponents of the theory that the conceptus, or the fetus, was a miniature adult whose every characteristic was set, either by virtue of the blood of its

---

[1] T. C. Schneirla and I were to have written a paper with the title "The Meaning of the Cryptohomunculus" for a conference on the biopsychology of development (Tobach *et al.*, 1971). The plan was never carried out because of his untimely death on August 20, 1968. Many of the ideas expressed herein stem directly from his protean theories, and my debt to him is evident. My use of the revised title is meant as a tribute to him and I hope that the deficiencies in the paper do not detract from my expression of honor and gratitude.

ancestors, or by some mystical force. Hildegard, the premedieval German cleric–scholar, described the conceptus as formed by the union of the seeds of the parents, a godhead-derived soul and a devil-derived corruption. The last two produced the character, and the perfection of the seeds determined the physical state of the child (Singer, 1917).

There are many modern formulations of preformationistic schemes. These schemes vary from each other in some ways, but they are similar in their insistence on the controlling, limiting, determining role of an "essence" that is in the genome. The discussion of these formulations in the various national scientific societies of our nation and in the public forums available to scientists (Crow *et al.*, 1967; *Scientific Research,* 1968; Bloom, 1969; David, 1971; *Nature New Biology,* 1971; Darlington, 1971; Ad Hoc Committee, N.A.S., 1972) has engendered much concern about the scientific, social, and ethical implications of the seemingly "purely scientific" concept of the genome.

To carry these discussions further, I propose an analysis of at least three aspects of the cryptanthroparion concept and their pertinence to the scientific problem, to the problem of research "logistics" (see below), and to society.

The Formulation of the Scientific Problem

I have been assigned to discuss "Gene–Environment Interactions and the Variability of Behavior." I propose that the concept of interaction as it is usually applied to genetic expression is inadequate to analyze the relationship between genetic processes and ontogenetic experiential processes that leads to behavioral individuality and variability. I therefore propose an alternative set of concepts that may be more useful for understanding that relationship.

Genes do not function in vacuo; the evolution of the first biochemical molecule was inextricably involved with the milieu in which it achieved an entity of its own, different from any other entity in the surround. One cannot discuss genes without stating the context or milieu in which their function is expressed. Genes function on a biochemical level. If the biochemical characteristics of the milieu are inappropriate, the function of the gene will be inhibited, changed or prevented immediately or at a later time.

All configurations of living matter have some developmental relationship to genes. All organisms at all levels of organization do not function in vacuo, but are in ongoing energy-transformations with the surround. During these energy-transformations, the internal state of the organism is changed, along with the surround.

All activities of an organism as a holistic entity in relationship with the surround may be termed behavior. Such holistic activities are distinguishable from the actions of the systems making up the whole. These systems are acting in relationship with their surround (usually internal) and such activities or functions may be conceptualized as taking place on the physiological level.

Behavior is an ordered set of phenomena derived from, based on, or related to physiological function and to the structural characteristics of the organism involved. Behavior, like all other biological processes, is a function of on-going energy-transformations with the surround. Behavioral phenomena, as well as the physiological and structural characteristics of the organism have a developmental relationship to genetic processes.

Behavioral phenomena may be analyzed in terms of patterns of component bits or subpatterns. To establish the validity and reliability of the analyzed component parts of the behavior pattern, it is necessary to demonstrate how the working relationship among the parts results in an integrated behavioral pattern. It is not appropriate to show a correlation between a component bit and the total pattern, because such a correlation may be fortuitous and biasing (Hayes, 1963). Rather, a functional relationship among the bits needs to be delineated.

Genes and environment are not contradictory opposites, as they have been traditionally defined. The molecular structure thought of as a gene is in fact a biochemical configuration in continuous activity. Its "environment" is a changing configuration of spatial energy relationships; that is, the neighboring nucleotide or sugar base of any one nucleotide may be considered "environment," just as the effects of light on the reproductive system in the bird is considered "environment." They are rather two aspects of the same process, whether it be replicative, directive, or relatively inactive. That process is the biological process-change in structure and function through continuous energy-transduction and transformation. However, knowledge of the expression of a biochemical process (genetic mechanism) in anatomy or physiology is not sufficient for the understanding of how the relationship between anatomical systems and their physiology is integrated to produce an activity by the whole animal.

The genome is not an anthroparion or a homunculus that simply changes in size when adult. "Behavior" is not in the genome. Between the biochemical expression of gene function and a behavioral pattern, there are many steps and functions on different levels of organization (Schneirla, 1972). Methods of population genetic studies, as well as biochemical and molecular investigations involving the isolation of particular genes that may have some direct or indirect relationship to a behavioral pattern or item, are valid approaches to initiating a program to understand the relationship between gene function and behavior. However, genetic processes are functional expressions of a particular biochemical process derived from a particular nucleotide configuration in a particular setting. These biochemical processes may be expressed as enzymes, structural proteins, ribosomal or transfer ribonucleic acid, and as regulatory substances. Only a very specific biochemical action or configuration can be traced to a specific gene or group of genes. To understand how the biochemical genetic mechanism "brings about behavior" also requires analysis on the physiological level, that is, analysis of gene expression in structure and function of systems. Before the relationship between genetic processes and behavior can be defined, the behavioral pattern or item must be analyzed in terms of its physiological or anatomical substrate. In addition, this analysis would require suitable investigation of the pattern under all relevant conditions of the organism (for example, hunger, reproductively active), in all the situations in which the pattern can be observed, as well as at all stages of the development of the pattern during the life history of the organism. In the case of nonhuman organisms, comparison of species, strains, and other subgroups would also be useful in fully defining the pattern being analyzed and in relating it to the anatomical and physiological processes which are more or less implicated in the behavior pattern.

It is always necessary, however, to acquire accurate information about the experience of the organism at all stages of development in order to make comparisons among members of the same species or among heterospecifics for the purpose of generalization. Experience is considered to be all effective stimulation (Schneirla, 1972) as well as all forms of energy-transformation, for example, nutriment ingested, chemical intake during respiration, or water circulation as in the case of aquatic organisms.

Thus, a complete analysis of the behavior pattern requires investigation on each preceding level, including that of the specific gene action. (Ewing, 1964; Rodgers, 1970; Schlesinger and Griek, 1970; Hirsch, 1962b; Ehrman, 1966). The synthesis of the processes uncovered by the analysis outlined above could now take place by integrating the different levels of organization during the development of the organism, in order to define the relationship between "a gene" and behavior.

Most often, the life process defined above is reduced to a conceptualization of two "components," gene and environment that interact. This concept is primarily derived from the statistical manipulations in the analysis of variance, which is a priori designed to discover the contribution of genes and environment to the variance in an additive fashion. As Erlenmyer-Kimling (this volume) and Morton (this volume) have pointed out, interaction does not have to be additive, but they still are concerned with the proportionate contribution of heredity and environment to behavior. It is important to point out that this interaction concept, as they and others use it, applies to populations. In the individual organism, the interaction cannot be additive. In the individual, the process of development makes such a separation meaningless, as each stage of development is derived from the preceding stage, which was formed by the fusion or synthesis of the temporally relevant effective energy input with the existing stage of development. All levels of intraorganismic organization play a role in this synthesis or fusion: biochemical (molecular, enzymatic), physiological, behavioral. The distinction between the application of the interaction concept to populations and the application to individuals is usually glossed over.

The suggestion has been made that I propose a redefinition of interaction so that the word continues to be used in the new sense, or that I propose another word. I am loath to do either; however, the only acceptable possibility would be to consider the existing stage of the organism at point X in time as the "thesis." The energy changes going on with time in the organism as it continues as an integral whole and in relation to energy input, constitute the "antithesis." The consequent state and stage of the organism at point $X + \ldots$ in time is the "synthesis." The concepts of thesis, antithesis, and synthesis are abstractable as separable entities for the purposes of analysis and integration only. In reality, they are not separable. I recognize that this results in a situation in which the synthesis and the thesis are the same. They become differentiated in the course of working with them: in the analysis of any process the thesis and synthesis continue to come into separate focus and to change into each other, depending on which question is asked when.

An example of the general usage of "interaction" in the statistical sense as in variance analysis is offered by the interesting research of Broadhurst and his colleagues (1971; and Rick *et al.,* 1971) who are by no means alone in such a conceptualization of interaction or in doing research based on the concept.

Despite the sophisticated and innovative combinations of experimental, behavior-analytic, biochemical, and statistical techniques designed to show that "genotype–environment interaction [is] . . . a much more dynamic situation than at one time envisaged [Broadhurst *et al.,* 1971, p. 3]," Broadhurst and his colleagues still conceptualize interaction as an additive mechanism in the statistical sense. In this article they state that environmental manipulation permits the behavior-geneticist to describe those genes which pull the phenotype in one or another direction. Consequently, change in behavior during experimental treatment is attributable preeminently to gene mechanisms (Broadhurst *et al.,* 1971, p. 6). For them, interaction is a mechanism whereby one component (environment) acting on the other component (gene) brings about the expression of the latter component. Only the parameters of gene expression have changed. Broadhurst's research and his interpretation of the results can be seen as an example of the impossibility of using an interaction approach to resolve the conflict between explaining all behavioral phenomena by genetic mechanisms (the majority view today) and relying on environmental procedures to explain, control, and predict behavior. The majority of theories in the latter vein consider genetic mechanisms as the ultimate limitation of environmental control of all behavior, particularly human intellectual and social performance (Tobach, 1972).

Broadhurst and his colleagues studied eight strains of rats in a systematic breeding experiment in which they defined the dominant–recessive gene ratio in each strain and succeeding generations of crosses. They studied conditioned avoidance behavior in all their animals. They then correlated the number of avoidances made by each strain with the ratio of the number of animals showing dominant characteristics to the number of animals showing recessive characteristics of each strain. They analyzed the conditioned avoidance data in six successive blocks of five trials each. They were thus able to observe the changing correlations as the animals continued to be conditioned. In the first block of five trials, there was a negative correlation between dominance and number of avoidances, that is, the higher the dominance–recessive ratio, the fewer the number of avoidances made. In the last five trials of the total of 30 trials, the correlation was positive; the higher the dominance ratio, the greater the number of avoidances made. These results were obtained only in rats that were stimulated in infancy. Rats that were not stimulated in infancy started out with negative correlations also. By the end of the 30 trials, however, the correlations only approached zero, and never became positive as in infancy-stimulated rats.

Clearly, the changes in the relationships between behavior and genetic characteristics cannot be explained by suggesting that the genes changed. Instead, the change in the relationship is attributed to the existence of two gene mechanisms, rather than one. The authors also note that stress in infancy may prime the individual to resist stress in adulthood and this may be related to the differences between the two groups of rats. Still, the effect of earlier stimulation is interpreted as functioning through the expression of a dual genetic mechanism.

It is noteworthy that stimulation in infancy leads to improvement in the ease of avoidance conditioning (Levine et al., 1956; Goldman, 1965; Goldman and Tobach, 1967). The change in the correlations in the stimulated group and not in the nonstimulated group may be attributed to this change in conditioning ability. If the infancy-stimulated rats made significantly more avoidance responses in succeeding blocks of trials, the correlations could change. If the nonstimulated animals did not make increasingly more responses, the correlations might not change as much. An explanation based on early experience may be more parsimonious in this instance than a dual genetic mechanism.

9. The Meaning of the Cryptanthroparion

The results of an interactionist conceptualization have done little toward resolving the environment–heredity dilemma. From the same data, one could argue that environmental manipulation, that is, appropriate earlier experience before training, is all that is necessary to bring about better avoidance conditioning, and thus change the relation between genetic characteristics and behavior. Or, one could argue that the change in avoidance conditioning depends on the expression of two genetic mechanisms that are expressed after environmental stimulation. The dilemma of choosing an "environmental" or a "genetic" explanation still exists.

The conceptualization of gene–environment interaction in this statistical way leads to a call to "optimize the environment for particular kinds of genotypes [Broadhurst *et al.*, 1971, p. 8] in the manner of Cronbach (1970) or Jensen (this volume). This is a rather static concept of interaction. The environment can only act within the limits of the genotype, which presumably has been defined by rigorous tests (see the following discussion about the extrapolation of population genetic characteristics to individual characteristics, Hirsch, 1970). The question of how much is contributed by genes, and by environment, about any behavioral pattern remains as always, despite the stated intent of Broadhurst and his colleagues to create a dynamic concept of interaction in which genotypic and environmental contributions are "fluctuating in relation to each other" and not "static" as "proportionate contributions of heredity and environment" are "often thought to be" (Broadhurst *et al.*, 1971).

I would like to suggest that the only way in which the dilemma can be resolved, and a dynamic interpretation of the relationship between experiential and biochemical processes (genes) conceptualized, is to use the developmental and levels approaches offered by Schneirla (1972).

The application of population genetic mathematics to ontogenesis of behavior is an example of typological thinking, as discussed by Dobzhansky, (1968b); Mayr, (1971) Hirsch, (1970). Typological thinking applies a fact about the characteristics of a group of organisms to an individual member of that group. Usually such group characteristics are defined in terms of frequencies and continua. Obviously, no individual can encompass such group parameters. Another instance of typological thinking is evidenced by the concept that human differences in development relate in some way to phyletic history or age.

The relationship between ontogeny and phylogeny has been widely discussed, particularly as a possible clue to understanding evolutionary processes. Several writers have commented on the hierarchical arrangement of phyla or species in a scale of increasing complexity possibly correlative with evolutionary age. Some have gone further (see Noble, 1969) and attempted to make the same type of correlation within species, particularly human beings.

It is possible to order phyla, or species, in terms of particular aspects of function such as behavioral plasticity of the nervous system. Within any particular phylum or subphylum or class, it may be possible through behavioral studies to hypothesize how these species may have been related in evolutionary history to some common ancestor. However, no contemporaneous species may be termed evolutionarily "superior" to another, since each species has survived through various evolutionary processes to occupy the ecological habitat in which the species is supported.

Great variation in developmental patterns is found not only when phyla are compared, but even when taxa below the phylum category, including closely related species, are compared. All physiological systems and structures do not develop at the same rate. As Nice (1962) has illustrated in her analysis of developmental patterns in birds, precociality or altriciality as a stage of maturation at hatching needs to be specified in regard to physiological or anatomical systems. Altriciality or precociality is not absolute, as different systems mature at different rates. This is also true in mammals, as for example, in the order Rodentia. In this group of animals, there are many types of developmental patterns, as evidenced by various stages of development of sensory or motor function at birth. The guinea pig, an outstanding example of precociality in this order is born with eyes and ears functional at approximately the adult level in most regards, and fully able to locomote. The laboratory rat is primarily altricial in regard to motor, visual and auditory systems, but apparently "precocial" in regard to gustation and tangoception. It is not possible to arrange rigidly all the families of rodentia in regard to behavioral plasticity and developmental pattern. In addition to the difficulty of defining situations which might be considered comparable for the purpose of defining a behavioral continuum (Hirsch, 1962b), there has been no resolution of the problem of deciding which systems shall be used as the basis for comparing development pattern and rate (Tobach *et al.*, 1971).

As Gottlieb (1971) has pointed out, a comparison of the pattern of sensory development can point to possible evolutionary relationships among various animal classes. It is also possible to arrange species hierarchically in regard to behavioral plasticity. But, no relationship has been demonstrated among the rate of development of motor or sensorimotor systems, the evolutionary history and the behavioral plasticity for a particular species.

In the order Primates, such a general hierarchical statement can be made comparing prosimians and anthropoids, but it is difficult to do so within the anthropoids. Certainly, it is not possible to do so in a correlational respect between motor development and general behavioral plasticity. A comparison between people and subhuman primates would seem to support the correlation between motor development and behavioral plasticity, but it is clear again that the comparisons are between suborders, not within species. One might attempt to correlate behavioral difference and evolutionary age, as in the case of aplacental mammals (Metatheria) and placental mammals (Eutheria), which present very different developmental levels of behavioral plasticity (such differences have not yet been determined fully) and which arose in different eras (Cretaceous and Eocene, respectively). To generalize from this about equivalent relationships between other suborders of mammals or genera is an example of a kind of "typological" thinking. The attempt to make such generalizations about the evolutionary relationship among different human populations in terms of motor development and "intellectual" development is another example of typological thinking. The designation of some ethnic groups as evolutionarily more "primitive" than other ethnic groups on the basis of differences in motor development is derived from cultural concepts, rather than from biological concepts of the evolutionary processes.

The experimental concepts used in developmental genetics are particularly applicable to the understanding of the relationship between gene function and behavior. The study of behavioral development has proceeded for the most part on the molar level, that is, changes in behavior with growth and maturation of the entire organism. When development is viewed as a process in which the total experience of the organism during its entire life history is the agent for change as seen in growth and maturation, as Schneirla (1972) did, there can no longer be an artificial separation between the biochemical functions attributed to the genome and other types of experience. Schneirla conceived of experience as subsuming all levels of organization and integration, including the biochemical level of genetic processes. His statement is only the beginning of the study of genes and behavior, however. His conceptualization is the basis for analysis and synthesis of fundamental processes in behavior suggested above. The fact that such an analysis and synthesis is difficult and complex for any organism is obvious.

In the case of the most complex type of behavior evidenced by the most complex organism known today, that is, cognitive function (including creativity) in human beings, the problem of analysis of a behavior pattern into its subunits that might be traceable to specific gene action is even more difficult. The integration of these traceable subunits into patterns of behavior on levels more distantly related to gene action is obviously necessary, and difficult. Human behavior is characterized by complex societal institutions. The importance of societal factors in the classroom learning–teaching situation has been demonstrated (Katz, 1968; Leacock, 1969; Gordon and Wilkerson, 1966). The attitudes of school personnel as they are expressed in testing and teaching situations affect all test results, even those that are designed to be impervious to any but "genetic traits." To assume that "personality" or "motivational" assessments under contemporary conditions would not be similarly affected is naive. Performance on a test is not "individual" behavior; rather it is a societal act and needs to be analyzed as such, requiring therefore analysis of performance under conditions in which societal factors are defined and their effects understood.

On the human level, physiology and psychosocietal phenomena, such as classroom learning, stand in a hierarchical relationship to each other. While physiological function is the *sine qua non* for all other functions (a sick human being cannot function in the same way as a well human being), human behavior is psychosocietal (Tobach and Schneirla, 1968). In addition, as with any behavior, one needs to consider that at early stages of development the physiological processes of growth and maturation derived from the fusion of experience at all levels of individual function, may play relatively more important roles than the psychosocietal, which may operate indirectly. Protein deficiency in a parturient woman, or protein deficiency during the early development of the parturient woman, may result in relatively permanent impairment of neural function in any children borne by that woman (Birch and Gussow, 1970). It is obvious, however, that the protein deficiency itself might be a psychosocietal factor as well as some enzymatic deficiency in the woman, related to genetic process during her own growth and development. In the context of the levels concept, the medical treatment of the protein deficiency carried out on the physiological level is within the psychosocietal context.

How are we to approach the problem of analysis of behavior, with its underpinnings of physiological and more basic phenomena, as well as societal processes? The behavioral scientist needs to identify the problem and formulate questions about the problem that are answerable by experimental investigation. It is extremely necessary, however, for the behavioral scientists to recognize the boundaries between levels. Some phenomena need to be analyzed in the context of societal processes with the appropriate societal techniques, principles, theories, and procedures. Others, on the individual level, need to be studied by means of biochemistry and other tools of the physiological armamentarium. Because the human behavioral scientist, by virtue of the subject matter, is always operating on the human societal level, the reasons for posing the questions for investigation must be made explicit.

Why do we ask questions about genetics and human behavior? The investigation and explication of genetic processes in all forms of life and their relation to all life processes are vital to human survival. The aims of research to understand the role of genetic processes in human behavior are like the aims of all scientific investigation: that is, the closer approximation of reality to permit human control of the relationship with environment. On the human level, however, it is not possible to avoid the societal value system that

9. THE MEANING OF THE CRYPTANTHROPARION

sets the aims of such investigations. The medical profession's use of concentration camp inmates for research without concern for the survival of the individual or the willing cooperation of the subject of the experiment is one example of value systems that may underlie scientific research.

Such questions might be asked for the purpose of improving the "national human quality" (Ad Hoc Committee Report, 1972). Whether this is to be done as Shockley (1971)[2] proposes is not always made explicit. It also may be that we ask these questions about genetic mechanisms in learning-behavior and performance on intelligence tests because we do not know how to solve our failures in the educational system (Leacock, 1969; Gordon and Wilkerson, 1966).

I propose that the "interaction" formulation put before this Conference is inadequate. I suggest that the question before the behavioral scientist is to understand the development and evolution of the behavior pattern under consideration. To do this, investigations need to be carried out on all levels of organization. No level is "more important" than any other. Extrapolation from one to another cannot be made superficially. Because of the complexity of the research proposed, the question of logistics becomes important.

The Logistics of Research in the Problems of Human Behavior and Genetics

Logistics is used here to refer to the deployment of scientific personnel and its support (facilities, equipment and supplies) as in the original meaning of the word in regard to the deployment of troops.

Intense societal concern has been generated about research in ethnic differences in performance on intelligence tests. The problem of logistics is forced on the Conference because of the contention that financial support is withheld from research that produces results that are possibly pejorative to some segment of the population of our country (Ad Hoc Committee, Na-

---

[2] "The First Amendment makes it safe for us in the United States to try to find humane eugenic measures. As a step in such search, I propose as a *thinking* exercise a voluntary, sterilization bonus plan.

Bonuses will be offered for sterilization. Income tax-payers get nothing. Bonuses for all others, regardless of sex, race, or welfare status, would depend on best scientific estimates of heredity factors in disadvantages such as diabetes, epilepsy, heroin addiction, arthritis, etc. At a bonus rate of $1000 for each point below 100 IQ, $30,000 put in trust for a 70 IQ moron of twenty-child potential might return $250,000 to tax-payers in reduced costs of mental retardation care. Ten percent of the bonus in spot cash might put our national talent for entrepreneurship into action.

A motivation boost might be to permit those sterilized to be employed at below minimum standard wages without any loss of a welfare floor income. Could this provide opportunity for those now unemployable [Shockley, 1971]?"

tional Academy of Sciences, 1972). I would like to pose some questions for discussion in this regard.

Is an individual's scientific program ever independent of the societal millieu? Are scientists free to pursue "truth" wherever it will lead them? Does the scientific "ambience" affect the course of an individual's research? Do considerations of economic restriction and societal conflict play a role in an individual's choice of a research question? Are scientists the ultimate decision-makers in research logistics?

The question posed by the scientist for investigation will determine the response of societal institutions, as the latter see their needs being fulfilled by the results of the investigation. The scientist needs to be fully aware of this, indeed is continually being reminded by the "justification" section of all grant applications. By consciously making the decision about the problem to be studied, the scientist becomes equally responsible with the societal institution. If the psychometrician accepts the assignment of determining the learning potential of a child before entering school, so that the societal institution can decide what type of schooling the child shall have, the amount of social effort, money, etc., the child is deemed worthy of, then the psychometrician asks research questions that will guarantee support by those societal institutions that want those answers. This relationship between scientist and society is a powerful factor in the logistics of research.

I would like to suggest that the problem might be defined as follows. People should be free to question all and any preconceived or established ideas. This is true in all parts of society, as well as in the scientific community. For example, it is the right of nonscientists to question what scientists believe and to question the priorities that scientists set for themselves. Who shall say how the national budget should be assigned in regard to research or services? Scientists also have the right to try to answer freely and without restraint any questions they may have. But scientists, just like others in society, should operate within an ethics system that governs all people, regardless of their type of work. When we say that the society has a right to question the act of any individual or group, we expect society to do so without endangering the lives of people, without circumscribing individual freedom, and without the dehumanization of people. Scientists similarly are constrained. The research done with captive populations by the Nazis, or by certain unscrupulous scientists today in our prisons, or in our "mental retardate" wards is to be criticized as we would criticize people who wish to change society without regard to an acceptable egalitarian nonexploitative ethic. It is questionable that only scientists should be involved in decisions in choosing research projects, when all of society is asked to support them. When the scientist's work becomes part of human culture as ideas, inferences, theories, and basis for societal action, the scientist is but one participant in the societal process of decision making. How shall scientists and nonscientists relate to each other in regard to the questions and formulations raised above?

It seems desirable to start from the premise that all people have the right to be told all that the scientist knows in order for society to define its priorities in line with its value system. It is equally necessary for the scientific community to inform all people about what it is not sure of, what it has doubts about, and of the controversies and disagreements that scientists have among themselves. The scientist may be as wrong in doing a particular piece of scientific work as any nonscientist may be in doing some other kinds of work. One important factor that operates in both cases (scientific and nonscientific work) is the directness of the negative feedback. One cannot devise an effective machine based on Boyle's Law, if Boyle's Law is incorrect. The engineer in that instance is immediately required to reinvestigate the basic premise, that is, the use of Boyle's Law. Similarly, if the nonscientist operates on a principle that defies or negates reality, at some point the basic assumptions underlying the behavior involved will have to be reexamined. In these instances the efficiency and accuracy with which these operations are carried out are a function of the many factors which are involved in behavior in general, either on a personal level or on a societal level. Profound mistakes in judgment may be made, and the processes by which these mistakes are made or avoided are complex and worthy of much discussion. The scientist at this stage of human history has as little basis for understanding those processes as has any other member of the human population.

It is important to note that the negative feedback check described above is most frequently missing in the behavioral sciences. Because human behavioral science involves communication of and working with ideas via language in a sociocultural setting, it is easier to convince each other about "reality" without empiric evidence (Eisenberg, 1972, in press.)

The Relationship of the Scientific Problem to Society

It is necessary to have a clear understanding of the relationships among the pursuit of scientifically valid research, the interpretation of the facts gathered in that way, and the use of the facts by society.

It has been suggested that certain groups in the United States are demanding that research results prejudicial to their group should not be published. This suggestion emanates most frequently from scientists who view all behavioral phenomena as ultimately explainable by genetic mechanisms. Other scientists who accept this explanation conclude that it is critical that gene mechanisms be understood, but genetic processes should be dismissed when the behavioral patterns of populations are being considered. Still others state that it is critical that genetics be understood so that steps may be taken to change the genetic pool of certain populations (Shockley, 1971; Herrnstein, 1971).

In these varied viewpoints, one sees the spectrum that can result from the same understanding of a "purely scientific" fact. Clearly, the value system of the scientist plays a significant role in the use to which the "fact" will be put. Scientists need to involve themselves in societal processes that directly affect their activities and in the development of value systems for society. It would appear to be imperative that scientists do this in order to bring about an effective and desirable relationship between their work and the use to which it is put by society.

Final Statement

What is the real meaning of the cryptanthroparion? The prefix "crypt" could refer to what is hidden in the little person (anthroparion); to the fact that the person is hidden in something; or that the person needs to be decoded in order to be understood. What is hidden in the crypthanthroparion? No one would deny the existence and the necessity of the material transmitted from progenitor to the progeny; that is, without the material being transmitted, there is no progeny. But, that is the *sine qua non* of all matter, animate and inanimate. Every phenomenon has a history. All matter is derived from other matter, though the form varies.

In a sense, the "little person" is "hidden" in the organism, at every stage of development. That is, what exists will always be represented to some extent and in some way in what is about to become. Also, what exists now has within it what went before. In the case of sexual vertebrate reproduction, the zygote "is" the sperm and the egg. Neither the egg nor the sperm is organized structurally or functionally to look or act like the zygote. The embryo, the foetus, the neonate, the juvenile, the adult, never look or act exactly like the individual will look and act at a later point in time. The rate of change will vary; the number of parameters changing will vary; the number remaining relatively stable will vary; but at no time is the individual the same as it was a moment ago.

But, one might say, there are some elements that stay the same from the stage of sperm and egg, and are always identifiable as the same both in structure and function. These are genes, which either endlessly replicate themselves in cells that are continually dying and becoming reborn, or genes which do not replicate themselves but remain structurally stable and continue to function as *de novo*. These stable systems are considered by some to be the hidden destiny of each organism.

However, as far as we know now, this apparent stability is dependent upon certain characteristics of the milieu remaining relatively stable, or changing appropriately to present the conditions under which genes can function.

The need to depend on the cryptanthroparion as the ultimate causal explanation actually stems from the philosophical view of a mind separate from a body, despite its apparent reliance on so physiological and corporeal a system as a gene. In this kind of thinking, the gene becomes a *deus ex machina* that creates a mental ability that does not reside in behavior; neither is it behavior. Rather it is some immaterial, vital force that one can only attempt to measure by inference.

Mental abilities and intelligence, when they are equated, presumably bear some relationship to "mind." As mental abilities, or intelligence, do not "reside" in or "consist of" behavior, it is difficult to see how one will be able to measure these if one can only do so through behavior. What then is the meaning of the cryptanthroparion?

Hidden in the little person is the future of the little person. This meaning of the cryptanthroparion is the critical one: to understand the organism—its structure and behavior—it is necessary to unravel its history, to determine how the various levels of organization were differently integrated in time to bring about what was observable at any point in time. While the current stage is being studied and identified, the "little being" is changing. It is there, but not completely so, as it were. The "little person" is the last stage before the next; it has within it the possibility of going on to the next stage. The "inside" and the "outside" necessary to bring it to the next stage is what has to be analyzed. The message is clear, but the solution is difficult.

Acknowledgments

This manuscript was prepared through partial support by NIMH 3KMH-21867.
I want to thank all my friends who discussed various aspects of the paper with me. They are, of course, not responsible for the errors therein. I particularly wish to thank the following for their critical reading: Judith S. Bellin, Lawrence Berlowitz, Alice Cottingham, Irving Crain, Lee Ehrman, Florence Halpern, Monica Hunt, Eleanor Leacock, Izja Lederhendler, Josephine Martin, Doris K. Miller, Cecelia Pollack, Ira Pomerance, Clara Rabinowitz, Martin Richards, Joel R. Seldin, Sylvia Scribner, Mark Smith, Salome Waelsch, Doxey Wilkerson. Most of all I wish to acknowledge the many theoretical discussions with Howard R. Topoff, which affected much of my thinking, and did much to improve this paper.

References

ABEELEN, J. H. F. van. (1968). Behavioural ontogeny of looptail mice. *Anim. Behav.* **16**, 1–4.

ABEELEN, J. H. F. van. (1970). Genetics of rearing behavior in mice. *Behav. Genet.* **1**, 71–76.

ABEELEN, J. H. F. van, SMITS, A. J. M., and RAAIJMAKERS, W. G. M. (1971). Central location of a genotype-dependent cholinergic mechanism controlling exploratory behavior in mice. *Psychopharmacologia,* **19**, 324–328.

Ad Hoc Committee on Genetic Factors in Human Performance. (1972). *Proc. Nat. Acad. Sci.* **69**, 1–3.

ALTMAN, J., DAS, G. D., and SUDARSHAN, K. (1970). The influence of nutrition on neural and behavioral development. I. Critical review of some data on the growth of the body and the brain following dietary deprivation during gestation and lactation. *Develop. Psychobiol.* **3**, 281–301.

ALTMAN, J., DAS, G. D., SUDARSHAN, K., and ANDERSON, J. B. (1971). The influence of nutrition on neural and behavioral development. II Growth of body and brain in infant rats using different techniques of undernutrition. *Develop. Psychobiol.* **4**, 55–70. (a)

ALTMAN, J., SUDARSHAN, K., DAS, G. D., McCORMICK, N., and BARNES, D. (1971). The influence of nutrition on neural and behavioral development: III. Development of some motor, particularly locomotor patterns during infancy. *Develop. Psychobiol.* **4**, 97–114. (b)

ANDERSON, E. N., Jr., and REED, T. E. (1969). Exchange of letters. *Science,* **166**, 1353.

BENEDICT, R., and WELTFISH, G. (1943). "The Races of Mankind." Public Affairs Pamphlet, No. 85.

BERTALANFFY, L. von. (1962). "Modern Theories of Development." Harper, New York.

BENNE, K. D. (ed.) (1965). The social responsibilities of the behavioral scientist. *J. Soc. Issues,* **21**, 1–84.

BIRCH, H. G. (1971). Levels, categories, and methodological assumptions in the study of behavioral development. *In:* "The Biopsychology of Development," (E. Tobach, *et al.,* eds.). Academic Press, New York.

BIRCH, H. G., and GUSSOW, J. D. (1970). "Disadvantaged Children: Health, Nutrition, and School Failure." Harcourt, New York.

BLOOM, B. S. (1969). Letter in the Harvard Educational Review in reply to A. R. Jensen, cited in ERIC, *Nat. Lab. Child. Educ.* **3**, #4.

BODMER, W. F., and CAVALLI-SFORZA, L. L. (1970). Intelligence and race. *Sci. Amer.* **223**, 19–29.

BONNER, J. T. (1971). The direction of developmental biology. *Current Topics in Develop. Biol.* **6**, xv–xx.

BROADHURST, P. L. *et al.* (1971). New lights on behavioural inheritance. *Bull. Br. Psychol. Soc.* **24**, 1–8.

BROWN, C. P., and KING, M. G. (1970). Developmental environment: Variables important for later learning and changes in cholinergic activity. *Develop. Psychobiol.* **4**, 275–286.

BURNHAM, D. (1971). Jensenism: The new pseudoscience of racism. *Freedomways,* 150–157.

CASPARI, E. (1971). Differentiation and pattern behavior in the development of behavior. *In:* "The Biopsychology of Development," (E. Tobach, *et al.* eds.). Academic Press, New York.

CATTEL, J. McK. (1906). A statistical study of American men of science. III. *Science,* **24**, 732–742.

CATTELL, J. McK. (1910). A further study of American men of science. *Science*, , 633–648.

COLLINS, R. A. (1970). Experimental modification of brain weight and behavior in mice: an enrichment study. *Develop. Psychobiol.* **3**, 145–155.

COLLINS, R. L. (1970). A new genetic locus mapped from behavioral variation in mice: audiogenic seizure prone (ASP). *Behav. Genet.* **1**, 99–109.

CRONBACH, L. J. (1970). Mental tests and the creation of opportunity. *Proc. Amer. Phil. Soc.* **114**, 480–487.

CROW, J. F., Neel, J. V., and STERN, C. (1967). Racial studies: Academy states position on call for new research. *Science,* **158**, 892–893.

DARLINGTON, C. D. (1971). The limited pool of human talent. The New York Times, August 10, p. 31.

DAVID, E. E. (1971). Quoted by "P.M.B." in *Science,* **171**, 875.

DAVIS, B. D. (1970). Prospects for genetic intervention in man. *Science,* **170**, 1279–1283.

DAVIS, K., DOBZHANSKY, T., GERALD, R. W., GLASS, H. B., HILGRAD, E. R., NEEL, J. V., SIMON, H. A., and TUKEY, J. W. (1972). Recommendations with respect to the behavioral and social aspects of human genetics. *Proc. Nat. Acad. Sci.* **69**, 1–3.

DELIUS, J. D. (1970). The ontogeny of behavior. *In:* "The Neurosciences, Second Study Program," (F. O. Schmitt, ed.) pp. 188–196. Rockefeller Univ. Press, New York.

DOBZHANSKY, T. (1963). Evolutionary and population genetics. *Science,* **142**, 1131–1135.

DOBZHANSKY, T. (1968). On Cartesian and Darwinian aspects of biology. The Graduate *J. Univ. Tex.,* 1968, **8**, 99–119.(a)

DOBZHANSKY, T. (1968). On genetics, sociology and politics. *Perspect. Biol. Med.* **11**, 544–554.(b)

DREGER, R. M. (1969). Letter on article by C. E. Noble. *Perspect. Biol. Med.* **13**, 117–121.

EDWARDS, H. P., BARRY, W. F., and WYSPIANSKI, J. O. (1969). Effect of differential rearing on photic evoked potentials and brightness discrimination in the albino rat. *Develop. Psychobiol.* **2**, 133–138.

EHRMAN, L. (1966). Mating success and genotype frequency in Drosophila. *Anim. Behav.* **14**, 332–339.

EISENBERG, L. (1971). Persistent problems in the biopsychology of development. *In:* "The Biopsychology of Development," (E. Tobach, *et al.* eds.). Academic Press, New York.

EISENBERG, L. (1972). The nature of human nature. *Science,* (in press).

EWING, A. W. (1964). The influence of wing area on the courtship behaviour of Drosophila melanogaster. *Anim. Behav.* **12**, 316–320.

FOREST, J. M. S. (1970). The effect of maternal and larval experience on morph determination in *Dysaphis devecta. J. Insect Physiol.* **16**, 2281–2292.

FOX, M. W. (1970). Neurobehavioral development and the genotype-environment interaction. *Quart. Rev. Biol.* **45**, 131–147.

FULKER, D. W. (1970). Maternal buffering of rodent genotypic responses to stress: a complex genotype-environment interaction. *Behav. Genet.* **1**, 119–124.

FULKER, D. W., WILCOCK, J., and BROADHURST, P. L. (1971). Studies in genotype-environment interaction: I. Methodology and preliminary multivariate analysis of a diallel cross of eight strains of rat. *Behav. Genet.* (in press).

GALTON, F. (1952). "Hereditary Genius." Horizon Press, New York.
GINSBURG, B. E. (1965). Book review of genetics and social behavior of the dog. (John Paul Scott and John L. Fuller, Univ. of Chicago Press, 1965). *Eugen. Quart.* **13**, 366–369.
GLASS, D. C. (ed.). (1968). "Genetics." The Rockefeller Univ. Press and Russell Sage Foundation, New York.
GOLDEN, M., and BRIDGER, W. (1969). A refutation of Jensen's position on intelligence, race, social class and heredity. *Ment. Hyg.* **53**, 648–653.
GOLDMAN, P. S. (1965). Conditioned emotionality in the rat as a function of stress in infancy. *Anim. Behav.* **13**, 434–442.
GOLDMAN, P. S., and TOBACH, E. (1967). Behaviour modification in infant rats. *Anim. Behav.* **15**, 559–562.
GORDON, E. W. (1971). Ethnicity, intelligence and education. Paper presented at Jefferson Medical College. February 3.
GORDON, E. W., and WILKERSON, D. A. (1966). Compensatory education for the disadvantaged. N. Y. College Entrance Board.
GOTTESMAN, I. I. (1968). Biogenetics of race and class. *In:* "Social Class, Race and Psychological Development," (M. Deutsch *et al.* eds.). Holt, New York.
GOTTLIEB, F. J. (1966). "Developmental Genetics." Reingold Publ., New York.
GOTTLIEB, G. (1971). Ontogenesis of sensory function in birds and mammals. *In:* "The Biopsychology of Development," (E. Tobach, *et al.* eds.). Academic Press, New York.
GRANT, M. (1916). "The Passing of the Great Race." Scribner's New York.
HARRINGTON, G. M. (1968). Genetic-environmental in'eraction in "intelligence." I. Biometric genetic analysis of maze performance of Rattus norvegicus. *Develop. Psychobiol.* **1**, 211–218.
HARRINGTON, G. M. (1968). Genetic-environment interaction in intelligence: II. Models of behavior, components of variance and research strategy. *Develop. Psychobiol.* **1**, 245–253.
HAYES, W. L. (1963). Statistics for Psychologists. Holt, New York.
HERRNSTEIN, R. (1971). I. Q. *The Atlantic,* **228**, 44–64.
HIMWICH, W. A. (1971). Biochemical processes of nervous system development. *In:* "The Biopsychology of Development," (E. Tobach, *et al.* eds.). Academic Press, New York.
HIRSCH, J. (1962). Discussion. *Amer. J. Orthopsychiat.* **32**, 891–892. (a)
HIRSCH, J. (1962). The contribution of behavior genetics to the study of behavior. *In:* "Expanding Goals of Genetics in Psychiatry," (F. J. Kallmann, ed.). Grune and Stratton, New York. (b)
HIRSCH, J. (ed.) (1967). "Behavior-Genetic Analysis." McGraw-Hill, New York.
HIRSCH, J. (1970). Behavior-genetic analysis and its biosocial consequences. *Semin. Psychiat.* **2**, 89–105.
HOLLOWAY, R. L., and INGLE, D. J. (1967). Exchange of letters. *Perspect. Biol. Med.* **10**, 679–682.
HOROWITZ, N. H. (1966). Perspectives in medical genetics. *Perspect. Biol. Med.* **9**, 349–357.
INGLE, D. J. (1964). Racial differences and the future. *Science,* **146**, 375–379.
INGLE, D. J. (1965). The 1964 UNESCO proposals on the biological aspects of race: a critique. *Perspect. Biol. Med.* **8**, 403–408.
INGLE, D. J. (1967). Editorial: The need to study biological differences among racial groups: Moral issues. *Perspect. Biol. Med.* **10**, 497–499.

INGLE, D. J. (1971). Causality. *Perspect. Biol. Med.* **14**, 410–423.
JAHODA, G. (1970). A cross-cultural perspective in psychology. *Advan. Sci.* **27**, 1–14.
JENSEN, A. R. (1968). Social class, race and genetics: Implications for education. *Amer. Educ. Res. J.* **5**, 1–42. (a)
JENSEN, A. R. (1968). Introduction to biogenetic perspectives. *In:* "Social Class, Race, and Psychological Development," (M. Deutsch, *et al.* eds.) Holt, New York. (b)
JENSEN, A. R. (1969). How much can we boost IQ and scholastic achievement? *Harvard Educ. Rev.* **39**, 1–123.
JENSEN, A. R. (1970). Race and the genetics of intelligence: a reply to Lewontin. *Bull. Atom. Sci.* **26**, 17–23. (a)
JENSEN, A. R. (1970). IQ's of identical twins reared apart. *Behav. Genet.* **1**, 133–148. (b)
KATZ, I. (1968). Some motivational determinants of racial differences in intellectual achievement. *In:* "Science and the Concept of Race," (M. Mead, T. Dobzhansky, E. Tobach, R. E. Light eds.). Columbia Univ. Press, New York.
KING, J. C. (1971). "The Biology of Race." Harcourt, New York.
KLOPFER, P. H. (1969). Instincts and chromosomes: What is an innate act? *Amer. Natur.* **103**, 556–560.
KUO, Z. Y. (1970). The need for coordinated efforts in developmental studies. *In:* "The Development and Evolution of Behavior," (L. R. Aronson, *et al.* eds.). Freeman, San Francisco.
LEACOCK, E. (1969). "Teaching and Learning in City Schools: A Comparative Study." Basic Books, New York.
LEDERBERG, J. (1966). Experimental genetics and human evolution. *Bull. Atom. Sci.* **22**, 4–11.
LEHRMAN, D. S. (1970). Semantic and conceptual issues in the nature-nurture problem. *In:* "The Development and Evolution of Behavior," (L. R. Aronson, *et al.* eds.). Freeman, San Francisco.
LEVINE, S., CHEVALIER, J. A., and KORCHIN, S. J. (1956). The effects of early shock and handling on later avoidance learning. *J. Personal.* **24**, 475–493.
LEWONTIN, R. C. (1970). Race and intelligence. *Bull. Amer. Sci.* **26**, 1–8. (a)
LEWONTIN, R. C. (1970). Further remarks on race and the genetics of intelligence. *Bull. Atom. Sci.* **26**, 23–25. (b)
LINDZEY, G., and THIESSEN, D. D. (eds.) (1970). "Contributions to Behavior-Genetic Analysis. The Mouse as a Prototype." Appleton, New York. (a)
LINDZEY, G., and THIESSEN, D. D. (1970). Genetic aspects of negative geotaxis in mice. *Behav. Genet.* **1**, 21–34. (b)
MacINNES, J. W., BOGGAN, W. O., and SCHLESINGER, K. (1970). Seizure susceptibility in mice: differences in brain ATP production in vitro. *Behav. Genet.* **1**, 35–42.
MASS, J. W. (1962). Neurochemical differences between two strains of mice. *Science*, **137**, 621–622.
MAYR, E. LETTER. (1971). *Perspect. Biol. Med.* **14**, 505–506.
MEAD, M. *et al.* (eds.) (1968). "Science and the Concept of Race." Columbia Univ. Press, New York.
MONROY, A. (1970). Developmental biology and genetics: A plea for cooperation. *In:* "Current Topics in Developmental Biology," (A. A. Moscona, and A. Monroy eds.), 5, xvii-xxi. Academic Press, New York.

MOTULSKY, A. G. (1968). Human genetics, society and medicine. *J. Hered.* **59**, 329–336.
MURPHY, E. A. (1966). A scientific viewpoint on normalcy. *Perspect. Biol. Med.* **9**, 333–348.
Nature New Biology, (1971) **232**, 99. Is behavior inherited?
NEEDHAM, J. (1956). "Science and Civilizations in China. Vol. 2. The History of Scientific Thought." Cambridge Univ. Press, London and New York.
NICE, M. M. (1962). Development of behavior in precocial birds. *Trans. Linn. Soc., N. Y.* **8**, 211 pp.
NOBLE, C. E. (1969). Race, reality and experimental psychology. *Perspect. Biol. Med.* **13**, 10–30.
OMENN, G. S., and MOTULSKY, A. G. (1971). Brain, genetics and behavior. Report of a meeting at La Jolla, California, supported by the National Genetics Foundation. *Science,* **173**, 1255–1256.
OTTINGER, D. R., and TANABE, G. (1969). Maternal food restriction: Effects on offspring behavior and development. *Develop. Psychobiol.* **2**, 7–9.
PASSAMANICK, B. (1968). Some sociobiologic aspects of science, race and racism and their implications. Presidential address, American Psychopathological Association, February 16.
PRZYBOA, E., ROGUSKI, H., SKOWRON, S., and SKOWRON-CONDRZAK, A. (1971). Problems of heredity in the light of developmental physiology. Public lecture before III Conference of the Polish Genetic Society, Sept. 17, 1970. *Folia Biol.* **19**, 107–118.
RABINOWITCH, E. (1970). Jensen vs. Lewontin. *Bull. Atom. Sci.* **26**, 25–26.
REED, C. F., WITT, P. N., SCARBORO, M. B., and PEAKALL, D. B (1970). Experience and the orb web *Develop. Psychobiol.* **3**, 251–265.
REED, T. E. (1969). Caucasian genes in American negroes. *Science,* **165**, 762–768.
RICHARDS, M. RICHARDS, K., and SPEARS, D. (eds.). (1972). "Intelligence and society. In Race, Cul ure and Intelligence." Penguin.
RICK, J. T., TUNNICLIFF, G., KERKUT, G. A., FULKER, D. W., WILCOCK, J., and BROADHURST, P. L. (1971). GABA production in brain cortex related to activity and avoidance behaviour in eight strains of rat. *Brain Res.* **32**, 234–238.
RIEGE, W. H. (1971). Environmental influence on brain and behavior of year-old rats. *Develop. Psychobiol.* **4**, 157–167.
RIESEN, A. H. (1969). Editorial: between simplistic nativism and extreme empiricism. *Develop. Psychobiol.* **2**, 1.
RIESENFELD, A. (1970). On some racial features in rats, dogs and men. *Homo.* **21**, 163–175.
RODGERS, D. A. (1970). Mechanism-specific behavior. An experimental alternative. *In:* "Contributions to Behavior-Genetic Analysis. The Mouse as a Prototype," (G. Lindzey, and D. D. Thiessen eds.), pp. 207–218. Appleton, New York.
ROSENZWEIG, M. R. (1969). Editorial: Human rights and social responsibilities in psychobiological research. *Develop. Psychobiol.* **2**, 131–132.
ROSENZWEIG, M. R. (1971). Effects of environment on development of brain and of behavior. *In:* "The Biopsychology of Development," (E. Tobach, et al. eds.). Academic Press, New York.
SARTON, G. (1927). Introduction to the History of Science." Volumes I, II and III. Williams and Wilkins, Baltimore.
SCHEIBEL, A. B., and SCHEIBEL, M. E. (1968). Editorial: The biologic uniqueness of the organism. *Develop. Psychobiol.* **1**, 223–224.

SCHLESINGER, K., and GRIEK, B. J. (1970). The genetics and biochemistry of audiogenic seizures. *In:* "Contributions to Behavior-Genetic Analysis. The Mouse as a Prototype," (G. Lindzey, and D. D. Thiessen eds.), pp. 219–258. Appleton. New York.
SCHNEIRLA, T. C. (1933). Some comparative psychology, *J. Comp. Psychol.* **16**, 307–315.
SCHNEIRLA, T. C. (1957). The concept of development in comparative psychology. *In:* "The Concept of Development," (D. B. Harris, ed.), pp. 78–108. Minnesota Univ. Press, Minnesota.
SCHNEIRLA, T. C. (1962). Psychology, comparative. *Encycl. Brit.* **18**, 690–703.
SCHNEIRLA, T. C. (1965). Aspects of stimulation and organization in approach/withdrawal processes underlying vertebrate behavioral development. In *Adv. Study. Behav.* **1**, 1–71.
SCHNEIRLA, T. C. (1972). "Selected Writings," (L. R. Aronson, E. Tobach, D. S. Lehrman, and J. S. Rosenblatt eds.), Parts III and IV. Freeman, San Francisco.
Scientific Research. Shockley finds proposal for research an uphill fight. pp. 17–19.
SHOCKLEY, W. (1968). Letter. *Science,* **160**, 443.
SHOCKLEY, W. (1969). Letter. The New York Times, November 8.
SHOCKLEY, W. (1970). Letter. The New York Times, January 1.
SHOCKLEY, W. (1971). Dysgenics—a social-problem reality evaded by the illusion of infinite plasticity of human intelligence? Paper prepared for a symposium on "Social problems: Illusion, delusion or reality." Amer. Psychol. Ass. Nat. Conv. September 7.
SHOOTER, E. M. (1970). Some aspects of gene expression in the nervous system. *In:* "The Neurosciences, Second Study Program," (F. O. Schmitt, ed.). Rockefeller Univ. Press, New York.
SINGER, C. (1917). "Studies in the History and Method of Science," Vol. I. Clarendon Press, Oxford.
STENT, G. S. (1970). DNA.*Daedalus* **99**, 909–937.
STERN, C. (1970). The continuity of genetics. *Daedalus,* **99**, 882–908.
STORRS, E. E., and WILLIAMS, R. J. (1968). A study of monozygous quadruplet armadillos in relation to mammalian intelligence. *Proc. Nat. Acad. Sci.* **60**, 910–914.
TOBACH, E. (1972). Social Darwinism rides again. Paper delivered at American Orthopsychiatric Association meeting, April 8.
TOBACH, E., and SCHNEIRLA, T. C. (1968). The biopsychology of social behavior in animals. *In:* "The Biologic Basis of Pediatric Practice," (R. E. Cooke, ed.), pp. 68–82. McGraw-Hill, New York.
TOBACH, E. *et al.* (eds.). (1971). "The Biopsychology of Development." Academic Press, New York. (a)
TOBACH, E., ARONSON, L. R., and SHAW, E. (eds.). (1971). "The Biopsychology of Development." Academic Press, New York. (b)
TYLER, P. A., and McCLEARN, G. E. (1970). A quantitative genetic analysis of runway learning in mice. *Behav. Genet.* **1**, 57–69.
WEHMER, F., PORTER, R. H., and SCALES, B. (1970). Prenatal stress influences the behavior of subsequent generations. *Comm. Behav. Biol.* **5**, 211–214.
WILLIAMS, R. J., and PELTON, R. B. (1966). Individuality in nutrition: effects of Vitamin-A-deficient and other deficient diets on experimental animals. *Proc. Nat. Acad. Sci.* **55**, 126–134.

DISCUSSION

ARTHUR R. JENSEN

University of California
Berkeley, California

Among the many topics brought up in Dr. Tobach's paper, the most challenging, perhaps, is the question of what Tobach calls the "logistics" of research in human behavior genetics. How can we go about finding out more than we already know? We want to advance our knowledge of the genetics of human behavior beyond mere statements to the effect that genetic factors are involved in many human behavioral characteristics.

There are three main problems. First, in research with humans, breeding experiments are not possible, or at least not feasible. We are forced to study the human material that nature provides. Second, complete control or randomization of environmental effects is unfeasible; we are more or less limited to studying persons in their "natural habitats." The best we can hope for at present is to seek out naturally occurring situations which more or less approximate experimental control of environmental effects, as in the case of adopted children, orphanage children, and identical twins reared apart. Third, most of the characteristics in which we are especially interested, such as mental abilities and personality traits, are continuously distributed in the population, at least at the gross phenotypic level of observation and measurement. Quite simple polygenic models fit these kinds of data very well, but about all they have permitted thus far is the estimation of heritability and in some cases the partitioning of the genetic variance into additive, dominance, and epistatic components, along with estimates of genotype × environment interactions and covariances.

If this is as far as we can go with the traits of greatest interest in human populations, the prospect does not appear very exciting or challenging to the researcher. We want to achieve more than a mere catalogue of heritability estimates and variance components of the lexicon of human traits. We want to get closer to the genes, as it were, and to hold out some hope for discovering the actual pathways from genes to behavior.

The question can be raised: Are some traits truly polygenic—is this actually the state of nature for, say, height or intelligence? The possibility that some traits may be due to a number of genes each with small, equal, and additive effects has not been ruled out. But perhaps it is best not to *begin* with the assumption that the particular trait we wish to study is irreducibly polygenic. If one begins with this assumption, then he is practically committed to remaining at the level of biometrical genetic analysis. It may be more heuristic (though also possibly more futile) to work on the hope that the trait in question, though seemingly polygenic, is potentially analyzable into a number of Mendelian characters. This is apparently the geneticists' summum bonum. If it is a false hope, we can find out only by trying. I know of no scientific laws or principles which a priori make it a false hope.

At this point I shall propose for consideration a set of steps which might be said to represent in general terms one possible logistics for human genetic-behavioral research on continuously distributed traits.

Establishing the Relevance of the Trait. This merely assumes that the class of behavior selected for study is of interest or importance for whatever reason. Few would question the relevance of individual differences in mental abilities, for example, while individual differences in walking gait may be viewed as trivial by comparison.

Establishing Reliability of Measurement. This is obvious. Unless there is a means of objectively and reliably identifying and measuring variation in the characteristic in question, it cannot be studied further. Stable, hardy traits seem preferable to ephemeral, capricious behavior as material for genetical analysis. In this respect, I list below some of the possible characteristics of psychological tests, particularly ability tests, which I think should be sought if they are to be considered for extensive genetical analysis:

1. *Tests that permit practice and repeated testing.* Some tests cease to yield meaningful scores if the subjects practice on the test materials, or practically all subjects reach a common criterion, eliminating variance, and so on.

2. Tests on which the subject's response is continuously graded, as compared with items on which the subject's response is either "right" or "wrong." The latter type of test is "reactive" on the behavior you are trying to measure. A sequence of responses perceived as wrong by the subject can lead to discouragement, thus dragging in other personality and motivational factors, which may contribute unwanted variance to individual differences in the trait you actually wish to measure, say, "intelligence."

3. *Tests which maintain the same meaning (that is, the same factorial composition, scale properties, etc.) over a considerable age-range, especially during the school-age years.* This is important for the use of sibling correlations, parent–child correlations, and for longitudinal developmental studies. Tests that are nominally the same but actually measure different things in different age groups have more limited usefulness in genetic research.

4. *Tests which "read through" sensorimotor variability.* The abilities of greatest interest are essentially independent of the sensory input and motor output channels. Unless we are specifically concerned with individual differences in sensorimotor capacities, those ability tests are regarded as best which reflect as little of the variance in sensorimotor skills as possible. A good criterion is whether the same ability as measured by one test also shows up the same individual differences on another test which calls for quite different sensorimotor capabilities. I have found, for example, that individual differences in memory span for numbers are the same whether the numbers are presented visually or aurally. In other words, either version of the test is measuring a central memory process and not individual differences in sensory abilities. Similarly, on the motor side, the Gesell Figure Copy test (in which the subject must copy 10 geometric forms of progressively increasing complexity) yields essentially the same results (essentially a mental age score) whether the subject makes the drawings with his preferred hand, his nonpreferred hand, or his foot. Again, some central process is measured, not just a peripheral motor skill. In general, it should be kept in mind that in ability testing (excluding testing of specific sensorimotor skills), behavior is the medium not the message. Intelligence, for example, is independent of any one sensory or motor modality. It does not *consist* of behavior and does not *reside* in behavior, and behavioral analysis per se will tell you very little about it. Very different behavioral phenotypes (e.g. number series test and block design test) can be highly correlated because they all reflect essentially the same central process, which is what we are really interested in when we study mental abilities.

In choosing tests, one must keep in mind a distinction between the usefulness of tests for practical prediction and for research purposes. A test that is good in the one case may not be good in the other. The Full Scale IQ yielded by the Wechsler Intelligence Scale, for example, has higher predictive validity for scholastic and occupational performance than the much narrower and more homogeneous Raven Progressive Matrices. But I would prefer the latter as a research instrument. The Wechsler score is a very complex composite with unknown weighting of all the various processes that enter into it. The Raven gets at a much more unitary kind of ability, though it is still more complex than one might like.

Heritability Estimation. A heritability study may be regarded as a Geiger counter with which one scans the territory in order to find the spot one can most profitably begin to dig for ore. Characteristics with low heritability are less likely to yield pay dirt. The reason, of course, is that all we have to work with, at least at the beginning of our investigation, is *variance,* and if what we are interested in is genetical analysis, we would like to know that some substantial proportion of the trait variance we are interested in is attributable to genetic factors. So we should not belittle heritability studies, but they should be regarded as only the beginning rather than as the goal of our efforts in genetical analysis.

Fractionation of Abilities. If a test score represents an amalgam of a number of psychological processes in each of which there are imperfectly correlated and genetically conditioned individual differences, we are stuck with our Fisherian polygenic model. Thus our aim should be to fractionate our ability measurements so as to get at smaller and more unitary components of ability. This is the province of the differential psychologist, but it requires also the methods of experimental psychology. Factor analysis alone is not the answer. A few behavioral geneticists have conjectured that abilities identified through factor analysis would correspond more closely to simpler underlying genetic mechanisms than abilities identified in other ways. This is probably a naive hope, which might hold true in a few cases only by coincidence. As it has generally been used, factor analysis has revealed common factors among tests which are already so complex as to guarantee nothing more than the fitting of a polygenic model. We more or less insure this outcome by seeking and constructing tests which spread subjects out continuously over a wide range of scores and preferably yield a normal distribution.

Factor analysis is based on correlations, and an important part of the question to the behavioral geneticist is: What is the cause of the correlation? If we do not attempt to answer this question, we keep mixed up a number of things which it should be our business to disentangle and we probably will not make much progress beyond simple biometric analyses until we make an effort in this direction. Psychological test scores on two or more tests can be correlated for three main reasons, or any combination of these. First, the two tests may be functionally related by calling upon some of the same psychological processes, or the ability needed in one test is a prerequisite for utilization of the ability called for in the second (i.e., hierarchical functional dependence). Second, the genetic basis of the two abilities may in principle be independent, and the two abilities may have no functional interdependence, but the genetic systems underlying each could have become associated through assortment. In fact, Robert Tryon, one of the pioneers of behavioral genetics, put forth the notion that the reason we find a general factor, or g, in our factor analyses of a large number of mental tests is because many genetically independent units of various abilities have become genetically assorted together as a product of social mobility in a socioeconomically stratified society. Third, two or more abilities may be intercorrelated because of pleiotropism, i.e., a single gene producing two or more phenotypically distinct effects. Factor analysis, therefore, in a sense lumps together abilities into common factors, which are thereby practically guaranteed to look polygenic. At this point I view the traditional use of factor analysis as a preliminary means of determining whether the test or ability on which one wishes to "zero in" is a part of any ability domain of real interest. Pure specifics are not of much interest to anyone. A test with a high g loading, on the other hand, seems worthy of further intensive psychological and genetic analysis.

The problem then becomes that of fractionating the test performance upon which we have decided to focus. The most dangerous pitfall here is apt to be the mistake of confusing the essential psychological processes involved with the peripheral behavior observed in the subject's performance. I have no simple prescription for avoiding this hazard. Psychometric and experimental ingenuity is all I can suggest. One prescription can be made with some confidence, however. One must get a grip on intraindividual variability. Unlike fingerprint ridges and the cephalic index, behavior fluctuates—it is an inherent characteristic of the process we wish to study. I have found that simpler, more unitary components of ability are more sensitive to intrasubject fluctuations (time of day, mood, fatigue, anxiety, etc.) than composite scores on omnibus-type tests, where intrasubject variability tends to cancel out, or on tests of past-acquired knowledge and overlearned skills, which are relatively insensitive to situational variables. So even if we discover a Mendelian unit of ability, because of intraindividual variability in performance the distribution of scores could look smoothed. Fortunately, if we use the right kinds of tests, repeated measurements make it possible to obtain mean scores for individual subjects which will show significant differences between nearly all possible pairs of subjects in a quite large sample. When we achieve this, we can then begin to look for discontinuities among siblings in the same families in hopes of finding evidence of segregation. When individual scores are too "smeared" by intraindividual variability, there is little or no hope in this search.

DISCUSSION

Experimental methods are needed to fractionate ability because some part of the between-subjects variance is more or less task-specific or a result of the interaction of procedural variables with subject variables. I have shown that even so simple an ability as memory span for digits is not unitary. A delay of 10 seconds between presentation and recall of the digit series significantly alters the rank ordering of subjects ranked for ability in immediate recall. Subjects who are the same in ability to recall 7 out of 7 digits perfectly can differ reliably when they are presented with, say, 10 digits (a supraspan series) and are required to recall as many of these in order as they can. In short, even memory span is bound to look polygenic as long as we fail to identify the subabilities that enter into it. Factor analysis can be useful here; we can factor analyze all these variations of memory span (or any other ability) performance under variations of the conditions of presentation, etc., and then obtain factor scores. Time does not permit my describing our findings when we have done this. I am now doing something similar with a test of intelligence (i.e., it correlates about .40 with Raven's matrices) based on a sensitive reaction time measure of information processing capacity. The procedure lends itself beautifully to fractionation of the subject's performance.

How far should one continue fractionation? There is no rule, except perhaps "as far as necessary" to find segregation. If there are "atoms" of ability, they will be those scores which segregate, that is, show single-gene inheritance. I can think of no purely psychological criteria for identifying "atoms" of ability, which I presume theoretically would mean a reliable source of individual differences which does not yield to any experimental efforts at further fractionation. But this concept of "atom" is outmoded even in physics and may be even less tenable in psychology. In general, it is probably better to fractionate too much rather than too little. In other words, situation A (Fig. 1) is preferable to C if one is seeking segregating units of behavior. C represents genetically complex abilities; B represents pleiotropism, and A represents single-gene units of ability (which when fractionated to this point may not even resemble abilities as the term is commonly understood).

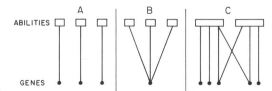

FIG. 1

Analysis for Segregation. Once one has measured what appear to be the fractionated components of an ability and can demonstrate satisfactory reliability, he can begin to do family studies to look for evidence of segregation. Linkage studies may also be possible, using a number of blood polymorphisms. It seems unlikely that linkages are apt to be found when the test score is too complex; the linked ability component will constitute too small a fraction of the total variance so that a linkage study will not be able to detect any given component through all the "noise" of other components. Initial scanning for possible linkages would consist of seeking correlations between narrow ability measures and a wide variety of blood groups. One would then carry out linkage studies for those abilities and blood groups that show correlations; not all such correlation will be due to linkage, of course. And one may question how promising linkage studies are apt to be in populations that are approaching genetic equilibrium. I leave this technical problem for the geneticists to worry about.

Finding Correlates of Mendelian Ability Units. Obviously, if one finds a segregating ability, he would be curious to know what other psychological tests and processes it enters into and how much of the variance it accounts for in larger composite measures of ability. I need not spell out the rather obvious statistical methodology for accomplishing this.

Chapter 10 Human Behavioral Genetics

N. E. MORTON
University of Hawaii
Honolulu, Hawaii

The population geneticist tries to make mathematics serve biology. The behavioral geneticist tries to make biology serve psychology. My training and interests prevent me from exploring this possible conflict, and I shall restrict my discussion to the use of mathematics to answer certain biological questions about human behavior.

What Are the Effects of Single Genes on Behavior?

The rationale for single-gene studies was given by Thiessen *et al.* (1970).

> The beauty of single gene analyses lies in the possibility of tracing the physiologic interactions from the initial gene alteration to the behavior in question. The alleles can be easily identified, manipulated as independent variables, and, in several cases, related to known metabolic pathways. It is as if only one letter of a word is allowed to vary at a time in order to study its special influence.

As in the mouse and *Drosophila,* many neurological mutants are known in man with gross effects on behavior, and probably a large fraction of other mutants modify behavior more discreetly. Considering that the academic interest of this material is reinforced by immediate prospects for better management of disease, surprisingly little research has been directed toward characterizing these effects, especially the more subtle and specific carrier signs. Some mutants, like total albinism, are similar to (and may be homologous with) genes in the mouse with known effects on behavior (including activity level and alcohol preference in the case of albinism). It is an interesting problem whether these effects are similar in mouse and man. Do the behavioral manifestations of phenylketonuria in homozygotes and carriers correspond to its dietary phenocopy in mammals? Answers to questions such as these would establish the comparability of human and mammalian behavior tests: for example, does alcohol preference in the mouse correspond to any behavioral trait in man (Henry and Schlesinger, 1967)?

The experimental design for such studies is extremely simple. Let

$$Y = 1 \text{ for a genotype to be tested,}$$
$$= 0 \text{ for a control genotype,}$$

and let X_i be the score on the ith test. Then the stepwise regression of Y on the X_i identifies tests which discriminate between the genotype and its control and assigns these tests efficient weights. If desired, some of the X_i may represent psychological factors defined as linear functions of the scores, in which case the regression provides a test of whether the factors are efficient discriminants of genetic differences. If the controls are paired, the regression may be performed within pairs.

One objective of such studies might be to select a battery of mutants which, in their characteristic expression, or more subtly in carriers, have highly specific behavioral effects. This would lead to an inversion of the current procedures for test validation: instead of circularly correlating a new test with old ones which are presumably better characterized, the test would be screened for its ability to differentiate a selected battery of genotypes.

The intraclass correlation for the discriminant is a measure of the behavioral effect relative to the unitary genetic difference: it may be called the *heritance* of the behavioral effect. Such an index of genetic determination of behavior is much more specific than the *heritability* defined on the general population, and its estimation is not fraught with any serious problems. If there remain psychologists who believe that behavior has no genetic component except at the neuropathological limit, and if such a position requires any answer but the mortality table, the heritance of single gene effects may be an appropriate rebuttal, since it simultaneously tests for a behavioral effect of a genotype recognized by other criteria and measures the strength of genetic determination.

A similar approach can be applied to relatives of probands for complexly inherited traits, like schizophrenia, to detect a possible carrier state, using the coefficient of relationship as dependent variable. The likelihood of heterogeneity is greater than for recognized single-gene traits.

There have been several attempts to detect behavioral effects of polymorphisms. Cohen and Thomas (1962) reported an excess of blood groups B and AB in nonsmokers or occasional smokers. Cattell *et al.* (1964) found an association between tender-mindedness and blood group A in tests of 14 personality factors. Parker *et al.* (1961) reported an excess of group O among manic depressives, which was not confirmed by Tanna and Winokur (1968). The other claims seem not to have been retested, and the evidence is far from convincing. A clear association between duodenal ulcers, group O, and ABH nonsecretors has been shown, with no indication that it is psychologically mediated (Roberts, 1959). The latter studies have used sib controls and (with much greater power) tests of association within ethnic groups as replicates to avoid stratification errors.

What Are the Effects of Chromosome Aberrations on Behavior?

As with single genes, little use has been made of chromosome aberrations in human behavioral genetics, the principal exception being the work of Shaffer (1962) and Money (1963) on specific space–form perception deficit in Turner's syndrome (both chromatin-positive and chromatin-negative). This suggests many questions; for example, do XO mice share this specific deficit? If so, homologous tests of space-form perception are thereby defined in the two species. Do XXX females differ in the opposite direction, in analogy with Lejeune's concept of antitrisomy? Are the defects developed before puberty, when hormonal infantilism and dwarfness complicate the picture? Are the behavioral effects of sex chromosome aneuploidy large by comparison (through the intraclass correlations) with effects on finger-ridge count, an almost perfectly heritable trait?

XXY and XYY males are prone to various psychological dysfunctions, which are poorly defined. Does the chromatin-negative Klinefelter share XXY behavior? Does the XXY tortoise-shell cat show homologous disturbances? Are the mental deficiency and schizophrenia occasionally associated with XXY of a specific type? What is the role of an extra Y on anti-social behavior, and how is that behavior characterized?

Many autosomal aberrations have gross effects on behavior by comparison with the X-chromosome anomalies, which rarely fall in the psychopathological range. However, even in the case of Down's syndrome (trisomy-21), mosaics and partial trisomics provide nearly normal material for study. Recent advances in cytogenetics detect small deletions, duplications, and inversions, and offer the hope of localizing genes which alter behavior without a concomitant morphological effect.

The method of discriminating behavioral correlates is the same as for single genes, and the same measure of heritance applies.

How Can Behavior Whose Transmission Is Unknown Be Screened for Sensitivity to Genetic Differences?

This question is the most difficult so far, and the theoretical and practical importance of an answer is less obvious, but it has attracted much attention as the "nature–nurture controversy." The alternative of investigating heritance of behavior for simple genetic differences should be considered.

Another possibility is to concentrate on predominantly environmental factors, such as regional school expenditure per child (Spuhler and Lindzey, 1967), parental income, and parental socioeconomic status. After allowing for test unreliability, the behavioral traits most sensitive to these effects (measured as a multiple correlation) may be least sensitive to genetic differences, but the possibility of gene–environment covariance makes quantitation suspect.

Since environmental effects on behavior are complex, the genetic model should be simple if it is to be of any use. Dominance, epistasis, and the genetic component of assortative mating must be neglected as unmeasurable where environment is not randomized. The test of Fisher and Gray (1938) provides some check on these simplifying assumptions. Let Y be a score for individual behavior, X be the midparent score, and Z be the product of maternal and paternal scores. Then the regression

$$Y = a + bX + cZ$$

provides a test of the hypothesis that $c = 0$, in which case certain nonadditive effects (both genetic and environmental) are negligible. Of course this test does not detect all deviations from additivity, but except for comparison of mean and variance between identical twins, it is the only method applicable to nonexperimental data. Note that it may be used for interracial crosses.

Gene–environment interactions are best studied by comparisons of relatives living together and apart. If each genotype selects its environment in a characteristic way, the more closely related are the members of a pair, the less will be the effect of separating them. For example, Berry *et al.* (1955) and Gartler *et al.* (1955) noted that the amino acid excretion pattern of

TABLE 1 Average Intra-Pair Variances for Twins Living Separately and Together[a]

	Alanine	Glutamine	Glycine	Threonine
Dizygotics				
Separated	124.0	54.0	652.0	67.0
Separated (Excluding #37)	50.4	17.9	60.3	10.25
Together	9.7	5.0	22.5	9.3
Monozygotics				
Separated	21.8	21.0	42.4	1.9
Together	12.8	19.6	151.6	1.6

[a] From Berry et al. (1955). Reproduced by permission.

identical twins was less affected by separation than the pattern of fraternal twins, and suggested that identical genotypes select similar diets and environments (Table 1). If this is a general phenomenon, as suggested for domestic animals by Robertson (1950), the limits of behavioral genetics are wide indeed: one would not ordinarily consider amino acid excretion a behavioral trait. Gene–environment interactions of this selective type vitiate heritability studies. More attention should be directed to them. Note that the members of the pair need not be separated for long periods of time, it being sufficient to compare performance together and separately.

Another kind of interaction has been reported in the *aussenvertreter* effect, whereby identical twins take opposite roles (Woodworth, 1941), increasing the within-pair and total variances. Eysenck and Prell (1951) have published data on neuroticism score which indicate that single monozygous twins are more variable than dizygous twins, due to an increase in the among-pair variances (Table 2). This large difference renders meaningless the comparison of intraclass correlations and calculations of heritability. One wonders if their result is repeatable: other workers have not noticed such an effect.

Cattell (1953) has developed models for relatives reared together and apart in terms of genotype–environment correlations, interactions being ig-

TABLE 2 Mean Squares on Neuroticism Scores[a]

Source	Within pairs V_W	Among pairs $V_W + 2V_a$	Individuals $V_w + V_a$	Intraclass correlation
Identical twins	13.68 (25)[b]	172.67 (24)	93.18	.853
Like-sexed fraternal twins	32.48 (25)	50.83 (24)	41.66	.220

[a] From data of Eysenck and Prell (1951). Reproduced by permission.
[b] Number of degrees of freedom shown in parentheses.

nored. Formally such a correlation is equivalent to shared environment, and it simplifies the notation to so consider it.

As an illustration of these principles, consider a trait Y measured on various relatives subjected to different commonness of environment. It is convenient to array the data as for the old method for calculating an intraclass correlation (Fisher, 1950, p. 214), each measured group of k relatives generating $k(k-1)$ ordered pairs. Let Y_1 be the first member of such a pair and Y_2 the second member, let R be the coefficient of relationship, let S be a vector of nongenetic variables (like age, income, socioeconomic status) whose linear effects are to be eliminated, and let C be a vector of common environment. For example, we might take

$C_1 = 1$ for individuals reared in the same household, as sibs or parent-offspring;
$ = 0$ otherwise;
$C_2 = 1$ for twins reared together;
$ = 0$ otherwise;
$C_3 = 1$ for individuals reared in the same household as parent-offspring;
$ = 0$ otherwise.

Then the regression

$$Y = a + \Sigma\, b_i S_i + \Sigma\, b_j C_j + BR$$

yields an adjusted variate

$$Y' = Y - \Sigma\, b_i (S_i - \bar{S}_i) - \Sigma\, b_j (C_j - \bar{C}_j) - B(R - \bar{R}),$$

from which the linear effects of the S_i, C, and R variables have been eliminated. Let us suppose that this is done for Y_1 and Y_2, using the coefficients calculated from the former (in which each member of a group of k relatives is repeated $k-1$ times). We may also correct Y for attenuation (i.e., test unreliability), but there seems little point in this.

The regression

$$\ln[(Y' - \bar{Y})^2] = a + \Sigma\, b_i S_i + \Sigma\, b_j C_j + BR$$

tests for differences in variance among groups. Only in the absence of such differences can the analysis proceed, and so the data must be partitioned into sets with homogeneous variance.

Assuming that this has been done, we make the regression

$$(Y'_1 - Y'_2)^2 = A + \Sigma\, b_j C_j + BR.$$

Under the model the heritability is

$$h^2 = B/A,$$

and for values of b_j significantly different from zero, the ratio b_j/A is the fraction of variance due to the jth type of common environment. The remaining fraction $1 - [(B + \Sigma\ b_j)/A]$ of the variation is due to unexplained environmental differences, errors of measurement, and interactions.

Data are almost never reported in a way suitable for this kind of analysis. Usually only the intraclass correlations are given, not adjusted for S variables. On the doubtful assumption that variances are the same for all groups, we may make the regression

$$\rho = A + \Sigma b_j C_j + BR,$$

where ρ is an intraclass correlation that may be weighted by the number of observations. Then $E(A) = 0$, and the heritability is $h^2 = B$, the significant values of b_j estimating the fraction of variance due to the jth type of common environment. The fraction $1 - B - \Sigma\ b_j$ of the variation is due to unexplained environmental causes.

As an illustration of this approach, it has been applied to median correlations reported by Erlenmeyer-Kimling and Jarvik (1963) from a literature survey of intelligence tests (Table 3). Weighting by the number of studies we find that

$h^2 = .675$, due to genetic differences;
$C_1 = .139$, due to common environment;
$C_2 = .016$, due to common environment specific for twins;
$C_3 = 0$, the difference between common environment of sibs and children;
residual $= .170$, due to random environment.

Errors of measurement due to unreliability of the tests depress h^2 by an amount which can be calculated. Other errors are irreparable, stemming mostly from failure to characterize relevant common environment. Thus twins or sibs reared apart share prenatal and some part of the postnatal environment, which tend to overestimate h^2. It is sometimes stated that arteriovenous anastomosis and mirror-imaging in monozygous twins depress heritability, without noticing that such an effect must increase the variance of single twins. If no such increase is detected, no claim of underestimation should be made. The selective type of gene–environment interaction, whereby close relatives choose similar environments, raises a philosophical problem. If such interactions are important, they tend to overestimate h^2, unless

TABLE 3 Intraclass Correlations for General Intelligence[a]

Relationship	Number of studies	Correlation ρ	Relationship R	Common environment C_1	Specific twin common environment C_2	Specific parent-offspring common environment C_3
Random pairs	4	−.01	0	0	0	0
Foster sibs	5	+.23	0	1	0	0
Foster parent-child	3	+.20	0	1	0	1
True parent-child	12	+.50	.5	1	0	1
Sibs reared apart	2	+.42	.5	0	0	0
Sibs reared together	35	+.49	.5	1	0	0
DZ twins, opposite sex	9	+.53	.5	1	1	0
DZ twins, same sex	11	+.53	.5	1	1	0
MZ twins apart	4	+.75	1.0	0	1	0
MZ twins together	14	+.87	1.0	1	1	0

[a] From Erlenmeyer-Kimling and Jarvik, 1963; after Spuhler and Lindzey, 1967. Reproduced by permission.

we wish to include the selected environment as heritable in some casuistic sense.

My conclusion is that measures of heritability when the environment is not randomized are fraught with uncontrollable difficulties. Instead of asking the geneticist to develop a better method of estimation, the psychologist should perhaps reconsider his reasons for wanting to estimate heritability when no selection experiment is envisaged.

Before leaving this subject, I would like to recall the most careful analysis yet performed, an exercise in path coefficients by Wright (1931) which seems unknown to behavioral geneticists. Burks (1928), in her classical study of foster children, had applied path coefficients to correlations corrected for attenuation, and had concluded that heritability of general intelligence is .75 to .80. Wright pointed out that her model (Fig. 1) is genetically unacceptable, since both parental IQ and child's IQ are resultants of parental genotype and environment. To develop a better model he had to make a number of assumptions, as follows:

1. The environment relevant to general intelligence is perfectly measured by a score for material and cultural advantages of the home;
2. dominance and epistasis are negligible;
3. heredity and environment are additive;
4. the ratio of residual to genetic determination is the same for parents and children, but environmental determination is greater for parents;
5. the correlation between midparental and child's genotype, allowing for assortative mating, is .78 (where .71 would be expected under panmixia).

Psychologists may be surprised that Wright made the first assumption more lightly than the second. His solution is given in Fig. 2. The heritability is $.71^2 = .50$ for children and $.56^2 = .31$ for parents, but this includes home

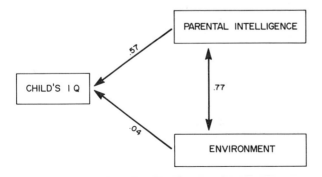

FIG. 1. Burks's model for inheritance of intelligence.

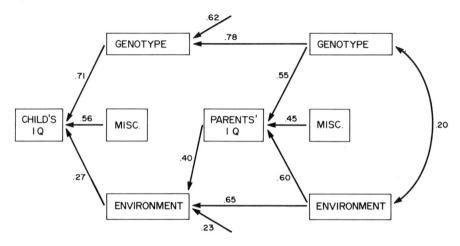

FIG. 2. Wright's interpretation of Burks's data.

environment not measured by the score for material and cultural advantages of the home. Essentially the same estimate is obtained much more simply from the values for sibs and foster sibs in Table 3:

$$h^2 = 2(.49 - .23) = .52,$$

which is less than the estimate from monozygous and dizygous reared together:

$$h^2 = 2(\rho_{mz} - \rho_{dz}) = 2(.87 - .53) = .68.$$

The first estimate assumes that environment is as similar for foster sibs as for genetic sibs, the second estimate assumes that environment is as similar for dizygous as for monozygous twins. Both assumptions are probably false, and so the heritability of general intelligence may well be less than any estimate so far made.

Twin researchers often use Holzinger's measure of heritance,

$$H = \frac{\rho_{mz} - \rho_{dz}}{1 - \rho_{dz}},$$

which is sometimes miscalled heritability. From Table 1, we see that a simple assumption is

$$\rho_{mz} = h^2 + T,$$
$$\rho_{dz} = h^2/2 + T,$$

where T is the environment common to twins. Thus

$$H = \frac{h^2}{2 - h^2 - 2T},$$

and so H is greater or less than h^2 according as T is greater or less than $(1 - h^2)/2$. This confusing statistic has no merit, since on the same doubtful assumptions.

$$h^2 = 2(\rho_{mz} - \rho_{dz}).$$

If I understand the present state of behavioral genetics, the phase of trying to convince skeptics that behavior is in some degree heritable has now ended. In any case, no intelligent skeptic would be converted by a heritability estimate, which a geneticist finds unconvincing. Good indices of heritability comes from single-gene traits, chromosomal aberrations, animal experiments, and the experience of biometrical genetics that a trait heritable at its extremes has never been found to have zero heritability within the normal range.

The usefulness of heritability estimates in man seems therefore limited to selection of heritable traits for further study. For this point, the criterion

$$2(\rho_{mz} - \rho_{dz})$$

is satisfactory, even if the assumptions on which it is an estimate of heritability are questionable.

Parenthetically Bartlett (1951) showed how to maximize an intraclass correlation. The same method can be used to maximize ρ_{mz}/ρ_{dz}, and hence heritability, but this seems less useful than to maximize discrimination of defined genotypes.

How Can the Inheritance of Behavioral Attributes Be Studied?

Traditional twin studies used for attributes the heritance

$$H = \frac{C_{mz} - C_{dz}}{1 - C_{dz}}$$

where C_{mz}, C_{dz} are the concordances for monozygous and dizygous twins, respectively. Concordance is defined as the probability that the co-twin of a twin proband be affected, and can be conveniently estimated by the Weinberg formula:

$$C = \frac{\text{number of twin probands with affected co-twins.}}{\text{number of twin probands}}$$

No precise genetic meaning is attached to this measure of heritance.

Recently two other approaches have been made to the inheritance of attributes. One is the generalized two-allele single-locus model, according to which the risk for affection is z, $z + td$, and $z + t$ in the genotypes GG,

GG', and $G'G'$, respectively. If A is the population incidence, and $z \ll t$, then $x = z/A$ is the proportion of cases which are *sporadic:* i.e., seldom recurrent in sibships (Morton *et al.,* 1971).

Such a model, while seemingly restrictive, is actually so flexible as to be difficult to exclude. Given the population incidence, there are seven hypotheses of rank one (with one parameter estimated from the data):

no phenocopies, GG completely penetrant	$(x = 0, t = 1)$
G dominant, completely penetrant	$(d = 1, t = 1 - z)$
G additive, completely penetrant	$(d = \frac{1}{2}, t = 1 - z)$
no phenocopies, G additive	$(d = \frac{1}{2}, x = 0)$
G recessive, completely penetrant	$(d = 0, t = 1 - z)$
no phenocopies, G recessive	$(d = 0, x = 0)$
no phenocopies, G dominant	$(d = 1, x = 0)$.

Similarly, there are five hypotheses of rank two:

no phenocopies	$(x = 0)$
G dominant	$(d = 1)$
G recessive	$(d = 0)$
G additive	$(d = \frac{1}{2})$
GG completely penetrant	$(t = 1 - z)$.

Commonly, two or more hypotheses of the same rank fit about equally well, and so cannot be discriminated. Models with $d = \frac{1}{2}$ are difficult to distinguish from additive polygenic inheritance.

An alternative model of rank one (given the incidence A) is *quasi-continuous* (Grüneberg, 1952). In the best derivation of this model (Smith, 1970), genes are assumed to act additively on a scale of genetic liability, which determines affection through a sigmoid risk function dependent on a single parameter, the heritability h^2 (Falconer, 1965). Edwards (1967) presented an alternative model that allows the risk to increase exponentially beyond unity; despite this unreasonable assumption, the Falconer and Edwards models fit equally well to actual data. There are no significant advantages to Edwards' model, and Falconer's is more appealing.

All of these models for inheritance of attributes can be applied to pooled data on different degrees of relationship and, with more power, to segregating families (Morton, 1967; Morton, *et al.,* 1971). It turns out that a critical distinction in terms of likelihood ratio between quasi-continuity and the best single-locus hypothesis of rank one is difficult. Recessivity and a high ratio of recurrence risk to incidence favor the single-locus models. No case has yet been found where quasi-continuity fits much better than its single-locus alternatives of the same rank. Estimation of specific recurrence risks does not depend critically on the genetic model, a fact that should cheer genetic counselors and depress geneticists.

So far the modern methods of analysis have not been applied to behavioral attributes such as handedness, where inheritance is controversial. For the most critical test, data should be reported and analyzed by families, without pooling of sibships with different sizes and numbers of affected.

It is sometimes suggested that obscure and presumably complex traits like schizophrenia have been subjected to so much inconclusive genetic analysis that further investigation should be suspended until biochemical or other resolution is obtained. This point of view neglects the fact that recent advances in complex segregation analysis, which are capable of eliminating many genetic mechanisms, have not yet been applied to behavioral traits. Their utility should at least be explored before reaching a conclusion which may well be premature.

In application of these methods, it is desirable to define the liability by a discriminant function. For example, let

$Y = 1$ for an unaffected first degree relative of a proband,
$= 0$ for an unaffected control with no affected first degree relative,

and let X_i be the score on the ith psychological, social, or biochemical variate thought to be relevant to the disease. Then the discriminant formed by regressing Y on the significant X_i can be studied by the methods of the preceding section to estimate heritability, and by the models outlined here if dichotomized into normal and abnormal. A followup study can determine a sigmoid risk function $Q(Y)$, where $Y = \Sigma\, b_i X_i$ is the discriminant, and $0 < Q(Y) < 1$ is the probability that an individual with score Y develops the disease in a specified time after testing. Then genetic analysis and counseling would both be reduced to the manageable problem of predicting the probability distribution of a continuous variable in relatives, given the distribution of liability in probands. Elston (1971) has suggested that pedigrees of three or more generations may provide the most powerful test of complex genetic hypotheses: such a test is better for a discriminant than for a rare disease.

Do Psychological Factors Have Genetic Significance?

It is not obvious to a geneticist why precise discrimination of abnormalities has not played more of a role in behavioral genetics. Perhaps the reason is to be found in the traditional hold of factor analysis over psychologists, which seems to have arisen somewhat like this. Suppose two psychologists independently decide to study general intelligence, defined intuitively. For this they must compose a suitable battery of tests. Now imagine that their intuitions are similar with respect to most aspects of intelligence, and differ in one respect: the first notes that some idiots are tactile insensitive and includes a test of this in his battery, while the second considers this irrelevant. Then a factor analysis of the first battery will reveal a factor of tactile sensitivity absent from the second battery (see O'Connor and Hermelin, 1963). Clearly, factor analysis can justly claim to reveal "vectors of the mind"—the mind of the person who composed the test battery. Whether tactile sensitivity is in fact an aspect of intelligence remains logically and operationally undefined, except by a discriminant function.

In recent years, the development of admirable statistical methods and even more admirable computers has brought factor analysis within reach of everyone, obscuring the logical difficulties. Some psychologists have even supposed that the facets of performance identified as factors are in some sense unitary determinants of behavior. To a geneticist, accustomed to organelles and loci, this is incomprehensible simply because it is not mechanistic. A psychological factor cannot be a unitary determinant of anything unless it resides in a specific organelle or macromolecule. However, we must not let our incredulity pass for knowledge. Many biologists felt just as incredulous about unit factors in genetics until they were shown to have a mechanistic basis. Much earlier, Socratic dialogues (the classical introspective analog of factor analysis) were enormously stimulating to philosophy, if not to science.

The hypothesis that a linear function of variables, determined from a correlation matrix, is an optimum discriminant of a genetic difference is readily testable, as in the first section. There is no *a priori* reason why this should be the case, unless the binary dependent variable expressing the genetic difference were included in the matrix. Therefore criterion analysis, in which relations are sought between factors and diseases, seems merely an inefficient way to construct a discriminant function, unless the original data are simultaneously submitted to regression. The test of our question "Do psychological factors have genetic significance?" is so easy that someone should try it. Presumably the answer will be "Yes, but not as much as ad hoc discriminants," which should replace factors as measures of behavior.

To What Extent Are Group Differences in Behavior Genetic?

Considerable popular interest attaches to such questions as "Is one class or ethnic group innately superior to another on a particular test?" The reasons are entirely emotional, since such a difference, if established, would serve as no better guide to provision of educational and other facilities than an unpretentious assessment of phenotypic differences. Although without practical consequences, the question is interesting as a methodological problem, which unfortunately remains unsolved. Assuming as the most economical hypothesis for correlations between relatives that intelligence is in some degree heritable, does it follow that a difference in performance on intelligence tests between two groups in different environments is also in some degree heritable? Obviously not, Jensen (1969) notwithstanding.

To study group differences, we may concentrate on environmental variables which differentiate the groups, and show that they account for at least a substantial fraction of the phenotypic difference. This requires two steps, in the first of which we discriminate between the groups in terms of environmental variables, to determine the most relevant set: and in the second we regress behavior on these variables *within* groups. Substituting the group means in this second regression, we predict the performance of each group in the absence of any genetic difference between them. Such an approach may show that a large fraction (perhaps all) of the observed difference is nongenetic, but it is subject to at least two criticisms: (1) some of the variation within groups may be due to correlations between environment and genotype, and to that extent the environmental part of the group difference will be overestimated; (2) the relevant environment cannot be perfectly measured, and to that extent the environmental part of the group difference will be underestimated, just as we suspect that the effect of home environment was underestimated when represented by a score for material and cultural advantages of the home in Burks's study discussed above.

An alternative approach is to look for members of the two groups, or of hybrids between them, living in the same environment. Maternal half-sibs living together offer the best material. Foster children are another possibility, but prenatal and early postnatal environment may be different between the groups. Interracial crosses usually involve considerable environmental similarity among hybrids of different constitution (F_1, B_1, B_2, etc.), the residual differences perhaps being small enough to be controlled by covariance analysis. A curvilinear relation between behavior and proportion of admixture could in principle be due to dominance, but could equally well indicate environmental effects. Thus Klineberg (1928) argued convincingly from intelligence tests in American Indians and mestizos, which by linear regression predicted a low IQ in pure Caucasians, that low performance in his material was social and not genetic. A final possibility is to abandon established differences in performance for differences in novel situations, like rate of learning of a new game, but perhaps this is no more culture-free than the so-called culture-free tests. Diallel crossing is more powerful if combined with covariance analysis of environmental differences. While group differences in a structured environment are messy for genetic analysis, promising methods exist which are almost never applied in behavioral genetics (Morton, et al., 1967).

Studies of the decline of performance with inbreeding require painstaking controls from sibs or neighbors. Given these, or covariance analysis as a poor substitute, inbreeding depression can be used as an index of heritance (Schull and Neel, 1965). Such studies gain interest if combined with segregation analysis to determine whether rare recessive genes or increased variance of liability is responsible for the inbreeding effect (Adams and Neel, 1967).

Recent controversy about ethnic differences in behavior is based on two fallacies: first, that a reliable estimate of heritability can be obtained when the environment is not random; secondly, that heritability is relevant to educational strategy. We have seen on page 255 that estimates of heritability within groups are unreliable when environment is not randomized among families: this applies *a fortiori* to heritability of ethnic differences. An analogy may be helpful. Voltaire described a man who killed swine with an appropriate mixture of prayer and arsenic. To analyze this, we would apply the two treatments in a factorial design, including arsenic without prayer and the converse. If, however, we could not use such experimental controls, and prayer was always accompanied by an unknown amount of arsenic, we could never rigorously separate the two treatments, which could conceivably correlate or interact in arbitrary ways. If we substitute heredity and environment for prayer and arsenic (or vice versa, according to one's predeliction), and school performance for mortality in swine, our parable is clear.

The claim that heritability is material to educational strategy is equally fallacious. Each child approaches school with certain attitudes and abilities determined by his family and neighbors; some of these behavioral traits are presumably genetic to an undetermined degree. The educational establishment tries to optimize the school output and might reasonably be expected to diversify goals and content of instruction to accommodate individual differences. However, the extent to which these differences are genetic is completely irrelevant, both to educational strategy and the success of that strategy. Of course, the genetic determination of individual differences remains an interesting academic problem, which is insoluble except by randomizing the environment.

It is perhaps understandable that those who urge the American educational establishment to give more consideration to individual differences should seek support from genetics that the differences are partly innate and "therefore" to be respected. Yet this argument is completely illogical and would not be accepted in other contexts: The physical therapist treating a case of poliomyelitis is not concerned with the extent to which susceptibility to polio may be genetic. It is only the reluctance of educators to diversify their product according to individual differences in attitudes and abilities which has led the well-meaning, and basically correct advocates of multiple streams in education to appeal to genetic arguments which geneticists find unsound. The logical structure would be improved by replacing the words *genetic, hereditary,* and *ethnic* by *familial* wherever they appear in discussions of educational strategy. Hopefully the controversy would then generate more light than heat.

What Are the Effects of Behavior on Population Structure and Selection?

Two important variables in population genetics are migration and selection, both of which are to greater or lesser extent behavioral. Migration clearly depends on topology, transportation, and political barriers: intense isolation by distance is found in Melanesians, where a boy who goes courting in the next village may lose his head (Friedlaender, 1971). It is not known whether genetic variability affects the tendency to migrate—such studies demand experimental material.

Closely related to migration are the customs of incest taboo and exogamy. Some anthropologists have speculated that group selection may have favored the incest taboo, which almost certainly arose and was promulgated for nongenetic reasons. I know of no experimental work on whether the tendency toward litter-exogamy is marked, and if so, heritable.

There is a fascinating but uncollated literature on behavioral responses to single-gene differences which could affect their fitness. The sanctity of albinos in the San Blas Indians probably increases their fertility. Deaf mutes tend to marry assortatively. Populations with thalassemia or hemoglobin S can occupy regions of hyperendemic malaria closed to unadapted groups. Consumption of fava beans may be contraindicated in a population with a high incidence of G6PD deficiency, and there is even a suggestion that the Pythagorean prohibition of beans stemmed from the susceptibility of G6PD deficient males to favism! Lactase insufficiency (the basis of which is still unclear) may have selected against pastoralism.

At a genetically more complex level, adaptations to extreme environments of heat, cold, and high altitude involve behavior as well as physiology. So far, it is unclear whether any of these adaptations are genetic, in the absence of the kinds of evidence on group differences discussed in the last section.

Going still further from simple genetics, polygamy of dominant males is a kind of phenotypic selection which may have some genetic basis. Are there genetic determinants of social dominance? One would suppose that heritability must be low for a trait subject to such intense selection. No genetic methods seem applicable to the primitive human populations where this question has been raised, and we must look to laboratory mammals for a critical study.

Summary

Available methods are discussed for answering seven questions in human material:
1. What are the effects of single genes on behavior?
2. What are the effects of chromosome aberrations on behavior?
3. How can behavior whose transmission is unknown be screened for sensitivity to genetic differences?
4. How can the inheritance of behavioral attributes be studied?
5. Do psychological factors have genetic significance?
6. To what extent are group differences in behavior genetic?
7. What are the effects of behavior on population structure and selection?

References

ADAMS, M. S., and NEEL, J. V. (1967). Children of incest. *Pediatrics*, 40, 55–62.
BARTLETT, M. S. (1951). The goodness of fit of a single hypothetical discriminant function in the case of several groups. *Ann. Eugen.* 16, 199–214.
BERRY, H. K., DOBZHANSKY, T., GARTLER, S. M., LEVENE, H., and OSBORNE, R. H. (1955). Chromatographic studies on urinary excretion patterns in monozygotic and dizygotic twins. I. Methods and analysis. *Amer. J. Hum. Genet.* 7, 93–107.
BURKS, B. S. (1928). The relative influence of nature and nurture upon mental development. *Yearb. Nat. Soc. Stud. Educ.* 27th, 219–316.
CATTELL, R. B. (1953). Research designs in psychological genetics with special reference to the multiple variance analysis method. *Amer. J. Hum. Genet.* 5, 76–93.
CATTELL, R. B., YOUNG, H. B., and HUNDLEBY, J. D. (1964). Blood groups and personality traits. *Amer. J. Hum. Genet.* 16, 397–402.
COHEN, B. H., and THOMAS, C. B. (1962). Comparison of smokers and non-smokers II. The distribution of ABO and Rh(D) blood groups. *Bull. Johns Hopkins Hosp.* 110, 1–7.
EDWARDS, J. H. (1967). Linkage studies of whole populations. *In:* Proceedings of the Third International Congress of Human Genetics," (J. F. Crow and J. V. Neel, eds.), pp. 483–489. Johns Hopkins Press, Baltimore.
ELSTON, R. C. (1971). (Personal communication).
ERLENMEYER-KIMLING, L., and JARVIK, L. F. (1963). Genetics and intelligence: A review. *Science* 142, 1477–1478.
EYSENCK, H. J., and PRELL, D. B. (1951). The inheritance of neuroticism: An experimental study. *J. Ment. Sci.* 97, 441–465.
FALCONER, D. S. (1965). The inheritance of liability to certain diseases, estimated from the incidence among relatives. *Ann. Hum. Genet. London* 29, 51–76.

FISHER, R. A. (1950). "Statistical Methods for Research Workers," 11th Ed., (Section 49.2). Oliver and Boyd, Edinburgh.
FISHER, R. A., and GRAY, H. (1938). Inheritance in man: Boas's data studied by the method of analysis of variance. *Ann. Eugen. London* **8**, 74–93.
FRIEDLAENDER, J. S. (1971). The population structure of South-Central Bougainville. (Submitted to *Amer. J. Phys. Anthrop.*)
GARTLER, S. M., DOBZHANSKY, T., and BERRY, H. K. (1955). Chromatographic studies on urinary patterns in monozygotic and dizygotic twins. II. Heritability of the excretion rates of certain substances. *Amer. J. Hum. Genet.* **7**, 108–121.
GRÜNEBERG, H. (1952). Genetical studies on the skeleton of the mouse. IV. Quasicontinuous variations. *J. Genet.* **51**, 95–114.
HENRY, K. R., and SCHLESINGER, K. (1967). Effects of the albino and dilute loci on mouse behavior. *J. Comp. Physiol. Psychol.* **63**, 320–323.
JENSEN, A. R. (1969). How much can we boost IQ and scholastic achievement? *Harvard Educ. Rev.* **39**, 1–123.
KLINEBERG, O. (1928). An experimental study of speed and other factors in racial differences. *Arch. Psychiat.* **93**, 1–111.
MONEY, J. (1963). Cytogenetic and psychosexual incongruities with a note on space-form blindness. *Amer. J. Psychiat.* **119**, 820–827.
MORTON, N. E. (1967). The detection of major genes under additive continuous variation. *Amer. J. Hum. Genet.* **19**, 23–34.
MORTON, N. E., CHUNG, C. S., and MI, M. P. (1967). "Genetics of Interracial Crosses in Hawaii." Karger, Basel.
MORTON, N. E., YEE, S., and LEW, R. (1971). Complex segregation analysis. *Amer. J. Hum. Genet.* **23**, 602–611.
O'CONNOR, N., and HERMELIN, B. (1963). "Speech and Thought in Severe Subnormality, (an experimental study)." Pergamon, Oxford.
PARKER, J. B., THEILE, A., and SPIELBERGER, C. D. (1961). Frequency of blood types in a homogeneous group of manic-depressive patients. *J. Ment. Sci.* **107**, 450, 936–942.
ROBERTS, J. A. F. (1959). Some associations between blood groups and disease. *Brit. Med. Bull.* **15**, 129–133.
ROBERTSON, A. (1950). Some observations on experiments with identical twins in dairy cattle. *J. Genet.* **50**, 32–35.
SCHULL, W. J., and NEEL, J. V. (1965). "The Effects of Inbreeding on Japanese Children." Harper, New York.
SHAFFER, J. A. (1962). A specific cognitive deficit observed in gonadal aplasia (Turner's syndrome). *J. Clin. Psychol.* **18**, 403–406.
SMITH, C. (1970). Heritability of liability and concordance in monozygous twins. *Ann. Hum. Genet. London* **34**, 85–91.
SPUHLER, J. N., and LINDZEY, G. (1967). Racial difference in behavior. *In* "Behavior-Genetic Analysis," (J. Hirsch, ed.), pp. 366–414. McGraw-Hill, New York.
TANNA, V. L., and WINOKUR, G. (1968). A study of association and linkage of ABO types and primary affective disorder. *Brit. J. Psychiat.* **114**, No. 514.
THIESSEN, D. D., OWEN, K., and WHITSET, M. (1970). Chromosome mapping of behavioral activities. *In* "Contribution to Behavior-Genetic Analysis," (G. Lindzey and D. D. Thiesen, eds.), Chap. 7, pp. 161–204. Meredith, New York.
WOODWORTH, R. S. (1941). Heredity and Environment. A report prepared for the Committee on Social Adjustment. Social Science Research Council, New York.
WRIGHT, S. (1931). Statistical methods in biology. Papers and Proceedings of the 92nd Annual Meeting. *J. Amer. Stat. Assoc. Suppl.* **26**, 155–163.

DISCUSSION

P. L. WORKMAN

University of Massachusetts
Amherst, Massachusetts

Professor Morton has provided an excellent summary of methods that are of very general use for behavior genetic studies both in man and in experimental animal populations. There are other methods, more restricted in scope, which utilize specific features of population structure in order to answer certain questions about the genetical control of behavioral variation.

For example, it is generally not possible to determine the effects of a common household environment on the manifestation of a trait in members of the same household. Such a problem is especially important when we are interested in the degree of genetic determination of a trait whose expression depends upon both physical and cultural factors in the environment. As discussed by Roberts (1973), if observations can be obtained from sibs or half-sibs living in different environments, then the total variance in trait expression can be partitioned so as to reveal the extent of a component due to a common household environment. Such data are, of course, very difficult to accumulate in western societies. However, in many nonwestern societies, there is a high degree of polygyny and, often, each of a man's wives and her children will have their own household. Roberts cites a study by Billewicz

DISCUSSION

et al. (1970), who analyzed the heritability of the level of immunoglobins (IgG, A, M, and D) in families of this type living in two villages of Mandinko in Gambia, West Africa. Using standard regression analyses of offspring on father, the heritabilities were estimated to be .42, .60, .28, and .50 for the different classes of immunoglobins. Sib correlations were in good agreement, providing estimates of .37, .61, .28, and .50. However, when an analysis was carried out using observations on half-sibs living in different households, although the heritabilities of IgG and IgM were not much changed, those of IgA and IgD were estimated to be .00 and .08, suggesting that for these immunoglobins, the familial correlations were the result of a common household environment.

This study exemplifies the notion that we should view the diversity of social and biological structures in human populations as providing an experimental laboratory in which we choose a population for study because of its suitability for answering particular questions. Another example is provided by a method which allows inference about a genetic basis for group differences.

Theoretical work in quantitative genetics shows that knowledge of within-group heritabilities can be used to infer the heritability of group differences only when the genetic relationship between the populations is known precisely. Such information, in fact, appears to be available only for experimental plant or animal populations in which the groups to be compared are lines derived from a single parental population. In human populations, given that the genetical relationship between two populations cannot be determined exactly, and, since such populations invariably are subjected to different physical and cultural environments, there appears to be no procedure by which direct comparisons can be made. As Professor Morton has suggested, one might look for environmental variables which differentiate the groups and attempt to relate such differences to variation in the trait under study. Of course, the inability to find such environmental variables does not mean that between-group differences have any genetic basis. An indirect approach to this problem can be developed if there exists a discrete dihybrid population formed by intermixture among members of the two populations which we should like to compare.

If a dihybrid population is close to genetical equilibrium, then the ancestral contributions to individual genomes will be quite similar. However, if immigration from the parental populations occurs in each generation, or if only a few generations have elapsed since the establishment of the hybrid population, or, if there is sufficient assortative mating with respect to ancestral origins, then there may be considerable genetic disequilibrium. Under such conditions, the contribution from either ancestral population to an individual in the hybrid population can vary between 0–100%, and individual differences in ancestral composition can be considerable. MacLean and Workman (1973a,b) have developed methods which provide an estimate of the ancestral origins of individuals in such hybrid populations. The method requires a knowledge of the frequencies of alleles at polymorphic loci in the parental populations and the corresponding phenotypes of individuals in the hybrid population. Loci at which parental frequencies are similar provide no information and cannot be used. In addition, one cannot utilize loci for which estimates of intermixture appear to be deviant from estimates from the majority of loci. Such estimates may reflect the confounding effect of selection (Workman, 1968) or an inaccurate determination of the parental gene frequencies. Standard regression techniques can be used to relate observations on any quantitative trait varying among individuals in the hybrid population to the individual estimates of ancestry. The absence of any significant regression would suggest that, in the hybrid population, there is no relation between ancestry and the level of expression of the trait. In general, the slope of the regression line should indicate the degree to which differences in ancestry are related to variation in the trait under study. Moreover, the extension of the regression line to trait values corresponding to 0 and 100% ancestry (from a particular parental population) suggests the probable degree of trait expression of individuals from the parental populations in the environment of the hybrid population. Differences between predicted trait values at the end-points and values actually observed in the parental populations indicate the effect of the hybrid population's environment (physical or cultural) and hence, the existence of genotype–environment interactions. This method, of course, provides only an indirect approach to questions about the genetic basis for between-population differences, but it may be as close as we can get to an answer.

Some limitations of this method should be mentioned. For behavioral traits whose expression may be affected by the extent to which an individual's appearance indicates his ancestral origins, the technique must be modified. For example, any analysis of intellectual performance in an American Black population would have to hold constant any morphological variation (skin color, hair texture, etc.) which might be found to show an association with variation in trait performance. In hybrid populations not in genetic equilibrium, heritable anthropometric variation is associated with genetic variation since genes derived from each ancestral population will tend to cluster according to the degree of ancestry and the extent of disequilibrium. Therefore, holding anthropometric variation constant might reduce the amount of genetic information to such an extent that reliable estimates of ancestry could not be obtained. For such traits in American Blacks, this method might not be applicable until considerably more laboratory and field work provides representative gene frequency data on African populations for a large number of polymorphic loci. On the other hand, for traits such as hypertension, whose expression may not be so confounded with visual appearance, the method may prove to be extremely useful. There are also other hybrid populations that can be studied depending upon which parental populations we should like to compare. For example, the modern Chileans appear to be a mixture of about 40% Indian and 60% Spanish, and data on the parental populations appears to provide a sufficient basis for studying a wide variety of traits in the Chilean population.

In theory, phenotypic variation can be partitioned into genetic and environmental components of greater or lesser specificity. However, research designs generally are oriented toward determining the magnitude of the heritable component. Although the foregoing special approaches do tell us something about the effects of a common household environment or the possibility of genotype–environment interactions in the special case of a hybrid population, there appears to be no adequate methodology for ascertaining the effects of the prenatal environment. Only when there are clearly defined prenatal insults such as rubella are we able to delimit such prenatal contributions to variation. I should like to mention one approach which may come to be extremely useful in this regard. In order to distinguish the prenatal environmental factors, we need to find measures of prenatal stress not confounded by genetic variation. A variety of plant and animal studies have shown that for characters showing bilateral asymmetry, the degree of bilateral asymmetry is relatively independent of genetic factors (apart from inbreeding) and may be taken as reflective of variation in the degree of environmental stress to which a population is subjected. Using dental asymmetry, as measured by differences in the mesiodistal diameter of antimeric teeth, Bailit *et al.* (1970) showed that the degree of asymmetry in four human populations could be related to differences in the environments of the populations as indicated by patterns of diet and disease. Since the relative sizes of the permanent teeth appear to be determined during the fetal stage, their results appear to have developed one indicator of the prenatal environmental conditions. In man, it might be possible to develop composite indices of developmental status of individuals based on the degree of asymmetry in several characters including, for example, dermatoglyphic features. Such scores could then be related to other variables of interest (behavioral, physiological, etc.). Other characters which might be suitable as indicators of prenatal stress might be provided by the so-called intermediate optimum traits; that is, traits for which the intermediate expression appears to be related to optimum fitness, such as birth weight or head circumference. Included in a measure of developmental status would be the degree to which an individual's measurement deviated from the optimum for the population. In general, it might even be possible to consider any anthropometric character in terms of an individual's deviation from the population mean, and, since we are here primarily interested in environmental sources of variation, each trait would be inversely weighted by its heritability. Different characters, of course, are affected at different times during the prenatal or postnatal development, so that associations with traits most affected at a specific time of development would provide an indirect way to access more specific causes of developmental stress. Although this method is, at the present time, largely theoretical, it provides a focus for work on growth and development which has thus far been generally ignored by behavior geneticists.

Determinations of heritability do not provide any basis for public policy or educational planning; neither do they tell us anything about individuals. If, on the other hand, individual measures of developmental status based on prenatal or early postnatal environmental factors were found to be associated with aspects of motor, intellectual, or perceptual ability, then behavior genetics could make a substantial contribution to the more applied problems in contemporary society.

References

BAILIT, H. L., WORKMAN, P. L., NISWANDER, J. D., and MacLEAN, C. J. (1970). Dental asymmetry as an indicator of genetic and environmental conditions in human populations. *Hum. Biol.* **42**, 626–638.

BILLEWICZ, W. Z., McGREGOR, I. A., ROBERTS, D. F., and ROWE, D. S. (1970). Family studies of Ig levels. *Proc. Int. Congr. Neurogenet, 3rd, Brussels* (in press).

MacLEAN, C. J., and WORKMAN, P. L. (1973). Genetic studies on hybrid populations I. Individual estimates of ancestry and their relation to quantitative traits. Submitted to *Ann. Hum. Genet. London.* (a)

MacLEAN, C. J., and WORKMAN, P. L. (1973). Genetic studies on hybrid populations II. Estimation of the distribution of ancestry. Submitted to *Ann. Hum. Genet. London.* (b)

ROBERTS, D. F. (1973). Anthropological genetics: Problems and pitfalls. In: "Theories and Methods of Anthropological Genetics," (M. C. Crawford and P. L. Workman, eds.), in press. Univ. of New Mexico Press.

WORKMAN, P. L. (1968). Gene flow and the search for natural selection in man. *Hum. Biol.* **40**, 260–279.

Editors' Comment

OMENN: A special comment about interpretation of twin concordance data may be helpful. It is customary to compare concordance rates in identical or monozygotic (MZ) and in fraternal or dizygotic (DZ) twins, with the understanding that monozygotic twins have identical genes. For traits determined quantitatively by the additive and interactive effects of genes and for traits with threshold effect, certain defined exceptions should be noted.

Discussion

First, because of the phenomenon of Lyonization or random inactivation of one of the two X chromosomes in each cell of normal females, females are mosaics for heterozygous X chromosomal loci. Normally, about half of the cells have the product of one allele and half the cells have the other. However, the random nature of X inactivation allows for deviation from the 50:50 expectation, depending on the number of cells present at the time of inactivation for a given tissue. Gartler and his colleagues have shown that among women in Sardinia proved to be heterozygous for glucose-6-phosphate dehydrogenase (G6PD) deficiency (by having one normal and one deficient son), about 1% have little or no G6PD activity, instead of the expected 50% of normal activity. At the other extreme, some such women have activity nearly 100% of normal. From binomial probabilities, Gandini, et al. (1968) estimated that only about eight cells must be present in the hematopoietic system at the time of inactivation. The importance of this phenomenon is that, for any X-linked markers, monozygotic female twins could differ markedly in the quantitative level of enzyme activity and possibly in a quantitative behavioral trait determined by such markers. It would be interesting to compare MZ male and MZ female twin pairs for variability in a search for X-linked gene effects. Second, it is possible that similar inactivation or lack of activation occurs for other loci on other chromosomes. Thus, the immunoglobulins are produced by clones of cells, each of which makes only one type of immunoglobulin. Individuals heterozygous for various antigenic markers on the immunoglobulins produce only one or the other marker in a given cell line (Grubb, 1970). Dr. Caspari noted that analogous mosaicism might account for certain "intermediate" phenotypes of scale colors in Ephestia and in the color of feathers in the blue Andalusian fowl.

References

GANDINI, E., GARTLER, S. M., ANGIONI, G., ARGIOLAS, N., and DELL'ACQUA, G. (1968). Developmental implications of multiple tissue studies in glucose–6–phosphate dehydrogenase-deficient heterozygotes. *Proc. Nat. Acad. Sci. U.S.A.* **61**, 945–948.

GRUBB, R. (1970). The genetic markers of human immunoglobulins. "Molecular Biology, Biochemistry, and Biophysics," Vol. 9, pg. 152 Springer-Verlag, Berlin and New York.

Chapter 11   The Future of Human
             Behavior Genetics

S. G. VANDENBERG
*University of Colorado
Boulder, Colorado*

As a critic of science fiction has said, it is difficult to prophesize intelligently, especially about the future. In trying to guess what the future of behavior genetics is going to be, it is well to keep in mind that those who are in a position to influence the course of events make the most successful prophets. Some knowledge of the history of science also helps. Unpredictable serendipity has led to major breakthroughs in research, but the ability to exploit such events required considerable knowledge of the existing science.

In this overview a distinction will be made of what *should* happen, in terms of research that is central to behavior genetics, and what is actually likely to happen. We will also consider more peripheral or auxilliary research that is needed if behavior genetics is to advance. In all of this the approach will be a rather pragmatic one, which resembles much of modern psychology in its emphasis on techniques and empirical facts. Near the end a more theoretical problem will be posed. Finally, in an appendix, a test battery for use in cooperative studies will be suggested.

Let us first look at research that is somewhat less likely to happen but which should be encouraged.

The multiplicative value of multivariate analyses of carefully selected tests which measure distinctive but possibly related abilities has been advocated by me before (Vandenberg, 1965, 1966, 1968), but has only been applied with twin data. It should next be applied to parent–offspring data and genetic abnormalities.

In some studies of rare genetic diseases it has become a fairly common practice to combine data on patients seen in a number of locations or even for investigators to adopt a common set of diagnostic procedures in order to obtain a sufficient number of probands for a meaningful analysis. Behavior geneticists will also have to find a way of doing more cooperative research: either a number of them will have to agree to collect the same kind of data for a co-authored study, or they will have to find ways of reporting on a small number of cases in sufficient detail to permit future integration of a number of separate reports.

A good example of what can be accomplished in this way is the summary by Moor (1967) of the effects—on the global IQ—of various types of sex aneuploidies from which I have constructed the graph shown in Fig. 1.

If the individual investigators from which these cases were collected had used a common battery of short tests of different abilities and had also obtained data on the performance of parents and sibs on these measures, an even more informative analysis could have been made by predicting the patients' scores from the number of X and Y chromosomes as well as from the scores of siblings and parents.

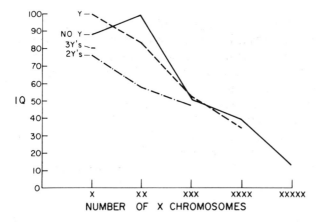

FIG. 1. Mean IQ of individuals with abnormal numbers of sex chromosomes (Moor, 1967).

This idea was suggested in part by a recent paper, Berman and Ford, (1970) who performed a study in which they predicted by a multiple regression equation the intelligence of children affected with PKU from IQ measures of parents and sibs. Then they related the difference between the predicted and observed IQ to blood phenylalanine levels. In children with truly elevated levels, there was a larger drop from the expected IQ, than in children with pseudo-phenylalaninemia.

Practical application of Ray Cattell's ingenious MAVA method (1953, 1960, see also Loehlin, 1965b), which calls for information about unrelated children raised in the same home, twins reared apart and other unusual situations, or the more conventional method of family studies involving more than two generations of interrelated nuclear families, will also require such cooperation. Still other examples are furnished by studies of the rarer types of aneuploidies such as XYY or XXY.

We need more *studies of adopted children,* and those in more detail than the one of Skodak and Skeels (1949), further analyzed by Honzik (1957).

It will be remembered that the correlation between the IQ of the adopted children and the adoptive parents' educational level was minimal, but it was substantial with the biological parent's education. It is often overlooked that for a number of cases the mother was actually tested. For these 63 cases, there was a difference of 20 IQ points between mean IQ of the adopted child and of the biological mother, in favor of the child as shown in Fig. 2. A future study may help us understand better how there could be such a general effect without any correlation with the socioeconomic or educational status of the adoptive parents.

While social agencies may be resistant to a single investigator mounting a frontal attack, perhaps a more personalized search for single cases by a number of individual behavior geneticists will encounter less organized resistance. Similarly we need *studies of children born to parents who were married more than once.* Again, an accumulation of cases by a number of separate investigators may be feasible. Perhaps a central organization could be set up to facilitate and coordinate such research.

Because there are only 23 pairs of chromosomes in man, the time has come to start routine *searching for linkage* between continuous variables and bloodgroups or other single gene markers as advocated by Thoday (1967). To make this more practical, there may also have to be a central facility which would provide serology laboratory services by airmail and computer facilities. The basic principles have been worked out and several computer programs for this purpose are now available from Elston.[1]

---

[1]More information may be obtained from R. C. Elston, Department of Statistics, University of North Carolina.

The method of *co-twin control studies,* which permits study of the influence of specific environment on a constant genotype, seems to have been completely abandoned. Even relatively small efforts, say with 10–15 pairs of identical twins, would be very informative. At best, the twins would attend a special nursery or kindergarten in which, for example, one of each pair was given number games and the other pre-reading games. Vandenberg explored this approach during one summer in Louisville and found it quite feasible. Some of the Headstart procedures could be checked out in this manner. The only study of this kind with which I am familiar is one from Sweden by Naeslund (1956), who compared the performance of 10 children taught reading by the phonics method with the performance of their MZ cotwins taught by the whole word or "sight" method. He found an interesting interaction with intelligence, i.e., the same method was not superior for all twins, but the brighter ones did better with the sight method and those of average ability did better with the phonic method.

Within and between Ethnic Group Comparisons

For theoretical reasons we need to study cognition crossculturally, if we are to arrive at biologically relevant generalizations about the species.

It is my considered opinion that attempts to estimate heritabilities in American Negroes, Mexican Americans or American Indians will be quite informative about heredity–environment interactions and will tend to show that heritability estimates on whites cannot serve as the basis for inferences about racial differences in ability. While this point should be obvious, it apparently is not widely understood and may need many more experimental demonstrations than the one small study of Vandenberg (1970), or the study of Scarr-Salapatek (1971), in which no individual zygosity determinations were possible.

If at this time it seems more expedient for political reasons not to do such studies in the continental United States, they could be done, perhaps also at less expense, on the various ethnic groups in Hawaii, or in Puerto Rico or Alaska, or even in Brazil.

There has been some talk about assignment of an index of white gene admixture to each of a number of Negroes in a study, using gene frequencies of ancestral African and white groups to arrive at the probability that a given allele is of white ancestry and weighting a number of these alleles to obtain for each person a total value (in the nature of a proportion of white genes in the total genome). This value can then be correlated with ability test scores. While there are at the moment too few well-established "African" frequencies for genetic markers to use this method (Reed, 1969), it will eventually be possible to do so.

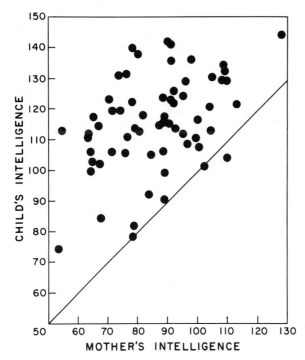

FIG. 2. Correlation between child's intelligence and natural mother's intelligence for 67 adopted children (after Skodak & Skeels, 1949).

Again I would not expect such a study to provide simple results which would give comfort to either racists or over eager equalitarians. If skin color and socioeconomic status were also measured, I would predict large interaction and covariance effects that may well outweigh additive genetic variance.

If this were the case, we would have the best scientific argument for the idea that social intervention needs to be tailored to the specific groups with which one is working.

Intergration of Behavior Genetics, Biochemistry, and Physiology

There is a good deal of research on animals in which techniques from biochemistry and/or physiology are combined with the methods of behavior genetics. However, we are still lacking in convincing demonstrations of the fruitfulness of this combination in studies of man, perhaps largely because such studies are expensive and therefore rare.

There have been many biochemical studies of schizophrenia, particularly, but so far these have not been productive (Kety, 1960; Rosenthal and Kety,

1968). In part this may be due to a reliance on psychiatric diagnosis, which may not only be inaccurate on occasion, but which could even be basically useless. The latter would be the case if there are several diseases with different modes of transmission but with somewhat similar behavioral effects. I, for one, do not see how one should proceed if one suspects that this is true. Nevertheless, this area continues to hold enormous promise for the eventual understanding of how genes influence behavior. After all, we often learn more about mechanisms and pathways from malfunctions than when everything works normally.

Although I earlier pleaded for more cooperative studies, this was mainly addressed to rather infrequent genetic anomalies of which the individual investigator can only hope to see a few. In the case of schizophrenia, large numbers may be a disadvantage. It may be better to investigate a smaller number of more similar cases, perhaps even with the same ethnic and socioeconomic background, in order to eliminate confounding factors.

More promising than psychoses may be drug addiction, alcoholism and reaction to medical drugs. Psychopharmacogenetics may be an apt name for this research area.

Most Likely Future Research

It is rather a safe bet to predict that there will be many more twin studies reporting on all kinds of variables. Such studies will in general not add much to our fundamental understanding, unless by chance or exceptional brilliance the authors discover some variable which is primarily controlled by a single gene or which demonstrates at least considerable bimodality. Even then, pedigree studies will be needed to prove the Mendelian nature of the trait. To be of any use at all, future twin studies should at least include a sufficient number of ability, personality or perceptual variables to permit a meaningful multivariate analysis of variance and covariance of the two types of twin data so that a contribution can be made to the unresolved question whether or not there is an important general hereditary component or whether there are a number of equally important independent hereditary components in cognition. If the latter is true, such studies can also begin to explore the precise nature of these components, both at the phenotypic and at the genotypic level.

It is a discouraging thought that, in a way, much of the research represented by ability factor analyses will have to be repeated with behavior genetic methods such as parent–offspring and twin studies, since in the conventional methods the effects of heredity and environment cannot be separated.

11. THE FUTURE OF HUMAN BEHAVIOR GENETICS 279

Besides twin studies there will undoubtedly be new parent–offspring studies. Because earlier studies did not use measures of special or "primary" mental abilities, one may hope that future parent–offspring studies will use such tests. In that case they can also contribute to the multivariate problem mentioned above.

A very worthwhile contribution can be made by combining the twin study method with the parent–offspring method. Elston and Gottesman (1968) have provided a method for obtaining refined heritability estimates from such data. This method is, in principle, capable of being extended to a multivariate model.

Without additional effort, such studies can also provide data for a study of assortative mating. There is no information about assortative mating for more modern, narrower and precise conceptions of special or "primary" abilities. Incidental to such work, it would be interesting to know how several of these abilities are distributed in both sexes at various (middle) ages. Other than in one study from Holland (Verhage, 1964), there are no such

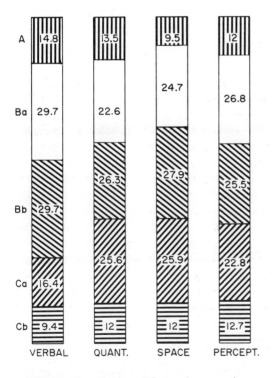

FIG. 3. Contribution of five socioeconomic groups (A, Ba, etc.) to the top 10% of children for four ability factors (after Nuttin).

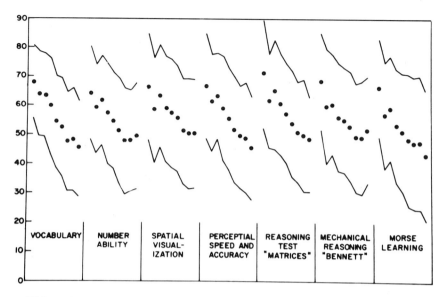

FIG. 4. Mean scores (and range of one standard deviation above and below the mean) for seven ability measures of Flemish recruits from nine socioeconomic levels (after Cliquet).

data. Even the distributions of these abilities in different socioeconomic classes are poorly studied.

Two interesting exceptions are studies by Nuttin (1965) and Cliquet (1963). Both studied the distribution of scores on a number of separate abilities for Belgian recruits from different socioeconomic backgrounds. Figure 3 shows the percentage of children from five socioeconomic classes who scored in the top 10% on four abilities, while Fig. 4 shows the mean scores (and standard deviation) on seven abilities for Flemish recruits from nine socioeconomic levels. It is again clear that sizable numbers from even the lowest group exceeded the mean for the highest group.

In all the preceding and following remarks, it should be noted that two parallel studies in rather different settings (or even different countries) would provide much more than twice the information. Perhaps it could be suggested to UNESCO that it would be worthwhile to organize multinational studies of twins or of parents and their children. Such studies could provide much information about the effect of different environments on heritabilities.

The third and final safe bet is that there will be many more reports of the psychological concommittants of diagnosed genetic anomalies, both single-gene substitutions and aneuploidies. As mentioned before, these will be of limited value by themselves, so that use of a common set of psychological variables that will permit comparisons across studies is to be recommended. In an appendix at the end of this report, an effort will be made to suggest some variables that would be useful common reference points.

Needed Ancillary Research

We now come to some less central problem areas in which progress is necessary if we are to avoid much inconclusive research with poor methods. As in all sciences, improvement in techniques should not be seen as merely tedious "development rather than research" oriented efforts. Human behavior genetics is not unique in having to rely on the available psychological tests. Unfortunately we seem to be going through a period in which work on such "applied" problems is regarded as second rate, hardly worth the efforts of ambitious scientists. It may be time to call a halt to the research dependent on poorly developed, ad hoc measuring techniques. The hope for quick solutions by instruments created for a single study is often accompanied by a rather contemptuous attitude toward the somewhat less glamorous efforts of improving existing tests. Factor analysis has been one very potent technique in such efforts. Unfortunately it has rarely used outside criteria. While conventional factor analysis continues to clarify the relationship between the many existing ability measures, many questions remain unresolved, partly because of its reliance on group administered tests and partly because a lack of concern for differential prediction or diagnosis. A few examples will suffice to amplify this point.

1. We still do not understand well the processes required in the performance on the subtests of the three Wechsler intelligence tests, although the studies by Cohen (1957, 1959) and by Saunders (1959) have given us some broad outlines. More studies are needed, such as the one by Davis (1956) in which the Wechsler subtests as well as a carefully chosen set of factorially relatively "pure" tests are administered.

2. We have only glimmerings of understanding about the relationship of success on the Piagetian tasks and their associated stages to conventional psychometric measures. Again, some beginnings have been made but much more work is needed, if possible on substantially larger samples without sacrificing the "clinical" quality of such investigations.

3. Research on the development of language will someday have to be integrated with the measurement of intelligence in young children. There is growing evidence that language development, performance on Piagetian tasks and the copying of simple geometric designs, mentioned by Dr. Jensen, all are proceeding in somewhat discontinuous fashion due to their dependence on biological maturation, so that even large amounts of practice have very little effect when the child is not ready for it. Perhaps it will be

possible to describe all these phenomena in a common terminology and use a small number of tasks as milestones for phases of the maturational process.

4. The relation between individual performance on various types of learning tasks and psychometric ability measures has received little attention, although a few promising studies exist.

The same problems exist in even more marked form in the area of personality, where there exists even less of a consensus about the relative merits of different approaches. The behavior geneticist is confronted with a large number of personality questionnaires and other tests, each supported by an increasing bibliography and advocated by devoted users. With the exception of a recent study by Sells *et al.* (1970), there has not been any major attempt to relate the various questionnaires to one another,[2] nor have there been systematic studies of test–retest correlations and attempts to understand lack of repeat reliability in terms of individual dynamic processes affecting responses to such tests on different occasions.

Cattell's efforts to develop "performance" measures of personality, just as similar work by Thurstone at an earlier date, have been largely ignored, nor has the possibility been explored that some of Guilford's very many ability measures may to some degree measure personality.

One promising way to make progress in the test construction area has been proposed by Loehlin (1965a) in his analysis of test items which showed high concordance for MZ twins but not for DZ. Tests specifically tailored for behavior genetics studies by this method or similar ones may well prove to be useful for other types of research as well.

The flip side, as disk jockeys say, of genetics, is environmental influences. The *assessment of environmental factors* influencing cognitive factors is a very difficult task that has perhaps too often been left to sociologists, because it is not easily brought under experimental control. The result is that there are many vague general statements but little hard knowledge. Perhaps the broad outlines of how such an assessment should proceed can be indicated, but little progress in refining these ideas has been made since Barbara Burks' (1928) paper, except for the very fine-grained analyses by scientists studying infantile perception of patterned versus nonpatterned stimuli, of the child's language environment, of mother–child interactions, or the more "impressionistic" formulations by cultural anthropologists.

Considering past efforts, some requirements can be specified.

Environmental assessment needs to take into account several levels or types of information.

[2] Lewis R. Goldberg is presently engaged in establishing "translation" equations (which will permit translating scores on one personality inventory into scores on another) for most of the better known personality tests.

1. Socioeconomic status. Warner's triplet: occupation, education and type of home still provides a good measure and up to date revisions are available (Reiss, 1961).

2. Size and composition of the family, plus ordinal position of a given child. These easily obtained data may not add much over and above that obtained from the first category except for within-family variance.

3. Psychological atmosphere in home:

   a. As indicated by more objective items such as number of books, types of magazines, membership of parents in clubs or other organizations, hobbies of child and parents.

   b. More "psychological" attributes that are more difficult to assess: Parental attitudes, expectations for the child's career and type of disciplinary control. Parent questionnaires may give mainly their perception of the currently fashionable child rearing practice. Some shrewd interviewers can do fairly well in getting below this surface impression. Some teachers may also be able to provide useful data.

Need for "Basic" Thoretical Formulation

On a much *more theoretical level,* we are lacking well worked out approaches to the structure of populations with respect to ability measures. While there are some large bodies of data that are relevant, most of these were collected without benefit of modern ideas about gene pools with restrictions on gene flow between these pools. We know next to nothing about factors controlling social mobility except for some highly visible, uniquely human attributes such as outstanding school grades, great beauty or social charm, and exceptional athletic or artistic gifts. Even these we know about mainly on an anecdotal or common sense basis. Purely theoretical work and computer modeling may help to advance our understanding of the very complex multidimensional processes governing the changing distributions of genes influencing psychological variables. It should be understood that few individuals are capable of undertaking worthwhile work in this area. An evolutionary perspective would have to be formulated which shows the subtle interaction of the personal motives of many individuals who mate and reproduce, and the often unintentional but sometimes serious ecological consequences of industrialization and continued expansion of human populations.

Such theories may soon be needed to combat with reasoned argument proposals to curtail reproduction or to impose economic penalties for producing retarded children. If it can be shown that different abilities are distributed differently in the population and that the lower half of the population distribution for a given ability contains many genes for high ability, no lower cut-off point can be defined such that there would be a noticeable reduction in retardation.

References

BERMAN, J. L., and FORD, R. (1970). Intelligence quotients and intelligence loss in patients with phenylketonuria and some variant states. *J. Pediatr.* **77**, 764–769.

BURKS, B. S. (1928). The relative influence of nature and nurture upon mental development; a comparative study of foster parent-foster child resemblance and true parent-true child resemblances. "Twentyseventh Yearbook of the National Society for the Study of Education" Public School Publ. Bloomington, Illinois.

CATTELL, R. B. (1953). Research designs in psychological genetics with special reference to the multiple variance analysis method. *Amer. J. Hum. Genet.* **5**, 76–93.

CATTELL, R. B. (1960). The multiple abstract variance analysis equations and solutions. *Psychol. Rev.* **67**, 353–372.

CLIQUET, R. L. (1963). Bijdrage tot de kennis van het verband tussen de sociale status en een aantal antrobiologische ken merken. (Contribution to the knowledge of the relation between social status and a number of anthrobiological characteristics). *Verh. Kon. Vlaam. Acad. Wetensch. Lett. Schone Kunsten Belg. Kl. Wetensch. No. 72,* Brussels.

COHEN, J. (1957). A factor analytically based rationale for the WAIS. *J. Consult. Psychol.* **21**, 451–457.

COHEN, J. (1959). The factorial structure of the WISC at ages 7-6, 10-6 and 13-6. *J. Consult. Psychol.* **23**, 285–299.

DAVIS, P. C. (1956). A factor analysis of the Wechsler Bellevue Scale. *Educ. Psychol. Meas.* 16, 127–146.

ELSTON, R. C., and GOTTESMAN, I. I. (1968). The analysis of quantitative inheritance simultaneously from twin and family data. *Amer. J. Hum. Genet.* **20**, 512–521.

HONZIK, M. P. (1957). Developmental studies of parent-child resemblance in intelligence. *Child Develop.* **28**, 215–228.

KETY, S. S. (1960). Recent biochemical theories of schizophrenia. *In:* "The Etiology of Schizophrenia," (D. D. Jackson, ed.) Basic Books, New York.

LOEHLIN, J. C. (1965). A heredity-environment analysis of personality inventory data. *In:* "Methods and Goals in Human Behavior Genetics," (S. G. Vandenberg, ed.) Academic Press, New York. (a)

LOEHLIN, J. C. (1965). Some methodological problems in Cattell's multiple abstract variance analysis. *Psychol. Rev.* **72**, 156–161. (b)

MOOR, L. (1967). Niveau intellectuel et polygonosomie: Confrontation du caryotype et du niveau mental de 374 malades dont le caryotype comporte un exces de chromosomes X ou Y. (Intellectual level and polyploidy: A comparison of karyotype and intelligence of 374 patients with extra X or Y chromosomes). *Rev. Neuropsychiat. Infant.* **15**, 325–348.

NAESLUND, J. (1956). "Metodiken vid den första läsundervisnigen. En översikt och experimentella bidrag." (Methods for the first teaching of reading. A survey and an experimental contribution). Almqvist and Wiksell, Stockholm.

NUTTIN, J. (1965). "De verstandelijke begaaftheid van de jeugd in de verschillende sociale klassen en woonplaatsen." (The intellectual ability of youth in different socioeconomic classes and urban and rural backgrounds). *Meded. Kon. Vlaam. Acad. Wetensch. Lett. Schone Kunsten Belg* Kl. *Wetensch.* 27, No. 7, Brussels.

REED, T. E. (1969). Caucasian genes in American negroes. *Science,* **165**, 762–768.
REISS, A. J. (1961). "Occupations and Social Studies." Free Press, New York.
ROSENTHAL, D., and KETY, S. S. (eds.) (1968). The transmission of schizophrenia. Pergamon, New York.
SAUNDERS, P. R. (1959). On the dimensionality of the WAIS battery for two groups of normal males. *Psychol. Rep.* **5**, 529–541.
SCARR-SALAPATEK. (1971). Race, social class, and IQ. *Science,* **174**, 1285–1295.
SELLS, S. B., DeMAREE, R. G., and WILL, D. P. (1970). Dimensions of personality: 1. Conjoint factor structure of Guilford and Cattell trait markers. *Multivariate Behav. Res.* **4**, 391–422.
SKODAK, M., and SKEELS, H. M. (1949). A final follow up study of one hundred adopted children. *J. Genet. Psychol.* **75**, 85–125.
THODAY, J. M. (1967). New insights into continuous variation *In:* "Proc. III Int. Congr. Human Genetics." (J. F. Crow and J. V. Neel, (eds.). John Hopkins Univ. Press, Baltimore.
VANDENBERG, S. G. (1965). Innate abilities, one or many? A new method and some results. *Acta. Genet. Med. Gemellol.* **14**, 41–47.
VANDENBERG, S. G. (1966). Contributions of twin research to psychology. *Psychol. Bull.* **6**, 327–352.
VANDENBERG, S. G. (ed.) (1968) "Progress in Human Behavior Genetics." Johns Hopkins Press, Baltimore.
VANDENBERG, S. G. (1970). A comparison of heritability estimates of U. S. negro and white high school students. *Acta. Genet. Med. Gemellol.* **19**, 280–284.
VERHAGE, F. (1964). Intelligentie en leeftijd (Intelligence and age). VanGorkum, Assen, (Netherlands).

Appendix 1

Suggestions for Ideal Body of "Core" Data to Be Collected in Cooperative Studies

Karyotypes preferably with the newer techniques of Caspersson *et al.* (1971), or of Drets and Shaw (1971), or if this is not possible, determination of Barr bodies.
Birthweight and data on subsequent physical growth to be compared to standards.
Height of father, mother and sibs.
Parental ages at birth of proband and ages of other children.
Fingerprints and palm prints.
Photos of proband repeated at following visits. (perhaps somatotype).
Sexual identity or gender role questionnaire and, when techniques become available, quantitative sex hormone assay.
EEG, especially kappa-waves.
Teacher ratings of aggressiveness, popularity, outgoingness or sociability. ("compared to all the youngsters you have known, how do you think x rates?")

If proband is capable of it, a personality questionnaire such as the one by Porter and Cattell (1968).

Social Competence. Vineland social maturity scale, or a more modern equivalent to be constructed. (Many persons classified as retarded when of school age seem to function adequately as adults, which suggests that at times an undue emphasis may be placed on verbal ability or other school-oriented skills) (see Nihira *et al.* 1970).

Intelligence. If possible at all a test should be administered which allows looking for patterning of abilities, such as the Pacific Multifactor tests (Meyers *et al.*, 1962, 1964) for ages 2–6, the Primary Mental Abilities Test (PMA) with five age levels: kindergarten and grade 1, grades 2–4, 4–6, 6–9 and 9–12, WPPSI (ages 4–6½), WISC (ages 5–15 yr, 11 mos), WAIS (ages 16 and up) or some European test battery such as the Snijders–Oomen test, which can be used with the deaf as well as with normal children (Snijders and Snijders–Oomen, 1959), the Intelligence Structure Test (Amthauer, 1955). A new battery is currently being developed in England. (Warburton, 1970). The McCarthy Scales of Children's Abilities for ages 2½–8½ measure verbal ability, memory, abstract reasoning, number ability, motor coordination, and lateral dominance.

For some of these tests, some information is available on the heritability of subtests. It is summarized in Table 2.

If more extensive testing of specific abilities is desired, the ETS kit of reference tests should be consulted (French 1951, French *et al.*, 1963) or Guilford and Hoepfner (1971).

If, because of time limitations, no individual testing is possible, it would

TABLE 1 Age Ranges and Testing Time Required for Several Batteries

Name of test	Age range	Test time
Pacific Multifactor Test[a]	2–6	1 hour
PMA[b] grades:		
K–1	5–7	1 hour
2–4	7–10	1 hour
4–6	9–12	107 min.
6–9	12–15	75 min.
9–12	15–18	74 min.
WPPSI[c]	4–6 ½	1 hour
WISC	5–16	1 hour
WAIS	16 and up	1 hour

[a] The Pacific Multifactor Test is not commercially available, but can be constructed from details furnished in Meyers *et al.* (1962, 1964).

[b] The PMA can be ordered from Science Research Associates, 259 East Erie Street, Chicago, Illinois, 60611.

[c] The 3 Wechsler tests and the McCarthy scales can be ordered from the Psychological Corporation, 304 East 45 Street, New York, New York, 10017.

TABLE 2 Percentages of Variance Due to Genetic Factors in the Subtests of Normal Batteries

Pacific Multifactor Tests	h^2 (Vandenberg et al., 1968)			
Motor ability	.40			
Perceptual speed	.57			
Language development	—[a]			
Reasoning	.08			
Memory	.24			
Number ability	.50			
Primary Mental Abilities Test	(Blewett, 1954)	(Thurstone et al., 1955)	(Vandenberg, 1962)	(Vandenberg, 1965)
Verbal	.68	.64	.62	.43
Space	.51	.76	.59	.72
Number	.07	.34	.61	.56
Reasoning	.64	.26	.28	.09
Wordfluency	.64	.59	.61	.55
Memory	not used	.39	.20	not used
Wechsler Intelligence Scale	(Block, 1968)			
Information	.74			
Comprehension	.55			
Arithmetic	.64			
Similarities	.45			
Digit Span	.35			
Vocabulary	.68			
Digit Symbol	.51			
Picture Completion	.33			
Block Design	.57			
Picture Arrangement	.43			
Object Assembly	.26			

[a] The 30 DZ pairs were more concordant than the 26 MZ pairs.

be valuable to check whether the proband has been tested in school and to obtain a copy of the results.

Intelligence of Siblings and Parents. Future analyses would greatly benefit from any data on the intelligence of siblings and parents because a given child's score may be considerably below the family mean and yet still be average.

Editor's Comment

OMENN: Collaborative studies are essential for psychological evaluation of inborn errors of metabolism. Any single center has too few cases of most of the interesting disorders, and comparable or identical testing materials must be applied in a standardized fashion after preliminary pilot studies have been undertaken. The list of tests must be shortened and refined to include only those with highest potential for factor analysis and with greatest ease of administration. Controls of children with other specific metabolic disorders matched for IQ and for socioeconomic status, and of children with undifferentiated mental retardation similarly matched are necessary. In addition, siblings and parents identified as heterozygous carriers should be tested, since even for a disease as rare as one per 40,000 individuals the frequency of heterozygotes is 1% in the general population. Thus, the sum of heterozygous carriers for various specific inborn errors is a substantial percentage of "normal" people. Enzymes known to be involved in the central nervous system, as opposed to the toxic mechanism of phenylketonuria and other disorders of liver metabolism, should be the focus of such studies.

References for Appendix

AMTHAUER, R. (1955). "Der Intelligenz Struktur Test." Hogrefe Verlag, Gottingen.

BLEWETT, D. B. (1954). An experimental study of the inheritance of intelligence. *J. Ment. Sci.* **100**, 922–933.

BLOCK, J. B. (1968). Hereditary components in the performance of twins on the WAIS, *In:* "Progress in Human Behavior Genetics," (S. G. Vandenberg, ed.), pp. 221–228. Johns Hopkins Press, Baltimore.

CASPERSSON, T. G., LOMAKKA, G., and ZECH, L. (1971). The 24 fluorescence patterns of the human metaphase chromosomes-distinguishing characters and variability. *Hereditas,* **67**, 89–102.

DRETS, M. E., and SHAW, M. W. (1971). Precise human chromosome identification by means of specific banding patterns (Abstract). *Genetics,* **68** Suppl., 16.

FRENCH, J. W. (1951). The description of aptitude and achievement tests in terms of rotated factors. *Psychometr. Monogr.,* **5**.

FRENCH, J. W., EKSTROM, R. B., and PRICE, L. A. (1963). "Kit of reference tests for cognitive factors," (Rev. Ed.) Educational Testing Service, Princeton.

GUILFORD, J. P., and HOEPFNER, R. (1971). "The Analysis of Intelligence." McGraw-Hill, New York.

MEYERS, C. E., ORPET, R. E., ATTWELL, A. A., and DINGMAN, H. F. (1962). Primary abilities at mental age 6. *Monogr. Soc. Res. Child Develop.* **27**, 1 (Whole No. 82).

MEYERS, C. E., DINGMAN, H. F., ORPET, R. E., SITKEI, E. G., and WATTS, C. A. (1964). Four ability-factor hypotheses at three preliterate levels in normal and retarded children. *Monogr. Soc. Res. Child Develop.* **29**, 5 (Whole No. 96).

NIHIRA, K., FOSTER, R., SHELLHAUS, M., and LELAND, H. (1969). "The Adaptive Behavior Scales," Manual. Washington, D. C. Amer. Ass. Ment. Def. (Revised 1970).

PORTER, R. B., and CATTELL, R. B. (1968). "The Children's Personality Questionnaire." Univ. of Illinois Press, Urbana.

SNIJDERS, J. T., and SNIJDERS-OOMEN, N. (1959). Non-verbal intelligence tests for deaf and hearing subjects. Wolters, Groningen (Netherlands).

THURSTONE, T. G., THURSTONE, L. L., and STRANDSKOV, H. H. (1955). A psychological study of twins. Report No. 4. Psychometric Laboratory, University of North Carolina.

VANDENBERG, S. G. (1962). The hereditary abilities study: Hereditary components in a psychological test battery. *Amer. J. Hum. Genet.* **14**, 220–237.

VANDENBERG, S. G. (1965). Multivariate analysis of twin differences. *In:* "Methods and Goals in Human Behavior Genetics," (S. G. Vandenberg, ed.), pp. 29–44. Academic Press, New York.

VANDENBERG, S. G., STAFFORD, R. E., and BROWN, A. M. (1968). The Louisville twin study. *In:* "Progress in Human Behavior Genetics," (S. G. Vandenberg, ed.), Johns Hopkins Press, Baltimore.

WARBURTON, F. W. (1970). The British intelligence scale. *In:* "On Intelligence, The Toronto Symposium on Intelligence," (W. B. Dockrell, ed.). Methuen, London.

DISCUSSION

BENSON E. GINSBURG

University of Connecticut
Storrs, Connecticut

Professor Vandenberg has attempted to wrestle with the very difficult problem of predicting future trends—if only those of the near future—from current indicators. Since a discipline usually evolves, its future would seem to be rooted in the findings, problems, techniques, needs and scientific dogmas of the present. A synthetic appraisal of where we are may, therefore, lead to an analytic imperative telling us where we must go.

We can learn by looking at the history of science. Darwin's introduction to the *Origin of Species* would have us believe that all of the component ideas and evidence needed to arrive at the theory of organic evolution were at hand before he wrote his book, and that one had only to pull it together —as Wallace also did. It would not have been difficult to predict, before Mendel, that the mechanism of hereditary transmission must one day be understood or, much more recently, that we would eventually come to know the physical gene and to understand its coding and self-replicating properties. If one reads *The Double Helix,* a point was reached when it clearly became a downhill race, and one knew where to look for the route map down the hill (Watson, 1969). Vandenberg's admonition to look to the history of

DISCUSSION

science as we address ourselves to the question of what needs to be done is, therefore, much more than rhetorical. Some of us here, or perhaps our students, may be closer to the mother lode than we think.

Let me now offer a few commentaries on the specifics of the paper and follow these by my own synthesis, which assimilates Vandenberg's reasoning and data to a different point of view.

My major premise is that behavioral genetics is a discipline with unifying concepts and that these belong neither to the empiricism of a technique-oriented psychology nor to biometry, which must serve as a tool rather than a guide. Thurstone's (1938) search for primary mental abilities is a case in point. These primary abilities were conceived to be distinguishable, independent attributes that were rooted in genetics. The biometrical techniques, such as factor analysis, applied to batteries of tests simply constituted a method for identifying intercorrelated clusters and differentiating one such cluster from another. Royce (1958) has combined these methods with those of Mendelian genetics, particularly with animal models, in an effort to demonstrate that the tests are getting at "real," biologically based phenotypes. There is no intuitive way to know that one is assessing a biological attribute, or what I would like to call a behavioral phenotype, unless one can show that it has a definite mode of inheritance and is based on a demonstrable biological mechanism. Our behavioral tests are not necessarily isomorphic with any natural phenotype. For example, males and females seemed to respond quite differently to particular components of the Rorschach test. Now, perhaps as a result of changing cultural norms, these differences are blurring and what once seemed a reliable differentiator has become less so. Rosenthal (1968) recently made an analogous point about heritability, pointing out that in one of Nichols' studies, heritability was high for some occupational choices, but low when it came to the choice of a laxative! We should not over-interpret behavioral tests simply because they are in use. But the question remains as to what the criteria of evaluation of behavioral tests should be. Clearly some tests have more biological significance than others. That of the Rorschach may be minimal, but that for colorblindness represents a phenotype—or, rather, several phenotypes—that we accept as real. Why? Because it yields simple and definite genetic results, involves analyzable visual mechanisms, and has few degrees of freedom between genotype and phenotype.

I do not agree with Vandenberg's position on the relative uselessness of animal models. The infra-human mammalian neuroendocrine system has sufficient phyletic similarities to man's to make it highly relevant, particularly at the level of homology of mechanism, which can be tested. In inbred strains, one has an approximation to an assembly-line production of replicable genotypes in a mammal. In mutants on inbred strain backgrounds, one has access to simple models of the effects of single genetic substitutions on neuromorphology, neurophysiology, neurochemistry, and behavior. Further, the neuroendocrine system is available for manipulation and study. Our own experiments on susceptibility to audiogenic seizures in mice (Ginsburg et al., 1969; Sze, 1970) have shown that the developmental study of single-gene mutations on inbred stain backgrounds permits the identification of the processes that exhibit departures from normality and of the time in ontogeny when these departures first occur. It is our hypothesis that these particular mutants affecting glutamic decarboxylase activity and others like them act by regulating the expression of other genes. If true, it should be possible to create phenocopies of such mutants in normal animals by manipulating the products and substrates of the reactions in brain at the time that the mutants act. Such manipulations, in fact, have produced animals whose behavior is as different as if they had the mutant gene after the critical time in development (Ginsburg et al. 1969). Complementary physiological studies have been reported by Henry and Bowman (1970), in which early "priming" by exposure to sound modified the inherited susceptibility to audiogenic seizures.

Let me cite another example from our own work involving chromosomal contributions to behavior. In reciprocal crosses between highly aggressive and nonaggressive strains of mice, it can be shown that the aggression scores are those of the stock from which the Y chromosome was derived, thus affording an opportunity to study the role of the normal Y in aggression (Ginsburg and Jumonville, in preparation). These experiments have also identified an autosomal component. Once the mechanisms have been elucidated in these animal models, they could be directly applicable to our own species.

OMENN: The hypothesis that an abnormal sex chromosomal complement (47,XYY) might be associated with criminal behavior in man has launched an interesting chapter in behavior genetics (see Omenn and Motulsky, 1972). XYY males were noted in the early 1960's among patients with gonadal abnormalities. A few years later, chromosomal screening in prison populations (Jacobs et al., 1965) turned up a surprising number of men with sex chromosome anomalies, including XYY and an even larger number of XXY (Klinefelter's syndrome) males. The XYY males were tall. The frequencies varied among studies, but 2–12% of prisoners over six feet tall in maximum-security prisons in the United Kingdom were XYY. Only then was an effort undertaken to define the frequency of this chromosomal abnormality in the general population. Chromosomal karyotypes had to be prepared, rather than screening buccal smears for Barr bodies, which are indicative of inactivated X chromosomes. About one in 800 liveborn males is XYY, so the vast majority of XYY males must exist in our "normal" population. The interest in the XYY phenotype was greatly enhanced by a psychological study by Price and Whatmore (1967) claiming that XYY individuals were "black sheep" from otherwise good families. They were unlike their prison buddies, who typically came from broken homes and had sibs and other relatives with arrest records. Also, they committed their first crimes earlier and directed their attention to offenses against property, rather than against people. Geneticists were stimulated to think that a particular behavioral entity in the large field of criminality was associated with the particular chromosomal abnormality. However, attempts to confirm that psychological study have failed to show such differences between XYY and XY male prisoners. In sum, individuals with the XYY karyotype (or XXY or XXYY) seem to have an enhanced risk of getting into trouble with the law, but the absolute risk is probably quite low and certainly influenced by social policies. Hopefully, better understanding will result from followup of the 47,XYY newborns identified in several large screening programs.

GINSBURG: As Vandenberg pointed out, simplistic analyses are also possible in humans. If we obtain serological and other easily detectable markers for all the human chromosomes, then it becomes feasible to attempt, through linkage studies, to identify particular genes that make contributions to complex human attributes. One example of preliminary data in this field is the postulated association between a major gene for verbal intelligence and the Rh blood group locus (Bock *et al.,* 1970).

The development of the language function is another area upon which Vandenberg would like to focus greater attention. Recent advances permitting the application of dichotic listening techniques to young children have opened the possibility of studying the lateralization of the decoding function in relation to the development of speech and reading. Genetic pathologies are another obvious area for further research, as are longitudinal and parent–offspring studies. However, the issue of how to ask the questions— and which questions to ask of these materials—still remains. One could, if one is considering global planning, also do a certain amount of innocuous experimentation by means of coordinating human artificial insemination programs.

JENSEN: Preliminary studies we carried out several years ago in the Berkeley area indicated that artificial insemination programs did not provide sufficient variation among donors or among husbands of recipients. Donors represented a fairly homogeneous group of high IQ and scholastic achievement—mostly medical students and staff. Recipients were also mostly professional people with a narrow range of above-average IQ. Families were cooperative in allowing studies to be planned, but no further studies were done. We would regard any plan to "experiment" by inseminating with genes of possible deleterious effects to be altogether unethical clinically and unjustifiable scientifically.

GINSBURG: Returning to my central thesis—that there are unifying concepts in behavior genetics—I would like to try my hand at framing the key questions. In order to do so, it will first be necessary to review the concepts and principles behind them. These concepts assume that the search for biologically natural units of behavior can be planned. Any behavioral difference that is genetically based may serve as a starting point. The next step must be an elucidation of the mechanism. It is possible to approach the mechanism by means of animal models and to further predict aspects of the phenotype which should be a consequence of these variations in morphology and process. When these variations are understood in a developmental context, it may be possible, as in the case of our own experiments with seizure susceptibility in mice, to create a normal phenocopy by reprogramming the genes through regulatory phenomena rather than waiting for the more utopian solution of restructuring the DNA molecules. The basic principles of genetics must apply. Individuals may exhibit the same phenotype despite getting there by different genetic routes or via different departures in underlying physiological mechanisms. That is, individuals may be isophenic but heterogenic, and ethnicity may play a large role, as in the distribution of favism, sickling anemia, and Tay–Sachs disease. Complementarily, individuals may be alike for the genes in question, but phenotypically unlike, due to environmental factors and differences in genetic background. Here again, ethnicity may be a factor. For example, the identical genome could turn out quite differently, behaviorally speaking, under various paradigms of nutrition and rearing (Ginsburg 1968, 1969). By the same token, another genome in the same species would have a different repertoire of interactions with the same spectrum of environmental changes (Ginsburg and Laughlin, 1971).

The implication of these principles is clear. A search for natural phenotypes and genetic mechanisms involved in the determination of behavioral capacities can and should be carried out using identifiable genes. Unless an identity of mechanism can be established, one set of schizophrenic twins, for example, may not be like another, particularly if they are drawn from different gene pools. The use of genetic "lesions" should serve to identify relevant processes from among the complex of neuromorphological, neurochemical, and neurophysiological correlates of behavior. Developmental considerations are also important and can lead to understanding of the degrees of freedom inherent in particular genotypes, making possible the creation of phenocopies that alter the further course of behavioral development. Finally, the problem of ethnic differences must be viewed in the context of developmental genetics. It is plausible that many of the key ethnic differences, including those reported by Jensen (1969), may have more to do with early normative conditions that activate the potential inherent in each genotype than with the variations in the genotype that undoubtedly exist on an ethnic basis.

References

BOCK, R. D., VANDENBERG, S. G., BRAMBLE, W., and PEARSON, W. (1970). A behavioral correlate of blood-group discordance in dizygotic twins. *Behav. Genet.* **1**, 89–98.

GINSBURG, B. E. (1967). Genetic parameters in behavior research. *In:* "Behavior-Genetic Analysis" (J. Hirsch, ed.). McGraw-Hill, New York.

GINSBURG, B. E. (1968). Genotypic factors in the ontogeny of behavior. *Sci. Psychoanaly.* **12**, 12–17.

GINSBURG, B. E. (1969). Genetic assimilation of environmental variability in the organization of behavioral capacities of the developing nervous system. *In:* "Foetal Autonomy," (G. E. W. Wolstenholme and M. O'Connor, eds.). Churchill, London. (a)

GINSBURG, B. E. (1969). Genotypic variables affecting reponses to pos'natal stimulation. *In:* "Stimulation in Early Infancy" (A. Ambrose, ed.). Academic Press, New York. (b)

GINSBURG, B. E. (1969). Developmental genetics of behavioral capacities: The nature-nurture problem re-evaluated. *Merrill-Palmer Quart. Behav. Develop.* **17**, 59–66. (c)

GINSBURG, and JUMONVILLE (in prep.)

GINSBURG, B. E., and LAUGHLIN, W. S. (1971). Race and intelligence. What do we really know? *In:* "Intelligence: Genetic and Environmental Influences" (R. Cancro, ed.). Grune and Stratton, New York.

GINSBURG, B. E., COWEN, J. S., MAXSON, S. C., and SZE, P. Y. (1969). Neurochemical effects of gene mutations associated with audiogenic seizures. *In:* "Progress in Neuro-Genetics," (A. Barbeau and J. R. Brunette, eds.). Excerpta Med. Found., New York.

HENRY, K. R., and BOWMAN, R. F. (1970). Behavior-genetic analysis of the ontogeny of accoustically-primed audiogenic seizures in mice. *J. Comp. Physiol. Psychol.* **70**, 235–241.

JACOBS, P. A., BRUNTON, M., MELVILLE, M. M., BRITTAIN, R. P., and McCLEMONT, W. F. (1965). Aggressive behavior, mental subnormality, and the XYY male. *Nature* **208**, 1351–1352.

JENSEN, A. R. (1969). How much can we boost I.Q. and scholastic achievement? *Harvard Educ. Rev.* **39**, 1–123.

OMENN, G. S., and MOTULSKY, A. G. (1972). Intra-uterine diagnosis and genetic counselling: Implications for psychiatry in the future. *In:* "American Handbook of Psychiatry" (D. A. Hamburg and H. K. Brodie, eds.), 3rd Ed., Vol. 6 Basic Books, New York.

PRICE, W. H., and WHATMORE, P. B. (1967). Behavioral disorders and pattern of crime among XYY males identified at a maximum security hospital. *Brit. Med. J.* **1**, 533–536.

ROSENTHAL, D. (1968). The genetics of intelligence and personality. *In:* "Genetics: Biology and Behavior Series" (D. C. Glass, ed.). Rockefeller Univ. Press and Russell Sage Foundation, New York.

ROYCE, J. R. (1958). The development of foetal and analysis. *J. Genet. Psychol.* **58**, 139–164.

SZE, P. Y. (1970). Neurochemical factors in auditory stimulation and development of susceptibility to audiogenic seizures. *In:* "Physiological Effects of Noise" (R. L. Welch and A. S. Welch, eds.). Plenum, New York.

THURSTONE, L. L. (1938). "Primary Mental Abilities." Univ. of Chicago Press, Chicago.

WATSON, J. D. (1969). "The Double Helix." The New American Library, New York.

Chapter 12 Comments on School Effects, Gene–Environment Covariance, and the Heritability of Intelligence

BRUCE K. ECKLAND

University of North Carolina
Chapel Hill, North Carolina

The Search for Explanation: Do Schools Make a Difference?

In most discussions of the relative contributions of genetic and environmental factors in the explanation of individual differences, I believe geneticists rather consistently err by attributing the unexplained or unmeasured variance in their models to the sociocultural environment. This may be a healthy attitude, especially from the standpoint of public policy issues, since it implies that decision-makers have a lot of room in which to maneuver without tampering with the gene pool or taking heredity into account.

Most of my colleagues in the social sciences would warmly agree with this. At least in sociology, we never seem to run out of social and psychological hypotheses for almost any phenomenon. Yet, although our theories have much to recommend them, our tools and our findings leave much to be desired. In fact, when weighed in balance, the evidence on a number of points tends to push the search for explanation back to biology.

Let me be more explicit, with special reference to our educational institutions in this country. What, for example, are the effects of our schools and colleges in producing individual differences in cognitive performance? This question has been investigated in many different ways. For purposes of this brief review, it will be convenient to distinguish between the organizational, contextual, and interpersonal effects of educational environments.

Organizational components, such as school facilities, curriculum, teacher characteristics, education practice, and so forth, have been studied in numerous settings. While there is evidence that good teachers (as measured by verbal aptitude) make some difference, little else seems to matter. In the most massive survey to date, working under the auspices of the U.S. Office of Education, the investigators were unable to turn up any convincing evidence that what goes on in our elementary and secondary schools has much effect (Coleman et al., 1966). Rather, differences in student performance are found primarily to reflect the students' backgrounds or, in other words, the cognitive and social skills they bring with them to the classroom.

At the college level, similar results have been obtained in a set of studies sponsored by the Educational Testing Service, the American Council on Education, and the National Merit Scholarship Corporation. Although some colleges seem to be more effective than others (Rock *et al.,* 1970), most differences in academic achievement can be predicted from the aptitude of students at entrance (Astin, 1968; Nichols, 1964). Differences, like the number of books in library, expenditures, proportion of faculty with doctorates, etc., account for very little of the variances in performance both within and between colleges.

Differences in teaching technologies also do not seem to matter. In a recent review of four decades of research on this subject, the investigators concluded quite emphatically that the data "demonstrate clearly and unequivocally that there is no measurable difference among truly distinctive methods of college instruction when evaluated by student performance on final examinations [Dubin and Taveggia, 1968]."

Another way in which the influence of school environments has been studied has been to measure the effect on students of different academic "climates" or, in other words, the normative context in which educational values arise and are reinforced. For such purposes, aggregate measures of individual traits, like ability and socioeconomic status, are taken to represent the informal milieu of the school or classroom, and these measures are then correlated with the individual performance of students or with some other dependent variable. Much of this research in recent years has dealt with the contextual effects of racially integrated and segregated schools.

The only general conclusions that can be made from this line of research is that whenever individual attributes, such as sex, ability and individual level social status, have been partialed out, most of the contextual effects on student performance (and on aspirations, self-concept, and the like) disappear. Contextual analysis has been hotly debated in the literature (Hauser, 1970; Bowles and Levin, 1968; and Turner *et al.,* 1966). However, the issues do not deal with whether environmental influences of this kind explain small or large amounts of the variance, but focus on how the findings should be interpreted and whether or not they are simply spurious.

Educational effects also have been extensively studied at the level of interpersonal influences, that is, the interaction between role incumbents. Most of the research in this area has focused upon the manner in which teachers differentially effect student performance. Indeed they do. But not with striking results.

Perhaps the most infamous study in the field, one that many environmentalists eagerly quoted when it appeared, was Rosenthal's *Pygmalion in the Classroom* (Rosenthal and Jacobson, 1968). Due to its farreaching implications, many educators still accept the findings unquestioningly. Yet, *Pygmalion* no doubt will go down in the history of educational research much like the Piltdown Man did in physical anthropology.

The central idea was that teachers are unduly influenced by their knowledge of the ability of their students and, as a result, the student's performance reflects the teacher's expectations. In a double-blind experiment, Rosenthal misinformed a group of teachers about what to expect from their new students based on IQ tests and later found that the students, independent of their measured intelligence, performed at a level consistent with what the teachers had expected—a self-fulfilling prophecy. However, not only has Rosenthal's original work been subjected to severe critical review (e.g., Thorndike, 1968), but more recent and meticulous studies have failed to replicate his results (Fleming and Anttonen, 1971; José and Cody, 1971).

Not all research on student–teacher relationships fares so badly, yet none of the work in this area has been particularly conclusive and none has contributed very substantially to our understanding of individual differences in student performance.

When taken as a whole, the last decade of research in the social sciences strongly suggests that at least for some behavioral traits, like individual differences in cognitive development, we should look more closely at familial sources of variation in our search for explanation. There is very little evidence that schools, per se, are a primary source of discrimination, or in any significant way change the pecking order of students—although the order may change for other reasons.

For the social scientist, this will mean focusing more closely on early childhood socialization and on the subcultural characteristics that students bring with them to the schools. But for the behavior geneticist, it should mean continuing to look for the genetic mechanisms that may be involved in producing individual differences. My point is, do not expect "educational hothouses" to do any wonders. Their products appear to be largely dependent upon the kinds of human seedlings with which they have to work.

Prediction or Explanation: Gene–Environment Covariance

Statistical models allow researchers to predict the outcome of events, given some quantitative measures of the variables under consideration and some very simple assumptions about the nature of reality. Psychologists can predict who will succeed in college, and they can do so at a level of accuracy that makes their work useful to admissions officers. Sociologists, better than parole judges, can predict which convicts, if released early, are likely to violate parole. And behavior geneticists, with reasonable accuracy, can predict the maze performance of different strains of mice and rats, as well as the performance of relatively isolated breeding populations on IQ tests.

However, we must never equate prediction with explanation. In the search for explanation, we often rely very heavily upon prediction and correlation techniques. But the scientific method *demands* that we also search for real or causal relationships, which in the last analysis can only be defined by the adequacy of the theories and data we bring to bear on the matter.

Based on several actuarial studies, for example, prison authorities generally find that murderers make good parole risks, while check forgers and drug addicts are poor risks. In other words, recidivism is *correlated* with the type of offense for which the offenders originally were committed to prison, thus making it possible to predict the high- and low-risk cases. (The level of accuracy here is irrelevant.) To assume, however, that recidivism depends, in a causative sense, upon the nature of the prior criminal act may be a gross error. The correlation could be entirely spurious.

At one time in the city of Philadelphia (another example) it was discovered that most of the prostitutes were Episcopalian. The "correlation" had the clergy quite worried. Was this the unanticipated consequence of the teachings of the Church or for some mysterious reason had the Church drawn under its wing most of the town prostitutes? Actually, neither of these hypotheses was correct, although both were partially correct. First, it was learned that these prostitutes had in fact been reared in the Church. But, more importantly, it was pointed out that the Episcopalians operated most of the orphanages in town and that, apart from their religious affiliation, orphanages tend to produce prostitutes. The correlation between prostitution and religious affiliation was spurious.

In her paper on gene–environment interactions, it seems to me that Dr. Erlenmeyer-Kimling is warning us against just this type of error when she speaks of the observed correlations between the adequacy of the child care and the mother's and child's IQ. As she indicates, the intercorrelations between these variables do not tell us whether the causative agent in regard to the child's IQ is the parent's IQ or the environmental conditions of the home, since the latter may in fact depend upon the former. Given our present state of knowledge, we have no way of knowing whether inadequate child care "causes" low IQ or whether the correlation is simply spurious, meaning that both inadequate child care and low IQ are the result of the mother's low IQ or some exogenous factor not being measured.

The same problem arises when trying to interpret the results of some of the latest unpublished studies from California and the U.S. Office of Education, which indicate that when enough sociocultural and personality factors are treated simultaneously in a multiple regression analysis, one can account for a very large proportion (if not all) of the mean difference between minority and dominant groups in achievement on IQ or related tests. Some readers have jumped to the conclusion that these findings prove that group differences in intelligence are due wholly to environmental factors. Yet, it is not at all clear how the background variables in these studies are causally related to cognitive development or the extent to which they also have a genetic component. In controlling for some kinds of background variables, like socioeconomic status, one runs the risk of partialling out not only environmental but also genetic sources of the variance.

Our most sophisticated multivariate techniques and prediction equations will not resolve this kind of problem. On the other hand, explanatory models that force us to spell out the theoretical assumptions we are making whenever we infer causation would help us avoid erroneous conclusions and help us to focus on the kinds of questions that merit further study.

The Heritability of Intelligence and School Achievement

Dr. Morton raised a most interesting question in one of our discussions when he asked: What can the genetic contribution (heritability) of an input phenotype (IP) and its genotype (G) tell us, if anything, about the interrelationships between IP and a school output phenotype (OP) and the educational environment (EE)? The path models which he gave appear in Fig. 1, with the residuals (R) representing exogenous or unmeasured factors upon which the relationships also may depend but have not been included in the diagrams.

Morton argued that the left-hand model, even if we were able to precisely measure the relative effects of G and R upon IP, would not have any influ-

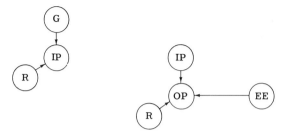

FIG. 1. Morton's models representing the relationships between genotype (G), input and output phenotypes (IP and OP) and the educational environment (EE).

ence upon the expression of the relationships outlined in the right-hand model; in other words, the effects of either *IP* or *EE* on *OP* would not depend on the magnitude of *G*. If we restrict the analysis only to gene and environmental interactions within the life cycle of a single generation, I suspect he may be correct.

The diagrams in Fig. 1 ignore intergenerational processes, however, and thus the full meaning of heredity. If the two diagrams were combined and the model extended to include the parental generation, I think a different conclusion would be drawn. That is, the genetic component of the input phenotype (*G* and *IP*) is indeed quite relevant to our understanding of the contributions of the input phenotype and the educational environment on school outputs.

Figure 2 illustrates the process I have in mind. The four new variables included in the model are the parents' genotypes and phenotypes for IQ and their own school achievement and attained socioeconomic status. The causal paths connecting these variables with each other and with the variables in Morton's model are identified by the arrows with broken lines. For purposes of this discussion, the child's input phenotype (*IP*) has been defined as IQ, and the output phenotype (*OP*) as school achievement. Let me briefly describe each of the new paths, beginning with those that extend from the parents' genotypes.[1]

The path from *PG* to *G* is the genetic parent–child correlation for intelligence. Under conditions of random mating, this path is .50. However, the more closely parents share the same genes for intelligence, the more closely their children will resemble them genetically. Thus, an additional component, one-half of the genetic assortative mating coefficient, must be added to this figure. If the coefficient for phenotypic assortative mating for intelligence is as high as .60, as some authors have suggested, the genetic

---

[1] For a more elaborate discussion of most of these paths and their consequences for intergenerational mobility, see Eckland (1971).

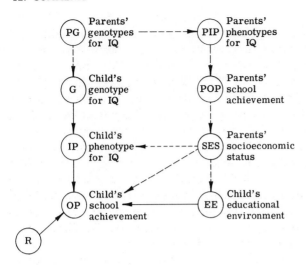

FIG. 2. An intergenerational model of the relationships between the genetics of intelligence and school achievement.

parent–child correlation may be as high as .65 to .75, depending upon the heritability of this trait.

The path from *PG* to *PIP*, like the path from *G* to IP, is the heritability of intelligence. The two differ only to the extent that the population variances of either the genotypes or the trait-relevant environments differ between the parents' and the child's generation. If such intergenerational differences are quite small, as I believe they are, the coefficients in these two paths should be essentially the same. Whether the values that we insert in these paths should be closer to .50 or closer to .90 (as Jensen has argued) also is incidental here. The point is that the contributions of *G* to *IP* and of *PG* to *PIP* are fairly strong.

The paths from *IP* to *OP* and from *PIP* to *POP* also are measuring the same thing, only in two different generations, i.e., the correlation between IQ and school achievement. The correlations here are generally found to vary from about .30 to .60, depending upon whether school achievement is being measured in terms of years of schooling, academic grades, or achievement tests. Owing to the rise of mass testing programs and the probable fact that educational selection has become increasingly meritocratic (Eckland, 1970), the correlation presently may be somewhat higher in the child's than in the parents' generation.

Next is the correlation between *POP* (the parents' school achievement) and *SES* (the parents' socioeconomic status). To the extent that our schools and colleges act as an avenue for upward mobility, which is the case in all large-scale open class societies, a relatively strong correlation will be found between educational and occupational attainment (or between education and any other measure of the general social status an adult has achieved).

From *SES*, three paths have been drawn, one to the child's phenotype for IQ (*IP*), another to the child's school achievement (*OP*), and another to the child's educational environment (*EE*). The first of these, *SES* to *IP*, can be justified on the common grounds that deprivation, as measured by the parents' social status, has an important effect upon cognitive development, whether due to inadequate child care or to a restricted learning environment. The path from *SES* to *OP* is included to indicate that, apart from the effects of the family's physical and cultural milieu on intelligence, the child's school performance may be similarly affected. When the child's IQ is held constant, academic achievement usually is found to still depend upon class background. This is particularly true during the high school years and in college-going decisions, although after students have entered college the independent effects of social class begin to disappear.

The educational environment (*EE*) or, in other words, where children go to school, is largely dependent upon social class (*SES*). We already have stressed the point that the correlation between academic performance and school quality in fact is largely due to the dependency of educational environments upon their socioeconomic character. Controlling for individual background variables, differences in educational environments have little measurable effect on student performance. Nevertheless, in order not to depart radically from Morton's basic model, we have retained the path from *EE* to *OP*, even though there is little evidence that much of the effects of *SES* on *OP* are mediated through the school environment. Such effects that *SES* has on academic performance apparently are far more direct and, for the most part, do not operate within the context of the school system.

The loop is now complete. Both sociocultural and genetic factors are shown to be implicated in school achievement (*OP*) *and in an interdependent manner,* which Morton's model does not take into account. Going back to his model in Fig. 1, the effects of *IP* and *EE* on *OP* are shown to be completely independent of the effects of *G* on *IP*. Our intergenerational model indicates otherwise.

First let us examine the relationship between the child's phenotype (*IP*) and his school achievement (*OP*). The observed or zero-order correlation between *IP* and *OP* has five separate components: the *direct* effect of *IP* on *OP*, plus four *indirect* effects, each of which involves a somewhat different set of variables or paths. The appropriate equation describing these paths is as follows:

$$r_{IP\text{-}OP} = p_{IP\text{-}OP} + p_{IP\text{-}SES}p_{SES\text{-}OP} + p_{IP\text{-}SES}p_{SES\text{-}EE}p_{EE\text{-}OP}$$
$$+ p_{IP\text{-}G}p_{G\text{-}PG}p_{PG\text{-}PIP}p_{PIP\text{-}POP}p_{POP\text{-}SES}p_{SES\text{-}OP}$$
$$+ p_{IP\text{-}G}p_{G\text{-}PG}p_{PG\text{-}PIP}p_{PIP\text{-}POP}p_{POP\text{-}SES}p_{SES\text{-}EE}p_{EE\text{-}OP}$$

All five components depend, in part, upon the relationship between G and IP, i.e., the heritability of intelligence. If the effect of G on IP is very large, the direct effects of phenotypic intelligence on school performance (IP on OP) may be substantially smaller than otherwise observed, particularly if SES is strongly involved in the process. This in fact appears to be the case, since the effects of SES on both IP and OP are partially determined by the genetic loop extending back to IP via the parents' background.

The relationship between the child's educational environment (EE) and school achievement (OP) is similarly affected, being composed of both direct and indirect effects, and all of which depend upon the size of the path from G to IP. These are:

$$r_{EE\text{-}OP} = p_{EE\text{-}OP} + p_{EE\text{-}SES}p_{SES\text{-}OP} + p_{EE\text{-}SES}p_{SES\text{-}IP}p_{IP\text{-}OP}$$
$$+ p_{EE\text{-}SES}p_{SES\text{-}POP}p_{POP\text{-}PIP}p_{PIP\text{-}PG}p_{PG\text{-}G}p_{G\text{-}IP}p_{IP\text{-}OP}$$

My conclusion is that the *measured* effects of both phenotypic intelligence and the educational environment on school achievement depend upon the heritability of intelligence. Other traits with significant heritabilities should be similarly affected. If intergenerational processes are ignored, of course, this connection will be overlooked.

If the model I have presented is basically correct (it is not substantially different from Wright's interpretation of Burke's data that Morton himself discusses in his paper), then the search for genetic sources of variation in the familial backgrounds of students is exceedingly relevant to understanding educational outcomes. Jensen indeed may be on target in suggesting that the failure of compensatory education (at least as applied to whites, who still make up a majority of America's poor) is *partially* the result of hereditary sources of variation that are associated with class distinctions in this country.

References

ASTIN, A. W. (1968). Undergraduate achievement and institutional excellence. *Science,* **161**, 661–668.

BOWLES, S., and LEVIN, H. M. (1968). The determinants of scholastic achievement—An appraisal of some recent evidence. *J. Hum. Res.* **3**, 3–24.

COLEMAN, J. S., *et al.* (1966). "Equality of Educational Opportunity." Washington: U. S. Government Printing Office.

DUBIN, R., and TAVEGGIA, T. C. (1968). "The Teaching-Learning Paradox." Univ. of Oregon Press, Eugene.

ECKLAND, B. K. (1970). New mating boundaries in education. *Soc. Biol.* **17**, 269–277.

ECKLAND, B. K. (1971). Social class structure and the genetic basis of intelligence. *In:* "Intelligence: Genetic and Environmental Influences," (R. Cancro, ed.), pp. 65–76. Grune and Stratton, New York.

FLEMING, E. S., and ANTTONEN, R. G. (1971). Teacher expectancy or My Fair Lady. *Amer. Educ. Res. J.* **7**, 241–252.

HAUSER, R. M. (1970). Context and consex: A cautionary tale. *Amer. J. Soc.* **75**, 645–664.

JOSE, J., and CODY, J. J. (1971). Teacher-pupil interaction as it relates to attempted changes in teacher expectancy of academic ability and achievement. *Amer. Educ. Res. J.* **8**, 39–49.

NICHOLS, R. C. (1964). Effects of various college characteristics on student aptitude test scores. *J. Educ. Psychol.* **55**, 45–54.

ROCK, D. A., CENTRA, J. A., and LINN, R. L. (1970). Relationships between college characteristics and student achievement. *Amer. Educ. Res. J.* **7**, 109–121.

ROSENTHAL, R., and JACOBSON, L. (1968). "Pygmalion in the Classroom." Holt, New York.

THORNDIKE, R. L. (1968). But you have to know how to tell time. *Amer. Educ. Res. J.* **6**, 692.

TURNER, R., BOYLE, R., MICHAEL, J., SEWELL, W., and ARMER, J. (1966). Comments on Neighborhood context and college plans. *Amer. Sociol. Rev.* **31**, 698–707.

Epilogue

GILBERT S. OMENN, ERNST CASPARI, and
LEE EHRMAN

Behavior Genetics and Educational Policy

Each of the authors has contributed to the primary theme of this book—the interaction or interrelationship of genetics and environment in behavior. Such an approach represents a rejection of both extreme positions in the age-old "nature–nurture" controversy. The genetic position of Galton led to claims of innate superiority of racial, social, or religious groups and rationalizations of special privileges for such subpopulations. In analogy to plant and animal breeding, some scientists advocated control of reproduction as a eugenic measure, to "improve" the gene pool for future generations. By contrast, the extreme behaviorist position, as represented by Watson and Skinner, claims that all "normal" individuals are equally responsive to behavioral manipulation. As summarized in Chapter 12 and in various reports of the Federal Office of Education and Office of Economic Opportunity in recent months, the environmentalist expectation that earlier and more intensive educational efforts would improve academic performance and overcome differences in ability of even quite young children appears to be meeting practical frustration. Thus, nowhere does recognition of genetic and environmental interaction have more important social consequences than in considerations of educational policies. In this final chapter, we wish to direct some summary comments to teachers and educational policy-makers.

The attribute most important for achievement in school is what both psychologists and laymen call "intelligence." Motivation, personality, and interpersonal relationships at home and in the school are also significant factors, affecting both achievement in school and performance on tests designed to measure "intelligence." Empirical correlations have established that such IQ tests do reflect ability to learn in school, at least in Caucasian population groups in Western societies. And it is apparent from their academic performance that school children vary in their capability for learning, especially abstract learning.

The variation in a quantifiable trait like IQ score can be studied with sophisticated methods to assess the relative role of genetic and environmental factors in the variation among individuals in any group (see Chapter 2). Comparisons of the IQ's of relatives, twins, and adopted children and their relatives indicate that high proportion of the variance observed within those Caucasian population groups that have been tested is genetic, usually 70–80% for IQ and even a significant figure for certain measures of personality. This does not mean that 70 points of an IQ score of 100 points are inherited. "Variance" refers to the range over which individuals of a specified population group score on a parameter like IQ. Part of this variance appears to be due to inherited differences in potential for learning, while the rest is due to differences in learning opportunities, in prenatal and postnatal physical conditions, and in the influences of parents, teachers, and peers on motivation and diligence in learning.

Much of the variance may reflect an interaction of genetic and environmental factors. For example, let us imagine that a group of individuals had a *genetic* potential for an IQ range of 85–115 in the usual school situations. With restricted learning opportunities, perhaps no one would attain more than an IQ of 100. Then the full extent of genetic variation would be underestimated. Poor nutrition during infancy also might impair the expression of the genetic potential and the response to later learning opportunities. Conversely, if learning methods which would enhance the performance of some of the children could be applied, then the total variance might become larger, producing an IQ range of 85–130, for example. The relative portion attributed to genetic factors could be either smaller or larger, depending upon the reasons for the improved results. The point to be emphasized is that the genetic–environmental interaction is dynamic. The fact that IQ has a large genetic component and that individuals differ in their genetic endowment should not discourage efforts to achieve the full potential of each individual.

One of the frustrating aspects of research in any field is the inability to specify just *how* certain factors operate. In genetics, we have learned that

DNA contains inherited information for the production of specific enzymes and other proteins that carry out vital functions of cells. The inheritance of some traits like red/green color blindness or diseases like cystic fibrosis or sickle-cell anemia is determined by a single gene on one of the 23 pairs of chromosomes. These patterns of inheritance can be analyzed just as Mendel analyzed the inheritance of color and size of pea plants almost 100 years ago. Individual genes that affect IQ adversely are also well known (Chapter 7). For example, phenylketonuria and galactosemia are metabolic disorders in which severe mental retardation occurs on a genetic basis, due to inability to handle the amino acid phenylalanine and the sugar galactose in the normal diet. Environmental manipulation—restricting these substances from the diet—prevents the toxic effects on the brain and allows affected children to develop normal IQ. Thus, these rare diseases are good examples of the interaction of genotype and environment.

However, most human traits and most common diseases are affected by multiple genes, as well as by external factors such as nutritional status, producing a pattern of polygenic or multifactorial inheritance. Analysis of these traits is complicated not only by the number of genes involved but also by the fact that the exact effects of each gene are usually not known. For example, blood sugar and blood cholesterol (lipid) levels, strongly influenced by genetic factors, are quantifiable traits that vary through a "normal" range and into the abnormal levels correlated with the diseases diabetes mellitus and coronary atherosclerosis, respectively. IQ may be considered an analogous quantifiable trait, having a large genetic component, but subject to influence by environmental factors. Dietary and drug management can predictably improve blood sugar and lipid values, but we are not yet aware of predictable ways of improving IQ scores (see Chapter 12). Also, it is known that abnormal blood sugar or blood lipid values may result from many different mechanisms, which require different, specific therapies. Thus, continuing with the analogy, different specific types of educational environments and teaching methods may be necessary to maximize the learning performance of different individuals. Constructive, innovative approaches to educational practice should be encouraged, but be subjected to objective evaluation. No single approach should be expected to be best for all students.

The frequencies of certain genes and the measures of quantifiable traits vary among different strains of animals (see Chapters 3,4,5, and 8) and among subpopulation groups of people. Human populations are commonly subdivided by socioeconomic status, geography, and ethnic or racial group. Certain measurable genes, like those affecting blood groups or hemoglobin, are known to have different frequencies in different subpopulations. Whether it is "better" or "worse" to have one form of a gene or another often depends upon the environment. Two examples have been discussed (Chapters 6 and 7). The sickle-hemoglobin gene became common in parts of the world where malaria was prevalent, since sickle-hemoglobin protects against malaria infections. Similarly lactose tolerance is associated with cultural use of milk by adults. Genetically-determined lactose intolerance leads to nausea, gas, and sometimes diarrhea, but only after consumption of milk. Some white children are lactose-intolerant and so are many black and oriental youngsters. Thus, teachers should be aware that milk as part of school lunch programs may be undesirable for some students.

More often, it is simply not known which form of the gene is more advantageous to survival, as with the ABO blood groups. But it is clear that tremendous genetic variation does exist among individuals and that the distribution of gene frequencies may be different from one subpopulation group to another. Thus, the genes that underlie normal intellectual development probably also differ in different populations. Nevertheless, two aspects of such genetic variation must be emphasized: first, the differences are characteristic of populations, not individuals, so that a broad range of genotypes exists within all subpopulations; and, second, a function so complex as cognition or intelligence undoubtedly involves a great many neuropsychological processes and many times more genes, so that we might expect genetic differences between groups going in a favorable direction for certain processes, in an unfavorable direction for others, and having no significant impact in yet other aspects of so global a phenomenon as intelligence. Thus, it is reasonable to expect that an individual's IQ score or the mean IQ score for individuals of any identifiable group might be different if the IQ test comprised different elements.

At the same time we must recognize, as have most teachers, that youngsters differ in their response to particular teaching approaches; some children thrive on discipline, while others seem to require greater freedom to explore some of their own interests. Individuals differ in their relative achievement in reading, vocabulary, mathematical, and other skills, so it may be advisable to place children in classes according to their ability in particular subjects, rather than primarily by chronological age or with the same children in each subject. Even for similar subject matter, it may be important to present the material with quite different teaching approaches for different children. The interrelationship of genetic and environmental factors in learning clearly constitutes a compelling challenge at both theoretical and practical levels to everyone involved in education.

Author Index

Numbers in italics refer to the pages on which the complete references are listed.

Abeelen, J. H. F. van, 202, *204*, 233, *234*
Adams, M. S., 262, *264*
Akrivakis, A., 182, *207*
Allison, A. C., 130, 143, *167*
Altman, J., *234*
Amano, T., 176, *177*
Amthauer, R. 286, *288*
Anderson, E. N., Jr., *234*
Anderson, V. E., 155, *167*, 174, *177*, *178*
Anfinsen, C. B., 131, *167*
Angioni, G., 271, *271*
Angst, J., 157, *167*
Anthony, E. J., 204, *204*
Anttonen, R. G., 299, *306*
Apgar, J., 133, *169*
Argiolas, N., 271, *271*
Armer, J., 299, *306*
Aronson, L. R., 38, *46*, 219, 226, *239*
Astin, A. W., 298, *305*
Attwell, A. A., 286, *288*
Augusti-Tocco, G., 176, *178*

Avis, O., 110, *122*
Ayala, F. J., 112, *120*

Baglioni, C., 136, *167*
Bailit, H. L., 269, *270*
Bakay, B., 154, *170*
Baker, P., 114, *120*, 125, *127*
Baker, W. K., 30, *43*
Barnicot, N. A., 137, *167*
Barondes, S. H., 137, 141, *167*, *168*
Barnes, D., *234*
Barranger, J. A., 177, *178*
Barry, W. F., *235*
Barthelmess, I. B., 69, *71*
Bartlett, M. S., 257, *264*
Bastock, M., 30, *43*
Bates, M. W., 29, *45*
Bateson, P. P. G., 36, *43*, 50, *52*
Bayless, T. M., 119, *122*
Benedict, R., *234*
Benikschke, K., 137, *169*

311

Benne, K. D., *234*
Bennett, E. L., 62, *66*
Bentley, D. R., 49, *52*
Benzer, S., 2, *4*, 68, *71*
Berman, J. L., 275, *284*
Berry, H. K., 250, 251, *264, 265*
Berry, R. J., 78, *96*
Bertalanffy, L. von, *234*
Bessman, S. P., 177, *178*
Bignami, G., 59, *65*
Billewicz, W. Z., 266, *270*
Birch, H. G., 228, *234*
Blewett, D. B., 287, *288*
Block, J. B., 287, *288*
Bloom, B. S., 220, *234*
Bock, R. D., 293, *295*
Bodmer, W. F., 88, *96*, *234*
Boesiger, E., 29, 34, *43, 45*
Boggan, W. O., *237*
Bonner, J. T., *234*
Borgman, R. D., 187, *207*
Borisov, A. I., 41, *43*
Boudreau, J., 2, *4*, 29, *44*
Bovet, D., 60, 61, *65*
Bovet-Nitti, F., 60, 61, *65*
Bowles, S., 299, *305*
Bowman, R. F., 292, *295*
Boyle, R., 299, *306*
Bramble, W., 293, *295*
Breese, E. L., 81, *96*
Brennan, J. G., 151, *169*
Bridger, W., *236*
Brittain, R. P., 292, *295*
Brncic, D., 30, *43*
Broadhurst, P. L., 76, *96*, 184, 188, 189, 193, 202, *204, 206*, 213, *214*, 223, 225, *234, 235, 238*
Brown, A. M., 287, *289*
Brown, C. P., *234*
Brown, J. C., 198, *204*
Bruell, J. H., 81, 82, 92, *96*
Brunton, M., 292, *295*
Bryan, A. L., 160, *167*
Bryan, J., 137, *167*
Buettner-Janusch, J., 109, *121*, 134, 137, *167, 169*
Bullock, T. H., 105, *121*, 161, *167*
Burks, B. S., 255, *264*, 282, *284*
Burnet, B., 69, *71*

Burnham, D., *234*
Burnham, L., 203, *206*
Burt, C., 210, 213, *214*

Callan, H. G., 135, *167*
Calvin, G. J., 133, *167*
Calvin, M., 133, *167*
Campbell, C. B. G., 110, *121*
Campenot, R. B., 196, *207*
Cardinal, R. A., 176, *178*
Carran, A. B., 60, *65*
Carter, J. E. L., 90, *96*
Caspari, E., 2, *4*, 75, *96,* 115, *121,* 184, 203, *204,* 215, *216, 234*
Caspersson, T. G., 285, *288*
Cattel, J. McK., *234, 235*
Cattell, R. B., 210, 213, *214,* 249, 251, *264,* 275, *284,* 286, *288*
Cavalli-Sforza, L. L., 88, 93, *96*
Centra, J. A., 298, *306*
Chadwick, L. E., 32, *46*
Chen, H. S., 146, *167*
Chevais, S., 37, *43*
Chevalier, J. A., 224, *237*
Child, I. L., 90, *96*
Childs, B., 161, *167*
Chilton, M-D., 140, *168*
Cliquet, R. L., 280, *284*
Chomsky, N., 158, 160, *167*
Christenson, J. G., 177, *178*
Chung, C. S., 10, *16,* 262, *265*
Cicero, T. J., 135, *167*
Clark, L., 199, *205,* 212, 214
Clark, P. J., 90, *96*
Clark, W. C., 198, *204*
Clausen, J., 112, *121*
Clow, C. L., 155, *170*
Cody, J. J., 299, *306*
Cohen, B. H., 249, *264*
Cohen, J., 281, *284*
Cohen, P. T. W., 144, 146, 148, 152, *167, 170*
Cole, J. M., 61, *66*
Coleman, J. S., 298, *305*
Collins, R. A., *235*
Collins, R. L., 60, *65,* 181, 198, *205*
Connolly, K., 69, *71*
Connor, J. D., 64, *67,* 154, *170*
Conway, J., 210, *214*
Cooper, R. M., 88, *96*

Author Index

Covington, M., 60, *66*
Cowan, W. M., 135, *167*
Cowen, J. S., 292, *295*
Cowie, V., 129, *171*
Cox, R. P., 182, *205*
Craig, W., 36, *43*
Crary, D. D., *46*
Cronbach, L. J., 225, *235*
Crow, J. F., 220, *235*
Curran, R. H., 38, *46*

Dahlstrom, A., 138, *169*
Dairman, W., 177, *178*
Dallaire, L., 155, *170*
Darlington, C. D., 220, *235*
Darwin, C., 30, *43*, 115, *121*
Das, G. D., *234*
David, E. E., 220, *235*
Davis, B. D., *235*
Davis, D. E., 88, *96*
Davis, K., *235*
Davis, P. C., 281, *284*
Dayhoff, M. O., 131, *167*
DeFries, J. C., 8, *16*, 90, *96*, 196, 197, *205*
DeLange, R. J., 134, *167*
Delius, J. D., *235*
Dell'Acqua, G., 271, *271*
DeMaree, R. G., 282, *285*
Deol, M. S., 78, *96*
DeSaix, C., 187, *207*
Diamond, I. T., 110, *121*
Dickie, M. M., 29, *45*
Dickerson, R. E., 134, *168*
Dingman, H. F., 286, *288*
Dixon, G. H., 131, *168*
Dixon, L., 196, *205*
Doane, W. W., 35, *43*, 70, *71*
Dobzhansky, T., 2, *4*, 27, 29, 30, 32, 33, 37, 39, *43*, *45*, *46*, 69, *71*, 87, *97*, 106, 112, 115, *121*, 200, 203, *205*, 210, *214*, 225, *235*, 250, 251, *264*, *265*
Dofuku, R., 152, *168*
Doyle, D., 3, *4*
Dreger, R. M., *235*
Drets, M. E., 285, *288*
Dubin, R., 298, *305*
Dunham, H. W., 186, *205*
Dutton, G. R., 137, *168*

Eccles, J. C., 166, *168*

Eck, R. V., 131, *167*
Eckland, B. K., 302, 303, *306*
Edwards, A. W. F., 93, *96*
Edwards, H. P., *235*
Edwards, J. H., 258, *264*
Ehrman, L., 29, 30, 39, 40, 41, 42, *43*, *46*, 52, *53*, 184, 205, 222, *235*
Eiduson, S., 173, *178*
Eisenberg, L., 231, *235*
Ekstrom, R. B., 286, *288*
Eleftheriou, B. E., 196, *206*
Elens, A. A., 30, 33, *44*
Eliasson, M., 177, *178*
Elston, R. C., 259, *264*, 279, *284*
Epstein, C. J., 131, 133, *168*
Erlenmeyer-Kimling, L., 29, *44*, 69, *71*, 129, *168*, 204, *205*, 253, 254, *264*
Erway, L. C., 176, *178*
Everett, G. A., 133, *169*
Ewing, A. W., 32, *44*, 49, 53, 222, *235*
Eysenck, H. J., 90, *97*, 251, *264*

Falconer, D. S., 6, 8, 9, *16*, 258, *264*
Fambrough, D. M., 134, *167*
Farmer, H., 61, 62, *66*
Faugères, A., 34, *44*
Feldman, D. S., 176, *178*
Felsher, B. F., 175, *179*
Ferguson, A., 116, *121*
Fessas, P. H., 182, *207*
Fine, A. E., 116, *121*
Fisch, R. O., 155, *167*, 174, *178*
Fisher, R. A., 13, *16*, 90, *97*, 217, *218*, 250, 252, *265*
Flatz, G., 116, *121*
Fleming, E. S., 299, *306*
Ford, R., 275, *284*
Forest, J. M. S., *235*
Foster, R., 286, *288*
Fox, M. W., *235*
Fox, R., 162, *168*
Fox, S. S., 139, *168*
Franklin, I., 130, 143, *168*
Fraser, A. S., 176, *178*
Fraser, F. C., 182, 203, *205*
Fraser, G. R., 182, *207*
Freedman, D. G., 155, *168*
Freedman, N. C., 155, *168*
French, J. W., 286, *288*
Friedlaender, J. S., 263, *265*

Fudenberg, B. R., 203, *205*
Fudenberg, H. H., 203, *205*
Fujimori, M., 139, *168*
Fulker, D. W., 6, *16*, 19, *21*, 76, *97*, 210, 213, *214*, 223, *235*, *238*
Fuller, J. L., 19, 20, *21*, 29, *44*, 64, *65*, 129, *168*, 181, 184, 193, 198, 199, *205*, *207*, 212, *214*
Fuxe, K., 138, *169*

Galton, F., *236*
Gandini, E., 271, *271*
Ganschow, R. E., 3, *4*
Garcia, J., 51, *53*
Gardner, B. T., 160, *168*
Gardner, R. A., 160, *168*
Gartler, S. M., 250, 251, *264*, *265*, 271, *271*
Gatenby, P. B. B., 116, *121*
Gauel, L., 215, *216*
Gaze, R. A., 49, *53*
Gazzaniga, M. S., 161, *168*
Geiger, P. J., 177, *178*
Gerald, R. W., *235*
Geschwind, N., 162, *168*
Gibbons, I. R., 136, *170*
Giblett, E. R., 146, 155, *167*, *168*
Gibson, J. B., 186, *207*, 210, *214*
Ginsburg, B. E., 181, 182, 192, 196, 198, 202, 203, *205*, *236*, 292, 294, *294*
Glass, B., 20, *21*, *235*
Glass, D. C., *236*
Glasser, S. R., 152, *171*
Glazer, D., 176, *178*
Golden, M., *236*
Goldman, P. S., 224, *236*
Gordon, E. W., 227, 229, *236*
Gottesman, I. I., 112, 113, *121*, 186, 201, *205*, *236*, 279, *284*
Gottlieb, F. J., *236*
Gottlieb, G., 226, *236*
Goy, R. W., 35, *44*, 151, *168*, *171*
Graham, G. G., 119, *122*
Grant, M., *236*
Gray, H., 250, *265*
Greulich, W. W., 112, 113, *121*
Griek, B. J., 62, *66*, 222, *239*
Gross, S. R., 175, *178*
Grossfield, J., 34, *44*
Grouse, L., 140, *168*

Grubb, R., 271, *271*
Grüneberg, H., 78, *97*, 258, *265*
Grünt, J. A., 35, *44*
Gudschinsky, S. C., 158, *168*
Guilford, J. P., 286, *288*
Gussow, J. D., 228, *234*
Guttman, R., 83, *97*

Hahn, W. E., 140, *168*
Haldane, J. B. S., 90, *97*, 135, *168*, 181, 186, 187, 190, *205*
Hall, W. C., 110, *121*
Haller, M. H., 200, *205*
Halsey, A. H., 210, *214*
Harding, R. S., 137, *171*
Harrington, G. M., 182, 189, *206*, *236*
Harris, H., 70, *71*, 125, *127*, 144, 155, *168*, 175, *178*, 203, *206*
Harris, R. E., 61, *65*
Harrison, G. A., 114, *121*
Harvey, S., 175, *179*
Hauser, R. M., 299, *306*
Hayes, C., 160, *168*
Hayes, K. J., 160, *168*
Hayes, W. L., 221, *236*
Hegmann, J., 197, *205*
Heinroth, M., 36, *44*
Heinroth, O., 36, *44*
Henderson, N. D., 19, *21*, 181, 189, 190, 191, 192, 193, 196, 198, 199, 202, *206*
Henry, K. R., 64, *65*, 248, *265*, 292, *295*
Hermelin, B., 260, *265*
Heron, W. T., 59, 60, *66*
Herrnstein, R., 232, *236*
Hess, E. H., 36, *46*
Heston, L. L., 90, *97*, 203, *206*
Hiesey, W. M., 112, *121*
Hilgrad, E. R., *235*
Hill, R. L., 134, *169*
Hillarp, N. A., 138, *169*
Himmich, H. E., 139, *168*
Himwich, W. A., *236*
Hinde, R. A., 50, *53*
Hirsch, J., 2, *4*, 29, *44*, 48, *53*, 69, *71*, 129, *169*, 184, 202, *206*, 222, 225, 226, *236*
Hockett, C. F., *169*
Hodos, W., 110, *121*
Hoepfner, R., 286, *288*
Hoffer, A., 29, *45*
Holley, R. W., 133, *169*

Author Index

Holloway, R. L., 236
Honzik, M. P., 275, *284*
Horowitz, N. H., *236*
Howard, M., 213, *214*
Howe, W. L., 76, 78, 79, *97*
Howell, R. R., 203, *206*
Hoy, R. R., 49, *52*
Hoyer, B. H., 139, *169*
Hsu, T. C., 33, *47*, 137, *169*
Hubby, J. L., 27, *44*, *45*, 70, *71*, 125, *127*
Hundleby, J. D., 249, *264*
Hurley, L. S., 176, *178*
Hutton, J. J., 175, *178*
Huxley, J., 29, *45*
Huzino, A., 177, *178*

Ikeda, K., 29, *45*, 49, *53*
Immelman, K., 50, *53*
Ingle, D. J., *236*, *237*
Ingram, V. M., 136, *169*

Jacobs, P. A., 292, *295*
Jacobson, L., 299, *306*
Jacoby, G. A., 29, *44*
Jahoda, G., *237*
Jakway, J. S., 35, *44*
Jarvik, L. F., 129, *168*, 253, 254, *264*
Jennings, R. D., 61, *66*
Jensen, A. R., 6, *16*, *237*, 261, *265*, 295, *295*
Jerison, H. J., 108, 109, *121*, 137, *169*
Jinks, J. L., 6, *16*, 19, *21*, 76, *97*, 184, 189, 202, *204*, 210, 213, *214*
Jolly, C. J., 137, *167*
José, J., 299, *306*
Jumonville, 292, *295*

Kaplan, W. D., 29, *45*, 49, *53*
Katz, I., 227, *237*
Katzin, M., 203, *206*
Kaufman, I. C., 193, *206*
Kaul, D., 32, 34, *45*
Keck, D. D., 112, *121*
Keele, D. K., 154, *170*
Kellogg, W. N., 160, *169*
Kerkut, G. A., 223, *238*
Kessler, C., 198, *207*
Kessler, S., 69, *71*
Kety, S. S., 149, *169*, 203, *207*, 277, *284*, *285*

King, J. A., 196, *206*
King, J. C., *237*
Klineberg, O., 261, *265*
Klopfer, P. H., *237*
Kluger, J., 212, *214*
Koelling, R., 51, *53*
Koehn, R. K., 136, *169*
Konishi, M., 50, *53*
Konopka, R. J., 68, *71*
Kooptzoff, U., 155, *170*
Korchin, S. J., 224, *237*
Koref Santibanez, S., 30, *43*
Kornetsky, C., 177, *178*
Krech, D., 62, *66*
Krechersky, I., 60, *66*
Kummer, H., 51, *53*
Kuo, Z. Y., *237*

Laird, C. D., 140, *168*
Langer, B., 30, 32, *46*
Laughlin, W. S., 294, *295*
Leacock, E., 227, 229, *237*
Lederberg, J., *237*
Lees, R. B., 158, *169*
Lehman, A., 29, *45*
Lehrman, D. S., *237*
Leland, H., 286, *288*
Lenneberg, E. H., 160, *169*
Lerner, I. M., 13, *16*
Levene, H., 33, *43*, 250, 251, *264*
Levere, R. D., 176, *178*
Levin, H. M., 299, *305*
Levine, P., 203, *206*
Levine, S., 193, *206*, 224, *237*
Lew, R., 258, *265*
Lewontin, R. C., 6, 10, *16*, 27, *44*, *45*, 70, *71*, 125, *127*, 130, 143, *168*, *237*
Li, C. C., 20, *21*, 105, 110, *121*
Lieberman, J. S., 176, *178*
Lindquist, E. F., 188, *206*
Lindzey, G., 59, 60, 64, *66*, *67*, 88, 89, 90, 91, 92, *97*, *98*, 181, 196, *206*, *237*, 250, 254, *265*
Linn, R. L., 298, *306*
Lisk, R. D., 152, *169*
Livingstone, F. B., 165, *169*
Loehlin, J., 59, *66*, 181, *206*, 275, 282, *284*
Lomakka, G., 285, *288*
Lorenz, K., 36, *45*, 112, *121*

Lubin, A., 188, 189, *207*
Lush, J. L., 8, *16*, 111, 118, 119, *122*
Lykken, D. T., 181, 196, *206*

MacLean, C. J., 268, 269, *270*
MacLeod, C., 182, *205*
Manning, A., 31, 32, 34, 35, *45*, 49, 50, 53, 69, *71*
Manosevitz, M., 59, *66*, 88, *97*, 181, 196, *206, 207*
Margoliash, E., 134, *169*
Markert, C. L., 136, *169*
Marks, J. F., 154, *170*
Marler, P., 50, *53*
Marqhisse, M., 133, *169*
Mass, J. W., *237*
Mather, K., 81, *96, 97*, 183, 187, *207*
Maxson, S. C., 292, *295*
Maxwell, J. D., 116, *121*
Mayer, J., 29, *45*
Maynard Smith, J., 34, *45*
Mayr, E., 29, 30, 31, 32, 33, 37, *45*, 110, 111, *122*, 225, *237*
McCarthy, B. J., 139, 140, *168, 169*
McClearn, G. E., 59, 60, 63, *66*, 91, *97*, *239*
McClemont, W. F., 292, *295*
McCormick, N., *234*
McCraken, R. D., 116, *122*
McDonald, G. S. A., 116, *121*
McGaugh, J. L., 61, *66*
McGregor, I. A., 266, *270*
MacInnes, J. W., *237*
McKenzie, J. A., 87, *97*
McKim, M., 213, *214*
McLean, P. D., 139, *169*
Mead, M., *237*
Medioni, J., 29, *45*
Mednick, S. A., 204, *207*
Melville, M. M., 292, *295*
Meredith, H. V., 113, *122*
Merrill, S. H., 133, *169*
Meyers, C. E., 286, *288*
Mi, M. P., 10, *16*, 262, *265*
Michael, J., 299, *306*
Minna, J., 176, *178*
Mirsky, A. F., 174, *178*
Money, J., 151, 154, *170*, 249, *265*
Monod, J., 135, 141, 157, *170*
Monroy, A., *237*
Moor, L., 274, *284*
Moore, B. W., 135, *167*

Morch, E. T., 113, *122*
Morley-Jones, R., 183, 187, *207*
Morrison, M., 136, *170*
Morton, N. E., 10, *16*, 258, 262, *265*
Motulsky, A. G., 130, 131, 143, 144, 146, 148, 149, 150, 153, 157, *167, 168, 169, 170*, 182, *207, 238*, 292, *295*
Murphy, E. A., *238*
Myers, K., 88, *97*

Naeslund, J., 276, *284*
Nagy, Z., 198, *207*
Needham, J., *238*
Neel, J. V., 164, *170*, 220, *235*, 262, *264, 265*
Nelson, P., 176, *178*
Neurath, H., 136, *170*
Newell, T. G., 196, *207*
Nice, M. M., 226, *238*
Nichols, R. C., 298, *306*
Nihira, K., 286, *288*
Nirenberg, M., 176, *177, 178*
Niswander, J. D., 269, *270*
Noble, C. E., 225, *238*
Nuttin, J., 280, *284*
Nyhan, W. L., 154, *170*

O'Conner, N., 260, *265*
Ohno, S., 135, 136, 152, *168, 170*
Ojemann, G. A., 162, *170*
Olian, S., 197, *207*
Oliverio, A., 60, 61, *65*
O'Malley, B. W., 152, *171*
Omenn, G. S., 140, 144, 146, 148, 149, 152, 157, *167, 168, 170, 238*, 292, *295*
Ornitz, E. M., 176, *178*
Orpet, R. E., 286, *288*
Osborne, R. H., 250, 251, *264*
Osmond, H., 29, *45*
Ottinger, D. R., *238*
Owen, K., 247, *265*

Paige, D. M., 119, *122*
Paigen, K., 175, *178*
Palay, S. L., 140, *171*
Papayannopoulou, T. H., 182, *207*
Pare, C. M. B., 157, *170*
Parker, J. B., 249, *265*
Parnell, R. W., 90, *97*
Parsons, P. A., 32, 34, *45*, 75, 76, 78, 79, 80, 82, 84, 86, 87, 90, 92, *97*
Passmanick, B., *238*

Author Index

Pavlovsky, O., 29, *43*
Peacock, J., 176, *178*
Pearson, W., 293, *295*
Pelton, R. B., *239*
Penhoet, E., 136, *170*
Penswick, J. R., 133, *169*
Perttunen, V., 29, *45*
Petit, C., 30, 31, 32, 34, 35, 39, 41, 42, *45, 46,*
Phoenix, C. H., 151, *171*
Piaget, J., 162, *170*
Ploog, D., 160, *170*
Polansky, N. A., 187, *207*
Polanyi, M., 213, *214*
Pollitzer, W. S., 125, 126, *127*
Porter, R. B., 286, *288*
Porter, R. H., *239*
Post, R. H., 119, *122*
Prakash, S., 70, *71*
Prell, D. B., 251, *264*
Price, L. A., 286, *288*
Price, W. H., 293, *295*
Przyboa, E., *238*

Raaijmakers, W. G. M., *234*
Rabinowitch, E., *238*
Rajkumer, T., 136, *170*
Rakic, P., 138, *170*
Ramsay, A. O., 36, *46*
Rasmussen, D. I., 136, *169*
Redeker, A. G., 175, *179*
Reed, C. F., *238*
Reed, S. C., 32, *46*
Reed, T. E., *234,* 276, *284*
Rees, L., 157, *170*
Reese, E. P., 50, *52*
Reiss, A. J., 283, *285*
Reitan, R. M., 155, *170*
Renaud, F. L., 136, *171*
Rendel, J. M., 33, *46*
Rensch, B., 130, 163, 165, *171*
Ressler, H., 197, *207*
Richards, K., *238*
Richards, M., *238*
Richards, O. W., 29, *46*
Richelson, E., 176, *177*
Rick, J. T., 223, *238*
Riege, W. H., *238*
Riesen, A. H., *238*
Riesenfeld, A., *238*
Riss, W., 35, 36, 37, *47*
Roberts, D. F., 266, *270*

Roberts, J. A. F., 249, *265*
Roberts, R. C., 7, *16*
Robertson, A., 251, *265*
Robertson, F. W., 34, *46,* 69, *71*
Rock, D. A., 298, *306*
Roderick, T. H., 62, *66*
Rodgers, D. A., 222, *238*
Roguski, H., *238*
Rose, A., 76, 80, 82, 84, 86, *97*
Rosenblatt, J. S., 38, *46*
Rosenblum, L. A., 193, *206*
Rosenzweig, N. S., *120,* 235, *127*
Rosensweig, M. R., 62, *66, 238*
Rosenthal, D., 129, *171,* 200, 203, *207,* 213, *214,* 277, *285,* 291, *295*
Rosenthal, R., 299, *306*
Ross, S., 198, *207*
Roth, N., 175, *178*
Rothenbuhler, W. C., 49, *53,* 183, *207*
Rowe, D. S., 266, *270*
Rowe, H. J., 136, *171*
Rowland, C. R., 196, *208*
Rowland, R., 165, *171*
Royce, J. R., 60, *65, 66,* 291, *296*
Russell, L., 116, *122*
Rutschmann, J., 198, *204*
Rutter, W. T., 136, *170*

Saengudom, C., 116, *121*
Sainsbury, M. J., 157, *170*
Sarich, V. M., 107, *122*
Sarton, G., *238*
Sato, G., 176, *178*
Saunders, P. R., 281, *285*
Sawin, P. B., 38, *46*
Scales, B., *239*
Scarr-Salapatek, 276, *285*
Schartzbroin, T., 61, 62, *66*
Scheibel, A. B., *238*
Scheibel, M. E., *238*
Schimke, R. T., 3, *4*
Schlesinger, K., 60, 62, 64, *65, 66,* 222, *237, 239,* 248, *265*
Schmitt, F. O., 141, *171*
Schneirla, T. C., 221, 222, 227, 228, *239*
Schull, W. J., 262, *265*
Schulsinger, F., 204, *207*
Schutz, F., 50, *53*
Scott, J., 198, *207*
Scott, J. P., 184, *207*
Scriver, C. R., 155, *171*
Searle, A. G., 78, *97*

Author Index

Searle, L. V., 59, 66, 211, *214*
Sebeok, T. A., 160, *171*
Selander, R. K., 70, *71*, 143, *171*
Seligman, M. E. P., 51, *53*
Sells, S. B., 282, *285*
Sewall, D., 69, *71*
Sewell, W., 299, *306*
Shaffer, J. A., 249, *265*
Shaw, E., 219, 226, *239*
Shaw, M. W., 285, *288*
Sheldon, W. H., 79, 89, 90, *98*
Shellhaus, M., 286, *288*
Shields, J., 129, *171*
Shih, Jean-Hung C., 173, *178*
Shockley, W., 229, 232, *239*
Shooter, E. M., 140, *171, 239*
Sidman, R. L., 138, *170*
Siegel, F. S., 155, *167*, 174, *177, 178*
Simon, H. A., *235*
Simons, E. L., 137, *171*
Simons, J. A., 176, *178*
Simoons, F. J., 116, *122*
Simpson, G. G., 130, 135, 163, *171*
Singer, C., 220, *239*
Sitkei, E. G., 286, *288*
Skeels, H. M., 275, *285*
Skinner, B. F., 60, *66*
Skodak, M., 275, *285*
Skowron, S., *238*
Skowron-Condrzak, A., *238*
Slater, E., 29, *46*, 129, *171*
Smith, B. J., 187, *207*
Smith, C., 258, *265*
Smith, C. U. M., 137, *171*
Smits, A. J. M., *234*
Snedden, D. S., *214*
Snijders, J. T., 286, *288*
Snijders-Oomen, N., 286, *288*
Snyder, S. H., 141, *171*
Solecki, R. S., 158, *171*
Sotelo, C., 140, *171*
Spassky, B., 2, *4*, 29, 39, *43*, *46*, 69, *71*, 210, *214*
Spears, D., *238*
Spelsberg, T. C., 152, *171*
Sperry, R. W., 161, *168*
Spevak, A. A., *214*
Spielberger, C. D., 249, *265*
Spiess, E. B., 30, 32, 33, 34, 39, *46*
Spiess, L. D., 30, 32, 34, 41, *46*,
Spieth, H. T., 33, *47*

Spuhler, J. N., 90, 92, *98*, 250, 254, *265*
Stafford, R. E., 287, *289*
Stamatoyannopoulos, G., 182, *207*
Steggles, A. W., 152, *171*
Stent, G. S., 172, *178, 239*
Stern, C., 120, *122*, 220, *235, 239*
Stevenson, R. E., 203, *206*
Storrs, E. E., *239*
Strand, L. J., 175, *179*
Strandskov, H. H., 287, *289*
Streisinger, G., 30, *43*
Sudarshan, K., *234*
Suntzeff, V., 135, *167*
Swadesh, M., 158, *171*
Swencionis, C. F., 196, *207*
Swin, P. B., 38, *46*
Symington, L., 61, 62, *66*
Sze, P. Y., 292, *295, 296*

Taddeini, L., 175, *179*
Tamura, M., 50, *53*
Tan, C. C., 32, *47*
Tanabe, G., *238*
Tanna, V. L., 249, *265*
Taveggia, T. C., 298, *305*
Tellegen, A., 155, *167*, 174, *178*
Tettenborn, U., 152, *168*
Theile, A., 249, *265*
Thibout, E., 34, *44*
Thiessen, D. D., 59 *66*, 91, *97*, 107, 111, *122*, 152, *171*, 181, 202, *206, 207, 237*, 247, *265*
Thoday, J. M., 186, *207*, 210, *214*, 275, *285*
Thomas, C. B., 249, *264*
Thompson, W. R., 59, *66*, 129, *168*, 197, 198, *207, 214*
Thomson, C. W. 59, 61, *66*
Thorndike, R. L., 299, *306*
Thorpe, W. H., 38, *47*
Thurstone, L. L., 287, *289*, 291, *296*
Thurstone, T. G., 287, *289*
Tinbergen, N., 115, *122*
Tobach, E., 219, 224, 226, 228, *236, 239*
Tolman, E. C., 59, *66*
Trout, W. E., 29, *45*
Truslove, G. M., 78, *98*
Tryon, R. C., 1, *4*, 29, *44*, 59, *66*, 211, *214*
Tukey, J. W., *235*
Tunnicliff, G., 223, *238*

Author Index

Turner, R., 299, *306*
Tyler, P. A., 63, 64, *66*, *239*

Udenfriend, S., 177, *178*

Vale, C. A., 182, 187, 191, 202, 203, *207*
Vale, J. R., 182, 187, 191, 202, 203, *207*
Valenstein, E. S., 35, 36, 37, *47*
Vandenberg, S. G., 213, *214*, 274, 276, *285*, 287, *289*, 293, *295*
van Martens, E., 215, *216*
Verhage, F., 279, *285*
Vesell, E. S., 182, *207*
Vicari, E. M., 62, *66*
Vitale, J. J., 29, *45*
Vogel, F., 156, *171*
Vogel, P., 203, *206*

Wade, P. T., 137, *167*
Walker, R. N., 90, *98*
Walsh, K. A., 136, *170*
Walsh, R. J., 158, *171*
Warburton, F. W., 286, *289*
Ward, A. A., Jr., 162, *170*
Washburn, S. L., 110, *122*, 137, *171*
Watson, C. J., 175, 176, *178*, *179*
Watson, J. B., 158, *171*
Watson, J. D., 290, *296*
Wattiaux, J. M., 33, *44*
Watts, C. A., 286, *288*
Wehmer, F., *239*
Weiner, J., 114, *120*
Weir, D. G., 116, *121*
Weir, M., 197, *205*
Welsh, J. D., 116, *122*
Weltfish, G., *234*
Wetterberg, L., 175, *179*
Whelan, D. T., 155, *171*
Wherry, R. J., 61, *66*
Whatmore, P. B., 293, *295*
White, M. J. D., 135, *171*

Whitney, G., 198, *207*
Whitset, M., 247, *265*
Wiesenfeld, S. L., 185, *208*
Wilcock, J., 223, *235*, *238*
Wilde, G. J. S., 212, *214*
Wilkerson, D. A., 227, 229, *236*
Will, D. P., 282, *285*
Williams, C. M., 32, *46*
Williams, R. J., *239*
Willits, V. L., 116, *122*
Willoughby, E., 116, *121*
Wilson, A. C., 107, *122*
Wilson, L., 137, *167*
Wimer, R., 60, 61, 62, *66*
Winokur, G., 249, *265*
Winston, H. D., 60, 64, *66*, *67*, 196, *206*, *208*
Winter, W. P., 136, *170*
Wirt, R. D., 155, *167*, 174, *178*
Witt, P. N., *238*
Wolf, H. H., 196, *208*
Wong, P., 29, *45*
Woodworth, R. S., 251, *265*
Workman, P. L., 125, *127*, 268, 269, *270*
Wright, S., 255, *265*
Wyspianski, J. O., *235*

Yang, S. Y., 70, *71*
Yates, F., 13, *16*
Yee, S., 258, *265*
Yeudall, L. T., 60, *65*
Young, H. B., 249, *264*
Young, W. C., 35, 36, 37, *44*, *47*, 151, *171*

Zamenhof, S., 215, *216*
Zamir, A., 133, *169*
Zech, L., 285, *288*
Zerbolio, D. J., Jr., 60, *67*
Zigas, V., 158, *171*
Zschiesche, O. M., 116, *122*
Zubek, J. P., 88, *96*

Subject Index

Accidental deaths, 165
Anatomical features of brain, 108, 137–139, 160–162
Advantage of the rare type, 39–42, 52, 184
Alcoholism, 278
Artificial insemination programs, 293–294
Assortative mating, 92, 126, 156, 164, 217, 279, 302
Audiogenic seizure susceptibility, 62, 181, 198, 292

Baboon social groups, 51
Behavior
    courtship, 30–34, 49
    and morphology, 79, 86, 89–90
    units of, 49, 150, 162, 164, 294
Behavior modification, 151, 157, 161, 166
Behaviorism, 56, 200
Biometrical analysis, 7–18, 81–83, 189, *see also* Methodology for human behavior studies

Brain
    anatomic features of, 108, 137–139, 160–162
    development, 150, 215–216
    protein synthesis in, 140, 152
    proteins, 134, 136, 141
    transcription of DNA in, 139–140
Brain waves, *see* Electroencephalograms
Breeding procedures, 58, 63, 224

Chromosomal abnormalities, 151, 154, 249, 292
    XO and space–form perception, 249
Chromosomal abnormalities, 151, 154, 249, 292
    XXY, 151, 249, 292
    XYY, 249, 292–293
Classroom interactions, 227–228, 297–300
    teachers' expectations, 299
Coadaptive gene complexes, 184
Color blindness, hunters and gatherers, 126

Subject Index

Cognitive processes
  development of 157–163, 227, 281–282, 304
  effects of schooling, 298–301
  fractionation of abilities, 243–246
Courtship behavior, *see* Behavior, courtship
Criminal behavior, 90, 292
Culture, 28
  cultural–biological interactions, 108–110, 114–115, 118, 126, 130, 162–166
  definition, 51
  evolution, 109–110, 114, 130, 132, 157–158, 162
  instinct and, 162–163

Dauermodification, 215
  burial practices, 158
Deaths
  accidental, 165
Developmental patterns, 226
  effects of protein under nutrition, 215–216, 228
  epistatic interactions, 183
  fixed-action patterns, 49
  ontogeny of learning, 211, 281
  sexual behavior, 34
  timing, 198–199, 270, 292
DNA (deoxyribonucleic acid)
  evolution of, 131, 133
  transcription of DNA in brain, 139–140
Dominance, *see* Gene–environment interaction
Drosophila species
  anatomical features, 31
  circadian rhythms, 68
  directional selection, 69, 72–73
  enzyme polymorphisms, 70
  extreme environments, 87–89
  geotaxis, 29, 72
  phototaxis, 29, 72
  rare types, 39–42, 184
  sexual selection, 30–37

Educational policy
  compensatory programs, 305, 307–310
  evaluation of schooling effects, 298–301
  gene–environment interaction, 297–305, 307–310
  racial groups, 260–263, 307
  school lunch programs and milk, 118–119
  variety of approaches, 127, 201, 262–263, 310
Electroencephalograms (EEG, brain waves) 155, 156
Environment, *see* Gene–environment interaction
Epistasis, *see* Gene–environment interaction, Developmental patterns
Ethics of behavioral research, 229–232
Ethology, 48, 198, *see also* Behavior, units of
Evolution
  by allelic genes, 133
  biological, 130–166, 220
  of brain, 57, 108–110, 124, 137–166
  cultural, 109–110, 114, 130, 132, 157, 158, 162
  of DNA, 131, 133
  by gene duplication, 135
  of human behavior, 57, 109–112, 120, 129–166
  of language, 158
  of proteins, 107, 112, 123, 133, 134, 173
  of RNA, 133, 135
Evolutionary trees
  linguistic, 158–159
  molecular, 107, 108, 134, 158
  paleontological, 132

Frequency-dependent selection, 39–42, 52, 126

Gene action, 129–130, 133, 139–141, 172, 222, 308–309
Gene duplication, 135
Gene–environment covariance, 184–187, 210, 250, 300
Gene–environment interaction, *see also* Learning
  adaptive potential, 114
  analysis of variance model, 188–191
  behavioral homeostasis, 80–94
  bird songs, 38, 50
  classification of types, 181–191

color blindness, 126
dietary effects
   galactosemia, 309
   general nutrition, 113, 294
   lactose intolerance, 116–119, 310
   phenylketonuria, 309
drug effects (psychopharmacogenetics), 156–157, 149
early life experiences, 38, 49, 191–197, 203–204, 213, 224, 294–295, 300–301
general, 1, 6, 93, 129, 181–204, 294
hormones, 35, 151–152
immunity to tropical diseases, 116
IQ, 186–187
mate selection in Drosophila, 28–34
mathematical treatment, 18, 19, 248–259
overcrowding, 34, 42, 88
porphyria, 175
reaction range concept, 122–114, 125–127, 201
school issues, 227, 250, 297–301, 307–310
socialization, 37, 50
Genes, allelic, evolution of, 133
Genes, major, for polygenic inheritance, 64, 126, 150, 241, 247–259
Genes/genotypes as variables, 65, 192–197, 248, 291–292
Genetic heterogeneity, 63, 174, 249, 278, 294
Genetic homeostasis, 73, 82
Geotaxis, 29, 72
Glottochronology, *see* Lexicostatistics
Glycolysis and energy production, 144–149, 173
Group differences, 260–263, 294
Guinea pigs, sexual behavior, 35–37

Hemoglobin, *see* Sickle hemoglobin and malaria
Heritability, 26, 73, 111, 184, 212, 243, 253–259, 262, 270, 279, 303–305, 308
   of group differences, 23–25, 260–263, 267
Heritance, 248
Heterozygotes for rare recessive disorders, 71, 154, 248, 288, *see also* Inborn errors in metabolism

Home environment, 282–283
Homo erectus, 108–109, 115
Homo neanderthalensis (Shanidar), 158
Homocystinuria, 154
Honeybees, 49, 183
Hybrid human populations, 100, 103, 113, 126, 261, 267–269
Hyperactive children, 157

Imprinting in birds, 36, 50
Inborn errors of metabolism, 152–155, 173, 287–288, *see also* Heterozygotes
Incest taboo, 263
Individuality
   behavioral, 64–65, 101, 110–111, 209, 262
   biochemical–enzymatic, 142, 144, 173
Innate versus acquired, 27, 36, 307
Intelligence, 59, 241–246, 253–256, 302, *see also* Cognitive processes
   factor analysis, 243–244, 259–260
   IQ scores, 255, 274–277, 308–310
   achievement and, 302–305
   heritability, 253, 262, 301–305, 308
   mother–child studies, 187
   sex aneuploidies, 274
   social class, 186, 210
   test design, 241–243
Isoenzymes (isozymes), 142–149, 173
Isozymes, *see* Isoenzymes

Language, 130, 157–163, *see also* Evolutionary trees, linguistic
   anatomical aspects, 160–162
   brain center, 161–162
   developmental aspects, 160, 281, 293
   evolution of, 158
   learned by chimps, 160–161
   relationships, 158–160
   vocalization, 160
   written, 163
Learning, 50, 59–65, 70, 101
   major gene effect, 64
   maze performance, 59, 193, 211–212
   measures of, 85, 197–198, 241–246
   multilevel interactions, 184
   neurochemical studies, 62
   ontogeny of learning, 211
   practice effects, 61
   specific abilities, 125, 243

Subject Index

strain differences, 59–65
Lesch–Nyhan syndrome, 152
  impulsive, destructive behavior, 152
Lexicostatistics (glottochronology), 158
Linkage studies, 126, 246, 275, 293

Malaria, *see* Sickle hemoglobin and malaria
Maternal and prenatal effects, 38–39, 191–197, 203–204, 213, 215–216, 269
Membrane macromolecules, 141, 176
Memory-storage processes, 141
Metabolism, *see* inborn errors of
Methodology in human behavioral studies, 89–91, 99–102, 142–157, 213, 241–246, 247–264, 266, 274–288, 290–295, *see also* Biometrical analysis, Chromosomal abnormalities, Human hybrid populations, Inborn errors of metabolism, Polygenic inheritance, Tests, Twins
  adopted children and half-siblings, 275
  cooperative research, 274–288
  intraindividual variability, 242–246
  non-Western societies, 266–267
  path coefficients, 255
  psychopharmacogenetics, 156–157
  sexual dimorphisms, 151–152
  single gene mutants, 247–249
Mice, various strains and hybrids
  behavioral measures, 76–87, 192–197
  early life treatment effects, 192–197
  extreme environments, 88, 101
  heterosis, 79–83, 94
  inducible enzymes, 175–176
  learning experiments, 60–64
  natural populations, 91, 100–101
  races and subraces, 91–92
  single gene mutants, 176, 248
  weight and skeletal differences, 76–79, 90
Milk drinking habit, 115–119, 185
  anthropological aspects, 117–119
  inheritance of lactase, 116–117
  intestinal lactase enzyme, 116

Neonates
  behavior, racial differences, 155
  maturation, 139
Nerve cells in culture, 176

Neurotransmitter agents, 62, 140, 149, 176
  related enzymes, 157, 176–177, 292

Pharmacogenetic analysis, 149, 156–157, 182, *see also* Psychopharmacogenetics
Phenylketonuria, 153, 155, 174, 183, 248, 275, 309
Phototaxis, 2, 29, 72
  adaptive value, 29–30
  mutants, 29
  variation, 29
Polygenic inheritance, 30, 146, 150, 174, 309, 120, 241
  major gene effects, 64, 126, 241, 247–250
  methodology for continuously distributed traits, 241–246, 146–150
  selective experiments, 72–73
Polymorphisms, of proteins and enzymes, 27, 70, 102, 124, 142–150, 155
Population crowding, 88
Primates, 101, 108–110, 123, 197, *see also* Baboon social groups
Protein
  brain specific, 134, 136, 141
  evolution, 133, 134
  synthesis in brain, 140, 152
Psychopharmacogenetics, 156, 157, 174, 278
Psychosexual orientation, *see* Sexually dimorphic behavior

Race
  definition, 93–94
  evolution, 110–115
Racial differences, *see also* Human hybrid populations
  gene frequencies, 93, 126, 268
  height, 112
  immunity to tropical diseases, 116
  IQ scores, 23–25, 260–263, 276
  lactose (milk) intolerance, 117
  neonatal behavior, 155
  polymorphisms, 125, 155, 268
  sensory traits, 92
  sociocultural analysis, 301
Rats
  early life treatments, 193

environmental manipulation, 88
learning experiments, 59–64, 211, 224
nutritional studies, 215–216
RNA (ribonucleic acid)
  evolution of, 133, 135
  transcribed from DNA, 139–140
  translation of RNA for protein synthesis in brain, 140, 152

Schizophrenia, 29, 90, 102, 153–154, 186, 203–204, 249, 278
Selection
  adaptability, 114, 211
  directional, 69, 81, 106
  experimental, 69, 72
  frequency-dependent, 39–42, 52, 126
  for lactose tolerance, 118–119
  natural, 106, 110–111, 123, 130, 134, 143, 163–165, 184
  sexual, 30–37
  social, 186, 263
Sensory-perception traits, 90, 92, 119, 176, 198
Sex hormones, effects of, see Sexually dimorphic behavior, 151–152
Sexually dimorphic behavior (male/female differences), 151–153
  effects of sex hormones, 151–152
  height differences, 113
  psychosexual orientation, 151
Sexual selection, 30–34
Sickle hemoglobin and malaria, 111, 130, 143, 165, 185, 310
Skin color, 116, 120, 126, 268

Social mobility, 283, 303–304
socioeconomic status, 210, 250, 301, 302–305
Somatotypes, 79, 89–90
Songs
  of chaffinches, 38–39
  of insects, 50
  of sparrows, 50

Teachers' expectations, see Classroom interactions
Technology, 130, 164–166
  tool use, 110, 124
Tests, 174, 241–243, 278–283, 285–287
  factor analysis, 243–244, 259–260, 278–279, 281
  multivariate analyses, 210, 213, 274–275
  Piagetian tasks, 281
  recommended battery, 285–287
Threshold, 124, 150, 174, see also Polygenic inheritance
Transcription of DNA in brain, 139–140
Twin studies
  behavioral tasks, 212
  discordance in monozygotic twins, 270–271, 276
  heritability estimates, 256–257, 279
  separated twin pairs, 250–251

Units of behavior, see Behavior, units of

X-linked traits, 68, 154, 271